JESSE RAMSDEN (
LONDON'S LE/
SCIENTIFIC INSTRUM

Science, Technology and Culture, 1700–1945

Series Editors

David M. Knight
University of Durham

and

Trevor Levere
University of Toronto

Science, Technology and Culture, 1700–1945 focuses on the social, cultural, industrial and economic contexts of science and technology from the 'scientific revolution' up to the Second world War. It explores the agricultural and industrial revolutions of the eighteenth century, the coffee-house culture of the Enlightenment, the spread of museums, botanic gardens and expositions in the nineteenth century, to the Franco-Prussian war of 1870, seen as a victory for German science. It also addresses the dependence of society on science and technology in the twentieth century.

Science, Technology and Culture, 1700–1945 addresses issues of the interaction of science, technology and culture in the period from 1700 to 1945, at the same time as including new research within the field of the history of science.

Also in the series

Phrenology and the Origins of Victorian Scientific Naturalism
John van Wyhe

John Herschel's Cape Voyage
Private Science, Public Imagination and the Ambitions of Empire
Steven Ruskin

Jeremiah Joyce
Radical, Dissenter and Writer
John Issitt

Jesse Ramsden (1735–1800)
London's Leading Scientific
Instrument Maker

ANITA McCONNELL

Routledge
Taylor & Francis Group

LONDON AND NEW YORK

First published 2007 by Ashgate Publishing

Published 2016 by Routledge
2 Park Square, Milton Park, Abingdon, Oxfordshire OX14 4RN
711 Third Avenue, New York, NY 10017

First issued in paperback 2016

Routledge is an imprint of the Taylor and Francis Group, an informa business

British Library Cataloguing in Publication Data
McConnell, Anita
  Jesse Ramsden (1735–1800) : London's leading scientific
  instrument maker. – (Science, technology and culture,
  1700–1945)
  1. Ramsden, J. (Jesse), 1735–1800 2. Machinists – Great
  Britain – Biography 3. Scientific apparatus and instruments
  – Design and construction – History – 18th century
  4. Scientific apparatus and instruments industry – England
  – London – History – 18th century
  I. Title
  681.7'5'092

Library of Congress Cataloging-in-Publication Data
McConnell, Anita
  Jesse Ramsden (1735–1800) : London's leading scientific instrument maker / Anita McConnell.
     p. cm. — (Science, technology, and culture, 1700–1945)
  Includes bibliographical references and index.
  1. Ramsden, J. (Jesse), 1735–1800. 2. Instrument manufacture—Great Britain—Biography. 3. Optical engineering—Great Britain—Biography. 4. Machinists—Great Britain—Biography. 5. Scientific apparatus and instruments—England—London—History—18th century. I. Title.
  TS511.G72L6676 2007
  681'.75092—dc22
  [B]
                                                                        2007005528

ISBN 13: 978-1-138-27835-6 (pbk)
ISBN 13: 978-0-7546-6136-8 (hbk)

Contents

List of Illustrations

List of Plates, Picture Credits

List of Plates

Picture Credits

Introduction 1, Introduction 2: © National Portrait Gallery, London.

1.1, 10.1, 10.2: ©Wellcome Library, London.

1.3, 1.4: © Westminster Archives Centre.

1.5, 4.1© The National Archives.

2.1, 7.1, 7.9: Sir Joseph Banks Electronic Archive, State Library of New South Wales.

2.2, 6.1, 8.2, 8.3, 8.7, 13.1, 13.9: By permission of the Syndics of Cambridge University Library.

2.3, 4.4, 7.4: © National Museums of Scotland.

3.1, 7.7a, 7.7b, 8.6: © H.A.L. Dawes.

4.2: © The Bethlem Art and History Collections Trust.

4.3: © Rijksarkivet, Stockholm.

5.1b, © Romualdas Svierdrys, reproduced by permission of Stasé Matulaityé.

5.2, 5.3a, 5.3b: © Romualdas Svierdrys.

5.4: Reproduced by courtesy of Museo La Specola, INAF-Astronomical Observatory of Padova, Italy.

5.5, 5.6: © Istituto Geografico Militare, Firenze.

6.2: Reproduced by courtesy of Mrs Mary Brück.

6.3: Reproduced by courtesy of Kai Budde, Landesmuseum für Technik und Arbeit in Mannheim.

6.5a, 6.5b: Reproduced by courtesy of Peter Brosche.

6.6: © Courtesy of Osservatorio Astronomico di Brera.

6.7: © Observatoire de Paris, Bibliothèque.

7.2: Courtesy of the Smithsonian Institution Libraries, Washington.

Plate 3, 7.8: © Science Museum/SSPL.

13.11: Courtesy of the Smithsonian Institution, Washington.

Plate 2, 7.3, 7.5, 7.6, 13.3a, b, 13.4, 13.5, 13.6, 13.8: Courtesy of Osservatorio Astronomico di Palermo Giuseppe S. Vaiana.

8.1, 8.4: © Royal Society.

13.2: Reproduced by permission of Hampshire Record Office.

8.5: Crown Copyright Ordnance Survey.

13.7: Courtesy of the Mariners Museum, Newport News, Virginia. Peter Ifland Collection.

13.10: © Patrick Marney Barometers.

Abbreviations for Manuscript Sources

AN	Paris: Archives nationales
ASF	Florence: Archivio di Stato
ASP	Palermo: Archivio di Stato di Palermo. Real Segretaria Viceregia
BCA	London: Birkbeck College Archives
BL	London: British Library
BOA	Milan: Brera Observatory Archives
BOD	Oxford: Bodleian Library
BPU	Geneva: Bibliothèque Publique et Universitaire
BPL	Birmingham Public Library, Boulton & Watt Collection
CNAM	Paris: Conservatoire national des arts et métiers
CRL	Copenhagen Royal Library
CUL	Cambridge University Library
CUM	Whitehaven: Cumbria Record Office
ERO	Chelmsford: Essex Record Office
Falkirk	Falkirk Council Cultural Services, Repository GB 558
HLS	Halifax: Halifax Local Studies Library
ICE	London: Institute of Civil Engineers
KCL	Kracow: Czartoryski Library
LGL	London: Guildhall Library
MHS	Oxford: Museum for the History of Science
MOL	London: Museum of London
NAS	Edinburgh: National Archives of Scotland
NHM	London: Natural History Museum Archives.
NLS	Edinburgh: National Library of Scotland
OIOC	London: British Library, Oriental and India Office Collections
PAO	Paris: Archives de l'Observatoire
RAS	London: Royal Astronomical Society
RGO	The National Archives, Cambridge University Library: Royal Greenwich Observatory Archives
RS	London: Royal Society
SML	London: Science Museum Library
TNA	Kew: The National Archives
UUL	Uppsala: Uppsala University Library
VUL	Vilnius University Library
WA	London: Westminster Archives
YRO	York: Yorkshire Record Office

Preface

During Ramsden's working life, England enjoyed constitutional stability with King George III on the throne and peace at home. Wars happened elsewhere: in North America, in the seas around Britain, and from time to time spilling across the Atlantic as far as the Caribbean. In Ramsden's last years British society watched events across the Channel with increasing alarm as the French Revolution took its horrific course.

The main impact of these major events on Ramsden's trade seems to have been the difficulty of transporting scientific apparatus across warring territory. Otherwise, as the physician Edward Jenner remarked in a letter to the Institut des Sciences, 'the sciences were never at war',[1] and even as the navies of France and England battled it out in the North Atlantic, their more distant scientific expeditions carried passports which kept them from attack. Indeed, their respective captains were often friends of long standing and members of each others' learned societies.

No personal papers or family letters have been located, which might have told us much more about Ramsden's background and character, his likes and dislikes, his religious or political sentiments. We know next to nothing of his boyhood education – did he arrive in London already provided with a competent grasp of mathematics and mechanics? Did he stay in touch with his relatives at Halifax, some of whom benefited at his death?

We have Robert Home's fine portrait in oils of Ramsden in his maturity; it is said that Ramsden objected to being shown wearing a fur-lined coat, declaring he had never owned such a garment. We learn that he was welcomed at the London and country houses of the Duke of Richmond and the Duke of Marlborough – men of the highest rank in political and social life – had he kept his Yorkshire accent? Was he the 'Mr Ramsden' seen on one occasion in 1789 with his acquaintance William Windham (1750–1810), then MP for Norwich, at a fencing display?[2] Was he among the vast crowd that gathered in September 1784, to watch Lunardi make the first balloon ascent in Britain?[3] We learn that he enjoyed sociable meetings with literary and scientific colleagues. Did he travel to Matthew Boulton's Soho Works outside Birmingham, and was the manpower organisation there an inspiration to him at Piccadilly or had he seen similar organisation in his home mill-town of Halifax?

At the lower end of the social scale, we hear of his frugality, and we have a legendary account of his fireside evenings with his apprentices – sharing a meal of bread and cheese and beer, and talking over the achievements or setbacks of the day. But who kept house for him after his marital separation? What of his surroundings? The weather? Local political disruptions? What caused his death, and where in Brighton did he die? At the end of the day, there are reams of official papers and a few illuminating letters between third parties. We know what the world thought of Ramsden's extraordinary skill, and also of his perpetual failure to meet promised delivery dates. But we are still unable to see into the soul of the man the Duke of Marlborough once referred to as 'The Gentleman of Piccadilly'.

I have been blessed by the proliferation of documents now available to the diligent researcher on internet, microfilm and typescript, and by the numerous kind friends and professional colleagues who shared valuable snippets of information. I hope that the whole is now greater than the sum of its parts, and that the reader, passing one day eastward along Piccadilly towards St James's Church, may be able to imagine Ramsden's now vanished houses and workshops and perhaps even sight the great man hastening towards an appointment for which he is undoubtedly late.

Anita McConnell
Combs, 2006

Notes

1   Cited in Beer (1960), xi.
2   Baring (1866), 1: 187. This seems entirely possible, as Windham was on the Select Committee of the House of Commons that in 1793 reconsidered the Board of Longitude's grant to the clockmaker Thomas Mudge for his marine watch. On p. 276 Windham mentions a troubled Ramsden coming to consult with him over the Committee's report. Windham was also a friend of Count Hans Moritz von Brühl, who figures in this story.
3   The report of this event in *Gentlemen's Magazine*, vol. 54/ii (1784), 711, is expanded in de Montluzin (2002), 134–7.

Acknowledgements

My warmest thanks to all those librarians and archivists at the Royal Society and Cambridge University Library who over the years have dealt with my requests competently and with good humour, and (in alphabetical order) Robert Anderson, Jim Bennett, John Brooks, Randall Brooks, Peter Brosche, Mary Brück, Willem Mörzer Bruyns, Peta Buchanan, Kai Budde, Ileana Chinnici and colleagues at Palermo, Gloria Clifton, Andrew Cook, Howard Dawes, Suzanne Débarbat and colleagues at Paris, Peter Delehar, Danielle Fauque, Brian Gee, Francisco Gonzales, Peter Hingley, Rod Home, Inge Keil, Agnese Mandrino and colleagues at Brera, Patrick Marney, Jean-Pierre Martin, Alison Morrison-Low, Luisa Pigatto and Valeria Zanini at Padua, Martin Rickenbacher, Simon Schaffer, Wolfgang Schaller, Mark Scudder, Tony Simcock, Allen Simpson, Romualdas and Aliucija Sviedrys, Arminas Stuopis and colleagues at Vilnius, Stuart Talbot, Liba Taub, Anthony Turner, Magda Vargha, Ruth Wallis, Michael Wright, Marina Zuccoli and colleagues at Bologna, and not forgetting the late John Millburn. I am also much obliged to the many archivists and librarians who kindly helped me to gain access to their search rooms, and then guided me in various languages through their cataloguing and ordering systems. Without their assistance I would have missed many of the treasures which they held.

In Memoriam

Derek Howse
(1919–1998)

Foreword

Lord Martin Rees*

Without scientific instruments, our knowledge of the natural world would not have advanced much beyond its level in Aristotle's time. Indeed the entire development of science has been driven by (and, in return, has driven) the invention of better instruments and novel techniques. Those who design and make these instruments deserve more acclaim than they normally receive: they tend to be overshadowed by those who thereby make discoveries, or who weave theories from what is discovered. That was true in earlier centuries, and is still, sadly, true today.

This book does something to redress that balance. It offers a fascinating insight into the life and times of a man who, though his craftsmanship and his innovation, made pervasive and lasting contributions. Jesse Ramsden was an innovator – especially in his designs of full circles rather than quadrants for astronomy, and in his elaborate theodolites that crucially improved the quality of surveys and cartography. He was an enterprising man, who benefited from the relatively liberal working practices among craftsmen in London, to get an edge over his counterparts on the continent. His workshop was large and versatile by the standards of his time.

Ramsden's instruments remained in use for many decades, and are still admired and cherished for their craftsmanship and elegance. They were commissioned and purchased not only in England, but throughout Europe. In Ramsden's time, there were few professional scientists, apart from surgeons and physicians – indeed the word 'scientist' had not yet been coined, and his artifacts were sometimes described as 'philosophical instruments'. But there were, throughout Europe, wealthy and discerning patrons – many of them conversant with the technology of the time and enthusiastic about the curiosities of the natural world. Such people – Sir Joseph Banks and the Duke of Marlborough prominent among them – were Ramsden's clients and customers.

Even those who have no special expertise in science will find much of interest in this book; it offers a fascinating perspective on the social texture of trade and craftsmanship in eighteenth century London, of Ramsden's business dealings across Europe, and of his social milieu. Anita McConnell has given us a work of meticulous scholarship that deserves wide readership.

* Professor Rees, Lord Rees of Ludlow, is Astronomer Royal and currently President of the Royal Society.

1   Jesse Ramsden by Robert Home, probably 1790.
     Oil on canvas, © The Royal Society

2 Astronomical circle made by Ramsden for Palermo Observatory.
Courtesy of Osservatorio Astronomico di Palermo Giuseppe S. Vaiana

3   Geodetic theodolite by Ramsden, 1787. © Science Museum / SSPL

4 © Hans Moritz, Count Bruhl, James Northcote, 1796. Petworth House, The Egremont Collection (acquired in lieu of tax by H.M. Treasury in 1957 and subsequently transferred to the National Trust). Oil on canvas (1270 mm × 1005 mm)

Introduction

The Instruments Trade[1]

The Instruments

When Ramsden came to London in 1756 and apprenticed himself to a maker of mathematical instruments he was entering into a craft which had been practiced in that city for about two centuries.[2] Mathematical instruments, those which serve principally to measure and compute lengths and angles, were everyday tools for 'mathematical practitioners', principally builders, shipwrights, land surveyors, cartographers, astronomers, military engineers and mathematicians.[3]

The origins of such instruments can be traced back to the astrolabes and other complex astronomical computing devices familiar to the medieval Islamic world from where their use and manufacture spread to Europe. Richard of Wallingford (d. 1336) is depicted using instruments to set out the gearing of his great astronomical clock, and fifteenth-century examples are known from Oxford Colleges.[4] From the mid-sixteenth century there is evidence of some Flemish engravers who had settled in London, supplementing their income by making mathematical instruments. Both crafts employed the same tools and skills to turn and polish metal, and to engrave divisions, lettering and figures. Brass resisted damp and insect damage and could be shaped and divided more accurately than wood, but until supplies became more readily available in the seventeenth century, it was a costly material. Rulers and other simple instruments were made from fine-grained woods. Wealthy customers commissioned pieces in ivory, silver or gilt.[5]

Spectacles had been made in Italy from the fifteenth century and examples were imported into England by Flemish pedlars. Around 1608 spectacle makers in the Low Countries discovered that a pair of concave and convex lenses set in a tube would bring distant objects more clearly into view. Soon afterwards it was found that the same two lenses, differently arranged, magnified small objects. During the seventeenth century the perceived military value of telescopes, and the rise of a leisured class interested in these new devices, brought into being a class of optical instrument makers who specialized in working clear crystal glass for lenses. By the eighteenth century the two skills had combined, with telescope lenses fitted to surveying and astronomical instruments, and microscope lenses arranged to read small divisions on their graduated scales.[6]

A third class, known as philosophical instruments, was designed to investigate elements of the natural world which could not be directly sensed or measured, namely magnetism, pressure, temperature and gravity. The manufacture of magnetic apparatus, barometers, thermometers, gravity pendulums and other meteorological and electrical instruments required a wide range of materials and skills, and called for a higher level of education among the craftsmen who constructed them.[7] A few instrument makers were involved with the skilled production of globes, an expensive business, requiring the cooperation of cartographer, engraver and globe-maker, not to say a wealthy patron.[8]

Advertisements developed from remarks in textbooks on the best source of instruments mentioned, to the flamboyant trade cards and bill-heads of the eighteenth century.[9]

The Trade

It is worth remarking here that, with a few famous exceptions, the makers of clocks and watches, and those who made weighing instruments, kept largely separate from the scientific instrument makers, having their own craft guilds and keeping their own shops.[10] Within the City of London these trades were controlled by these guilds, or companies, and there were also local guilds in some of the ancient chartered provincial cities. But as a man who had completed his apprenticeship might take his freedom in the guild of his master, or that of his father, who might be in an entirely different trade, scientific instrument makers are consequently found in almost all the London companies, the Spectaclemakers and Grocers Companies having the largest share. Freemen could trade publicly and take apprentices, but the freedom brought its obligations, including the payment of quarterly dues, and a large proportion of men preferred to work as sub-contractors or waged journeymen. Guild membership was not obligatory for those who traded in Westminster and in London's growing suburbs, and Ramsden, who worked exclusively in Westminster, did not belong to any company, and was able to employ foreign-born craftsmen without the need to obtain the consent of the Corporation of London's Common Council.

The Customers

Driving the demand for scientific instruments, besides the humble mathematical practitioners, were the professionals, such as mathematicians, architects and physicians; teachers, including the many peripatetic lecturers who toured Britain giving illustrated talks and demonstrations; the wealthy 'dilettanti' and 'virtuosi',[12] who devoted themselves to natural and experimental philosophy; and the nobility and royalty for whom the finest instruments were treasured possessions and status symbols, to be displayed alongside their globes and clocks, or to be presented as gifts to others of the same rank. The term 'amateur' used here to describe some of Ramsden's customers is in no way pejorative, but indicates that these men enjoyed the fruits of commerce, agriculture or industry, and had the learning, leisure and determination to make good use of their fine instruments – indeed many published their results, adding to the sum of human knowledge. If the Enlightenment blazed more brightly in Scotland and across the Channel, its ideas inspired the numerous informal groups who met in coffee houses to discuss new scientific discoveries. The weft underlying these groups was freemasonry; the same coffee houses hosted their lodges, and although no formal membership records of the London lodges survive, it is known that many freemasons were also Fellows of the Royal Society. Membership was strong in France and Germany, and several of Ramsden's visitors were freemasons. These formal and informal groups welcomed men of noble and common birth and of any religious affiliation, encouraging a more open society than existed in France. With due respect shown, a tradesman such as Ramsden could be accepted and respected by the nobility and titled men with whom he dealt.

The terms 'scientist' and 'scientific instruments' were not in used in Ramsden's day. There was no professional civil service, and few people were paid to pursue what we would now consider to be careers in science. In the second half of the eighteenth century, the government, when in need of advice on matters relating to science, usually turned to Sir Joseph Banks (1743–1820), President of the Royal Society from 1778 to 1820 (see Figure 1). Banks, a wealthy landowner with a taste for natural history and a very wide spectrum of acquaintances, consulted suitable persons for the task, sometimes organized in committees. These specialists were expected to provide their services gratuitously. The title of 'Le Ministre des affaires philosophiques' humorously bestowed on Banks by Lord Auckland was justified – he was the de facto minister of science.[13] His right-hand man in many matters was Sir Charles Blagden (1748–1820) (see Figure 2), a physician and experienced traveller, who served as Secretary of the Royal Society from 1784 to 1797.[14] Both men became well acquainted with Ramsden.

The Astronomer Royal was a salaried post created in 1675 with the founding of the Royal Observatory at Greenwich. In Ramsden's early days the Astronomers were James Bradley (1693–1762), then Nathaniel Bliss (1700–1764). From 1765, the formidable Nevil Maskelyne (1732–1811), was in charge.[15] Paid instructors at naval and military colleges taught the theory and use of instruments needed by hydrographers, navigators and surveyors, men who during their working lives were often expected to provide their own equipment.

University astronomy had a long history, but prior to the eighteenth century, was often undertaken at the whim of the professors of mathematics, by the simple means of carrying a telescope out onto a suitable roof when the weather allowed.[16] The construction of royal observatories in Paris in 1672 and at Greenwich in 1675 heralded the start of a programme of regular carefully-timed celestial observations for the benefit of navigation and cartography. Over the years both were furnished with excellent clocks and a suite of apparatus comprising small portable telescopes, large telescopes mounted on great quadrantal arcs so as to swing from horizontal to vertical, telescopes supported to point at or near the zenith, and telescopes mounted to move steadily in time with the rotation of the Earth.[17]

Astronomers from lesser observatories who visited these great institutions went home to ask for similar apparatus. Many European universities were run by the Society of Jesus which was willing to provide buildings and apparatus for astronomy.[18] The order was suppressed in one country after another during the 1760s and 1770s, but many of its institutions were taken over by secular rulers who, seeing a necessity for mapping their lands, kept the observatories supplied with new instruments for astronomy and surveying. In Ramsden's day the Dollond family were universally acknowledged to be the leading makers of telescope lenses,[19] but Ramsden was often called on to make the mounts for their

Sir Joshua Reynolds Pinx.   S. W. Reynolds Sculp.

Figure 1   Sir Joseph Banks Bart., by Joshua Reynolds, 1771–73. Oil on canvas. (1270 mm × 1015 mm). NPG, D621.

Figure 2   **Sir Charles Blagden, by Mary Dawson Turner, after Thomas Phillips, 1818. Etching. NPG, D14478.**

telescopes. Touring astronomers acted as travelling salesmen for Ramsden and Dollond. In addition to these 'official' customers Ramsden counted two wealthy private individuals: George Spencer (1739–1817), fourth Duke of Marlborough and the baronet Sir George Shuckburgh-Evelyn (1751–1804), both able to afford staffed observatories with major instruments.[20]

During both war and peace demand for all classes of instruments rose steadily throughout the eighteenth century, with multiple orders for such military items as gunnery instruments, small telescopes, octants and sextants, and away from the battlefield, theodolites and levels, microscopes, larger telescopes, barometers, thermometers, and much more. Orders and payments for these small items were negotiated through a reputable instrument maker, who probably kept a shop where he sold new or refurbished instruments to the passing trade. He had to buy the brass, wood and other components, and pay his sub-contractors weekly or by unit, carrying this debt until he was reimbursed by the purchaser. The sub-contractors laboured on their own lathes and workbenches in their homes or in small shared backyard workshops. In the years leading up to Ramsden's entry into the trade, the two George Adams, father and son, held much of this business.[21] The finest surveying and astronomical instruments were despatched to Greenwich and many European observatories from the workshops of George Graham, one of the few eminent makers of both clocks and instruments, and from the craft dynasty of Graham's former apprentice John Bird, Bird's former apprentice Jonathan Sisson, and Sisson's son Jeremiah.[22]

The Sissons were in competition with Parisian instrument makers but the latter were increasingly handicapped by their inability to raise capital for the purchase of materials, and by the restrictive and archaic Parisian guilds.[23] By the 1770s, with a great expansion in the size and number of European observatories, London was seen as the only place to purchase the larger and costlier apparatus, and Ramsden's products were the most eagerly sought after. With items commissioned to a novel design, the principal instrument maker had to be on hand lest modifications were needed as construction proceeded, while generous space and headroom were needed for the final assembly and testing. Nor was that the end of the matter, since the instrument then had to be partially dismantled and crated up for its long journey by sea and land to its destination. The whole process could take two years or more, from commission to delivery and payment.

Ramsden was perhaps more fortunate than most in the timing of his own development. In Britain the creation of a national topographical survey began in 1783 and burgeoned into the Ordnance Survey, while the desire to modernize apparatus at Greenwich Observatory was paralleled by the founding of numerous observatories in the small dukedoms and principalities of western Europe, principally with the intention of making cartographic surveys of their national territories. The astronomers appointed to establish these institutions usually travelled to Paris and London, where they made their way to Ramsden's workshop. Instrument makers of a similar competence were not to be found on the continent, nor could European glasshouses produce the good quality 'flint' glass needed for achromatic lenses.[24]

Soon after setting up as an independent tradesman, Ramsden made the transition from craft workshop to big business in order to meet this new demand. He acquired a spacious former carriage works, staffed with foremen overseeing a skilled labour force supplemented by apprentices or trainees, and a shopman. He provided the dividing engines and some of the machine tools. His establishment of 40 to 50 men was so unusual for its day that it was remarked on by many of his overseas visitors. To keep his men fully occupied, and to generate the flow of cash needed to pay their wages, they produced a variety of lesser instruments, standard and bespoke, with repairs and refurbishment of second-hand instruments. He developed a circular dividing engine which allowed him to divide the arcs of sextants, theodolites and protractors faster, and with a greater accuracy than hitherto; he contrived a series of improvements to the equatorial telescope and indeed to many other instruments which he was asked to make. By 1788 he was seeking to persuade his customers that circular instruments would be preferable to quadrants, large and small, and that the results would justify the extra expense. One great circle, planned for an observatory in Florence, failed to materialize; another, for an observatory near Dublin, was finished by his successor Matthew Berge. But his three great geodetic theodolites, the first made for William Roy's survey in 1787, and the famous Palermo astronomical circle of 1789, were masterpieces of design and construction. Yet by the early 1800s the concept that 'big is best' had given way to increased precision, much of it emerging from French and German workshops.

Ramsden – the Myth and the Reality

During his productive lifetime Ramsden was acknowledged by his European scientific contemporaries as a consummate artist while at the same time they bewailed his failure to meet delivery deadlines. They travelled to London in order to pressure him to accept their commissions, knowing full well that his insistence on perfection probably meant a serious delay before the instrument arrived and could be put to the task for which it was intended. At home, Ramsden's desire to improve both his machine tools and the instruments which he made, without thought of financial gain, led to his election to the Royal Society in 1786, a rare honour among instrument makers. Subsequently he received the Society's Copley Medal, its highest award.

Most of the information about Ramsden's life and works up to 1789 comes from a letter written by the Italian astronomer Giuseppi Piazzi to the French astronomer Joseph-Jerôme de Lalande, published in French that year and later translated into English.[25] This material, supplemented by a personal tribute contributed by his friend Louis Dutens, was combined in a memoir published in *Aikin's Biographical Dictionary* soon after Ramsden's death.[26] The lack of a subsequent detailed historical examination may be assigned first, to the absence of any personal or business papers; second, to the modest character of Matthew Berge, his long-time foreman, who continued Ramsden's business until his own death in 1819, and who also left no papers; and third, to Ramsden's eclipse by the products of French workshops, in respect of smaller instruments, and by those of the London maker Edward Troughton, and the German instrument and glass workshops, which by the nineteenth century were able to produce smaller astronomical and surveying instruments, graduated to the precision formerly attainable only in Ramsden's costly and unwieldy great circles, making the latter look increasingly old-fashioned.

A hundred years later, those Ramsden instruments which had survived this period of obsolescence which brought with it the hazards of war, and the predations of scrap merchants, began to be admired and valued for their antique beauty, and so they have continued to the present day. And yet, though Ramsden's works receive honourable mention in such instrument histories as Daumas's *Instruments scientifiques* …; Anthony Turner's *Scientific instruments, 1400–1800*; Allan Chapman's 'The accuracy of angular measuring instruments used in astronomy between 1500 and 1850'; on specialist works such as Jim Bennett's *The Divided Circle*, John Brooks's 'The circular dividing engine: its development in England'; Peter Ifland's history of sextants in his *Taking ths Stars. Celestial Navigation from Argonauts to Astronauts*, Alan Stimson's articles on Ramsden sextants, and many more, these focus on the instruments themselves, as they survive in national or private collections, somewhat detached from the time and circumstances of their construction and use.[27]

The only attempts at a 'life' have been that of Agnes Clerke in the old *Dictionary of National Biography* who relied on the earlier published sources, and a new memoir in the current *Oxford Dictionary of National Biography*, composed by Allan Chapman, drawing on many biographical details provided by the present author. Briefer memoirs, concentrating on his working years, include that by Rod Webster in the *Dictionary of Scientific Biography* and in various encyclopedias and biographical dictionaries published during the nineteenth centuries in Britain and Europe. These essays offer an assortment of dates for his birth, for the award of the Copley Medal, for the invention of his dividing engine (which one?) and for several other important manufactures. The repeated statements about his numerous employees have not hitherto enlightened us as to their identities or nationalities. Sweeping statements concerning his outstanding designs and manufactures overlook the bad ideas and second-rate products that emerged, some from his own hands, others at least bearing his signature.

It is now much easier to enter more fully into Ramsden's life and works. Discarding the old thrice-boiled sources, I have turned to contemporary correspondence, both to and from Ramsden, and that between third parties which is often more illuminating. I have toured most of the European observatories which he furnished, and unearthed documentation and invoices from those institutions, from the Greenwich Observatory, the National Archives of England and Scotland and from assorted local history libraries and provincial archive offices in Britain and Europe. I have endeavoured to set his activities within the political, economic and climatic events of his time while the *Oxford Dictionary of National Biography* illuminates the lives of his acquaintances.

A Note on Measures of Length and Coinage

Measures of length proliferated in eighteenth-century Europe with each small state setting its own standards but Ramsden made his instruments to British measure, and this was accepted by his foreign customers, although they sometimes reverted to their own measures when describing or listing them. The French measures mentioned in this text were the Paris foot ('pied du Roi'), which was divided into 12 inches, each inch divided into 12 'lines'. This duodecimal system was almost universal in Europe, but in England inches were more often divided into tenths. One Paris inch was equivalent to 2.7066 cm, somewhat longer than the English inch of 2.5399 cm.

Measure equivalents are set out in Horace Doursther, *Dictionnaire universel des poids et mesures anciens et modernes* (Brussels, 1840); monetary equivalents are given in Patrick Kelly, *The Universal Cambist*, 2 volumes (London, 1813). To convert eighteenth-century sterling to present-day values consult www. nationalarchives.gov.uk/currency.

Notes

1   Recent sources abound for this topic; the majority of those cited provide additional references. Sources for most named individuals may be found in the published (2004) or on-line versions of the *Oxford Dictionary of National Biography* (*OxDNB*).
2   Turner, A.J. (2003).
3   Taylor (1968).
4   *OxDNB*: J. North, 'Richard of Wallingford'; Turner, A.J. (1987), Chapters 1 to 3 (pp. 11–86).Turner, G.L'E (2000a).
5   See also *OxDNB* for memoirs on Tompion, Gemini, Ryther.
6   Clifton (1993a).
7   Turner, A.J. (1987), Chapter 4 (pp. 87–170.); Turner (1998). Porter *et al.* (1985).
8   Clifton (1993b).
9   Bryden (1992); Crawforth (1985).
10   Brown (1979a and b); Crawforth (1987); Clifton (1993a).
12   The 'dilettanti' enjoyed their own Society of Dilettanti, founded in 1734 and patronized by young men who had returned from the Grand Tour enthused by art and antiquities. By the later part of the eighteenth century 'virtuosi' and to a lesser extent 'dilettanti' had broadened their membership to include the experimental philosophers, whose collections included the apparatus and instruments of their field of interest: astronomy, physics, chemistry, or other aspects of the natural world.
13   Gascoigne (1998) uses this phrase several times, citing HMC Fortescue, II, 225, Lord Auckland to Lord Grenville, 6 November 1791, but it could have been truthfully said 15 years earlier.
14   Gascoigne (1998).
15   Howse (1998).
16   Gunther (1969a and b).
17   Forbes (1975); Howse (1975); Wolf (1905).
18   Udias (2004).
19   *OxDNB:* G. Clifton, 'The Dollond family'.

20 *OxDNB*: A McConnell, 'Evelyn, Sir George Shuckburgh' and S.M. Lee, 'Spencer, George, fourth Duke of Marlborough'.
21 Millburn (2000).
22 *OxDNB:* J.L. Evans, 'George Graham'; D. Howse, 'Jonathan Sisson' and 'Jeremiah Sisson'.
23 Turner, A.J. (1989); Turner, A.J. (1998).
24 Wolf (1905).
25 Piazzi (1778); Anon (1789); Piazzi (1803).
26 Aikin (1813).
27 Daumas (1972); Turner, A.J. (1987); Chapman (1993); Bennett (1987); Brooks (1992); Ifland (1998); Stimson (1985).

Chapter 1

Early Life

The greatest man who was born in Salterhebble. Halifax councillor T.W. Hanson, in a speech at Salterhebble, March 1934.

Halifax Courier and Guardian, 24 March 1934, 11.

Landscape and Family

The story begins in the West Riding of Yorkshire, in the parish of Halifax. The parish consisted of 26 'townships' or hamlets, spread over 100 square miles (259 km^2), an area slightly larger than that of the county of Rutland, with which it was often compared in size. On the Pennine flanks the upland rose to 1,500 feet (457 m) and over most of the region the slopes and thin soils precluded farming as a full-time occupation. Black oats, and more recently wheat, were the only crops, although a few domestic animals were pastured on level ground which was seasonally flushed by the many springs. The value of the land lay in the coal seams underlying parts of the parish, the good building stone to hand, and in the power taken from the many streams fed by the high rainfall.

The poverty of this environment had been relieved since at least the seventeenth century by the deliberate introduction of woollen manufacture. Raw wool was brought in and converted into yarn that was both woven and knitted. With its numerous fulling mills, the town of Halifax, along with Leeds and Bradford, was described in 1642 as rich and populous.[1] But prior to the construction of canals and toll-roads in the late eighteenth century, transport was difficult, with no navigable rivers and few roads. With meagre local harvests of grain, this basic foodstuff had to be brought from more fertile areas and, with other goods, was carried by packhorse. The parish church of St James and its outlying chapels served the Anglican parishioners but by the early eighteenth century, here as in other northern districts where there was a growing population, Anglicans were being outnumbered by non-conformists who supported the various dissenting chapels and meeting-houses.[2] Ramsden was a common name throughout the district. One line descended from John Ramsden, granted arms in 1575, but there is no evidence of Jesse Ramsden being closely related to this or other gentry families, nor is it known if his father was the Ramsden who served as a Trustee to Skircoat Workhouse in the 1740s. It has also not been possible to confirm from genealogical records the statement made in certain sources (such as Henry King's *History of the Telescope*, p. 162) that he was a great-nephew of Abraham Sharp (1651–1742) of Bradford. In his youth, Sharp, a mathematician and instrument maker, had worked under John Flamsteed, the first Astronomer Royal, at Greenwich Observatory.

Jesse Ramsden was born in September 1735 at Skircoat township, near Salterhebble, a small settlement south of Halifax.[3] His father Thomas Ramsden, then employed as a shoemaker, married Abigail Flather of Hip at the Northowram and Coley non-conformist chapel on 12 September 1734, and again on the same day at the parish church of St James, Halifax.[4] Such dual marriages were common, as only marriage in a parish church had legal force. By the time their first-born, Jesse, was baptized on 3 November 1735, Thomas was innkeeper at the Elephant and Castle at Skircoat.[5] From the age of nine to twelve Jesse was a pupil at the Free Grammar School of Queen Elizabeth at Heath, near Halifax, where he received the usual classics grounding.[6]

Myths, Legends and Chinese Whispers

From this point, we are offered three versions of Ramsden's early years: the earliest account, a letter, dated 1 September 1788, written to the Parisian astronomer Joseph-Jerôme Le François de Lalande (1732–1807) by Giuseppe Piazzi (1746–1826), the astronomer from Palermo who spent several weeks in London in 1787–88 urging Ramsden to complete an instrument for his observatory. This letter was published as 'Lettre sur les ouvrages de M. Ramsden, de la Société Royale de Londres adressée à M. de la Lande par le R.P. Piazzi, Professeur Royale d'Astronomie dans l'Université de Palerme' in the *Journal des savans* for November 1788, pp. 228–55. It also prefaces Lalande's translation of the description of Ramsden's dividing engine.[7]

A translation of Piazzi's text, without the opening address to Lalande, was published in the *European Magazine*, February 1789; much later, considerably reworded and with the opening paragraph to Lalande reinstated, it appeared in the *Philosophical Magazine*, 1803, accompanied by a stipple engraving, head and shoulders, by Charles Parsons Knight captioned 'taken from an original in the possession of Mr Colnaghi' – the art dealer Paolo Colnaghi (1751–1833) (see Figure 1.1). All the articles follow Piazzi in misinforming us that Ramsden was born on 6 October 1730. They continue:

> His earlier studies excited in him an extreme desire of dedicating himself to the pursuit of literature, and the favourite subjects which first struck his youthful mind were history and antiquities. Mathematics and chemistry next engaged his attention, but his father, who was a clothier, pressing him to follow some trade, he continued at home, in an employment which was not too well-suited to him, till he was one and twenty.
>
>   At this period young Ramsden came to London in quest of some occupation more worthy [of] his genius. Among other things, he applied himself to engraving, under the tuition of Burton.

At this point, a footnote in the *Philosophical Magazine* text identifies Mark Burton (w.1755–d.1786):

> Mr Burton was a thermometer and barometer maker, and divider of instruments. Instruments at this period were divided by means of a plate applied to them, and the divisions were in this manner marked off. Mr Burton was one of the best workmen of

Mr Ramsden

London Published Augst 31, 1803 by A Tilloch, Carey Street.

Figure 1.1 **Jesse Ramsden, by C. Knight after R. Home, 1803. Stipple engraving.**

his time, and worked for Short, Bird and other eminent artists. Mr Ramsden bound himself apprentice to Mr Burton for four years, and after his time was expired entered into partnership with Mr Fairbone, who lived afterwards in New Street, Shoe Lane. This partnership, however, did not long continue, Mr Ramsden opened a shop on his own account in the Strand, and, having married Miss Dollond, became possessed of a part of Mr Dollond's patent for achromatic telescopes. Mr Ramsden in the course of a few years removed to the Haymarket ...

From 'under the tuition of Burton' the *European Magazine* presses on:

In this situation a fortunate circumstance led him to the object for which Nature seems to have designed him – the improvement of astronomical instruments; for the invention and construction of which he undoubtedly ranks the first in Europe. In the course of his employment, mathematical instruments were frequently brought to him to be engraved. The more he examined them, the more he noticed their defects, and a secret instinct prompted him to wish to remove them. This wish was followed by a resolution to attempt it. He soon made himself master of the file and the lathe, and even of working glasses; and in 1763 made instruments for Sisson, Dollond, Nairne, Adams and others. About 1768 he opened a shop for himself in the Haymarket ...

The two unexplained names in the *Philosophical Magazine* footnote were those of James Short (1710–1768), maker of reflecting telescopes, and John Bird (1709–1776), maker of large astronomical apparatus. The four later names were of Jeremiah Sisson (1720–1783), John Dollond senior (1706–1761), Edward Nairne (1727?–1806) and George Adams senior (1709–1772). Together, these eminent craftsmen served the top end of the market in mathematical, optical and philosophical instruments and apparatus. We shall encounter them in later chapters.

So much for a story originally written while Ramsden was still alive. The next biography, composed shortly after his death by Thomas Morgan but sourced as from Piazzi's 'Life',[8] was published in John Aikin's *General Biography*, volume 8 (1813). This memoir can be found in Appendix 1.

Morgan expands on Ramsden's education: after his years at Heath school his father sent him to stay with an uncle in the North Yorkshire district of Craven. There:

he was tutored by the Revd Hall who had a reputation for the mathematical sciences. Under this gentleman's tuition young Ramsden became proficient in geometry and algebra and was proceeding with delight in studies for which his genius was particularly suited, when his father sent for him home, and put him an apprentice to a clothier in Halifax. After he had followed this occupation three years, he was placed in the capacity of clerk with another manufacturer in the same town, in whose service he continued till he was about twenty years of age. At this period of his life he went to London, where he became clerk in a wholesale cloth warehouse. This situation he retained for two years and a half, when his inclination for the sciences revived; and as he possessed at the same time a strong mechanical turn, he resolved to qualify himself for some business which should prove suitable to the bent of his mind. With this determination he bound himself an apprentice for four years to Mr Burton, who lived in Denmark Court near the Strand and was one of the best workmen of his time in making thermometers and barometers, and in engraving and dividing mathematical instruments. Not long after the expiration of his apprenticeship, he became a partner with a workman of the name of Cole, under whom he was at first a journeyman, with no higher wages than twelve shillings a week. This partnership, however, did not long continue; and after its dissolution Mr Ramsden opened a workshop on his own account ...

Yet another truncated story was published long after his death by the author and historian George Lillie Craik, who immortalized Ramsden among the 'Professors of optical discovery'.[9] One last detail, that Ramsden's father apprenticed him to a local man, probably a presser, comes from a column in a Halifax newspaper, repeated in the article with an engraved portrait celebrating the bicentenary of his birth.[10] The most recent local biography drew on these earlier texts.[11]

Out of the shadows

Among this tangle of names and dates, there lurks a plausible biography, some elements of which can be confirmed. We left Jesse Ramsden, born and baptized, in the care of his parents, and given his first schooling at Heath. The elusive Reverend Hall, 'famous in the mathematics' but not an alumnus of Oxford or Cambridge, and who does not figure in clerical directories, may have been a dissenting minister.[12] Young Ramsden comes across as a lad inspired by his schoolteachers at Heath, and then relishing his growing mastery of mathematics gained under Hall's tuition. During these years Ramsden is said to have spent his spare time in studying what was then known as 'natural philosophy' and reading all he could find about the precision instruments that were now casting light on so many aspects of what we now regard as 'science'. A clerk for a Halifax clothier, he would have been aware of the New Style calendar change. In 1752 Great Britain adopted the Gregorian Calendar, whereby, by the omission of 11 days, 3 September was reckoned as 14 September, and the beginning of the new year was moved from 25 March to 1 January. This brought Britain in line with most of western Europe, which had changed to the Gregorian Calendar in or shortly after 1582. The loss of 11 days brought headaches to clerks, bankers, accountants, shopkeepers and suppliers generally, when they calculated wages and interest.

Unfortunately no will has been located for Thomas Ramsden, which might have indicated a modest inheritance to fund Ramsden's new life when, aged 21, he set out for London, a three-day journey by coach. His first employment there was in a wholesale cloth warehouse, a position probably arranged prior to his departure and perhaps with a relative. Ramsden worked as a clerk for about two and a half years, before binding himself apprentice to Mark Burton of Denmark Court, Strand, for four years from 1756 at a fee of £12.[13] Burton was well known as a mathematical instrument maker, and advertised on a trade card, post 1760, that he made and sold all sorts of mathematical, philosophical and optical instruments at the Euclid's Head, near St Mary's Church in the Strand.[14] One arm of a set-square covered with trial marks and crudely engraved 'J Ramsden 1757' and again '1757' (see Figure 1.2), must represent a trial piece made under Burton's supervision.[15]

Figure 1.2   Set-square with Ramsden's trial engravings. National Museum of Science and Industry. Inv. 1954-326.

Ramsden's spell as journeyman to Cole, is most likely to have been with Benjamin Cole senior (1695–1766) or his son, also Benjamin Cole (1725–1813), both living and working at 'The sign of the Orrery', 136 Fleet Street. 'Mr Fairbone of 20 New Street, tube-drawer' is mentioned in the notebook of an unnamed surveyor now in the National Library of Scotland.[16] Again, there are several Fairbones, who were both timber merchants and mathematical instrument makers, at this address.

Westminster, the extensive area north of the Thames, upriver from the City of London, embraced the City of Westminster itself, with its two ancient parishes of St Margaret and St John, plus the larger area of the Liberties of Westminster, which met the City of London at Temple Bar. Its southern parishes included St Martin-in-the-Fields, St James Piccadilly, St Mary-le-Strand, St Paul Covent Garden and St George Hanover Square; to the north lay the parish of St Mary-le-Bone. At its completion in 1750, Westminster Bridge was the sole alternative river crossing to London Bridge. This was the district in which Ramsden entered as an apprentice and within which he spent his entire working life. Although home to the venerable abbey of St Peter (Westminster Abbey) and the location of Parliament and the offices of government, Westminster had become so insalubrious that it took an Act of 1765 to instigate improvements. John Noorthouck, in his *New History of London* (1773), described the former situation, remarking that although the high streets had raised pavements on each side for pedestrians, these were not kept in repair; that they were divided from the road – equally in disrepair – by lines of thick posts which diminished the width of the pavement and at night were dangerous obstructions. In older streets the house-roofs drained into spouts which simply poured rainwater onto the passers-by, while in all the streets, sign-boards – as large as the shopkeeper could afford – hung on frames projecting from the houses, obscuring the view and creaking horribly in windy weather. The Act banished the projecting shop-signs, laid flag-stones on the footpaths and edged them with kerb-stones. The carriageway was cambered and provided with channels each side, and surfaced with squared granite setts, brought by sea from Aberdeen. Westminster had brought in a lamp rate in 1761; the Act of 1765 required the oil-fuelled lamps to be placed at equal distances. Noorthouck relates that 'the expedition with which this grand design was executed, was as remarkable as the elegant appearance of the streets when finished'.[17]

Independence and marriage

Ramsden's own premises are first identified in the Westminster rate books for Exchange Ward, Strand, paying rent of £18 in 1763, and in 1766 more particularly for premises between Curle Court and Marygold Court, this entry being overwritten 'Midsummer ... Ramsden in the Haymarket'.[18] He took his first apprentice on 28 February 1763. James Ware from Christ's Hospital was bound to serve Ramsden for five years, but on 8 June 1764 he was turned over into the shoemaking trade.[19] Ramsden's friend, the Huguenot writer Louis Dutens (1730–1812), tells us that Ramsden had already made improvements to the sextant and had invented, though not brought to perfection, his first dividing engine.[20] It is not known how the commission for the transit telescope at Vilnius came his way (see Chapter 5 and Figure 5.2) but it was possibly passed on by Jeremiah Sisson whose work was already appreciated in Europe.

Marygold Court was adjacent to the premises of John Dollond (1706–1761), silk weaver turned optician, by this time joined by his sons Peter (1731–1820) and John (1733–1804). The Dollonds were renowned for the quality of the doublet lenses fitted to their large achromatic refracting telescopes.[21] Ramsden is said to have spent his evenings in the Dollonds' house where he learned about optical instruments, and he became attached to John Dollond's youngest daughter Sarah (1743–1796). They were married on 17 August 1766 at St Martin-in-the-Fields. Several memoirs, notably that of Agnes Clerke in the *Dictionary of National Biography* (1896) and that of Roderick Webster in *Dictionary of Scientific Biography* (1975) state, without giving any source, that Sarah brought as her dowry half the rights in John Dollond's patent for the achromatic lens, which would have been a valuable income for a young man starting out on his own. (We may discredit the statement in the *Halifax Guardian* of 10 January 1857 that John Dollond bequeathed half the share to Ramsden since he died intestate.) John Dollond had obtained this patent in 1758,[22] income from the rights being shared with Francis Watkins (1703–1791) who had contributed the fee. When Peter inherited his father's half-share, he paid Watkins £200 for the remainder and began court actions against those opticians whom he claimed were infringing this patent. If the story of the dowry is true, Peter then gave the Ramsdens a half-share in the rights. However, the litigation, which seems not to have involved Ramsden at the time, throws doubt on this legend.

The Ramsdens moved into their own premises in Haymarket, still a flourishing market for hay (see Figure 1.3.). A circa 1774 plan locates his frontage of 21 feet on the east side, two doors south of the theatre,[23] where Ramsden set up his sign showing 'The Golden Spectacles'[24] (see Figure 1.4). He paid £54 rent as a tenant of the Hon. Edward Russell, who in turn held a lease from the Crown Estates.[25] Ramsden immediately took out a Sun Insurance Policy to insure £200 worth of household goods, £60 worth of wearing apparel and £540 worth of utensils and stock.[26] Four children (six, according to a Dollond history,[27] or seven according to the article in *Philosophical Magazine*[28]) were born in the Haymarket: Sarah, who lived only from 22 November to 1 December 1767, John, born 15 September 1768, who as we shall see entered the East India Company Navy in 1780; Jessica, born 31 March 1770 and Dolland [*sic*] born 27 May 1771. All were baptized at the church of St Martin-in-the-Fields.[29] No marriages or burials have been found for these last two, nor are they mentioned in later accounts of Sarah Dollond's family and it seems likely that they died young. Sarah was no stranger to the workshop: she and the apprentices worked at Ramsden's first dividing engine, which was sent to France in 1775.[30] On 9 November 1768 Ramsden was elected to the Society of Arts, Manufactures and Commerce. He had been proposed by a Dr Irwin, but the Society's record notes 'no subs' and it appears that he took no part in the Society's activities.[31]

Figure 1.3   The Haymarket, City of Westminster, in the eighteenth century. WA, Box 44 (31b)

Figure 1.4   A restored eighteenth-century shop front in Haymarket. WA, F138 (35).

During this period, it seems that Ramsden and the Dollond brothers cooperated on the construction of optical instruments. The astronomer Jean Bernoulli III (1744–1807), who visited London during the summer of 1769, was surprised by Peter Dollond's lack of theoretical knowledge and was somewhat critical of the quality of apparatus bearing the Dollond name:

> Dollond's workshop was quite large; from it came achromatic telescopes, excellent gregorian telescopes, and generally almost all the optical and astronomical instruments

used in England, but the most valuable of these latter have to be ordered. We should note, however, that brass instruments are made more or less by the dozen and that those outside England make a big mistake if they suppose that an astronomical instrument that bears the name Dollond must be excellent in every respect. Should an instrument of this quality be received, that is a sign that it had not been finished by one of Dollond's workmen; often, to maintain his reputation, he has the mountings, the divisions etc, made by his brother in law Ramsden who passes for one of the best craftsmen in London of this type.

This Ramsden also sells telescopes with a good reputation. Maybe brothers in law assist one another; whatever the situation, Ramsden's telescope mounting gave me much pleasure; it is firm, and better than that which Dollond generally gives his achromatic telescopes. It suffices to tighten or slacken a spring to give the instrument a fast or slow movement.[32]

For this criticism of Dollond and his remarks on both craftsmen, Bernoulli was taken to task by the Portuguese carrier of news John Hyacinth Magellan (1723–1790) (on whom, see Chapter 2). Writing in April 1772 Magellan informed Bernoulli that he was certainly mistaken about Dollond and Ramsden. Magellan was annoyed that Bernoulli had not got to know Ramsden personally, and that he had been so outspoken about Dollond's theoretical ignorance, which had also upset Dollond.[33] But Bernoulli was not alone in his judgment; the Danish astronomer Thomas Bugge (1740–1815), visiting London in 1777, noted that 'none of the Dollond brothers seem to have any theoretical knowledge'.[34]

Bernoulli's apology appeared in 1772, where he admitted that he had made several mistakes in his account of Dollond and Ramsden. He had not intended to utter any warning against them and he urged anyone going to England to say nothing without getting to know both men or having spoken to their friends, adding that the Dollond whose undoubted learning and skill he admired was Peter, and that John was the name of his younger brother.

Louis Dutens, in his personal tribute appended to the article in John Aikin's *General Biography*, gives us a good idea of Ramsden's good looks and the charm he exerted on his numerous personal and business acquaintances (see Plate 1).

In person he was above the middle size, slender but extremely well made, and to a late period of life possessed of great activity. His countenance was a faithful index of his mind, full of intelligence and sweetness. His forehead was open and high, with a very projecting and expressive brow. His eyes were dark hazel, sparkling with animation but without the least fierceness. His nose aquiline and very handsome. His mouth rather large, but in speaking it had an expression of cheerfulness and a smile the playful benevolence of which will not easily be forgotten by his friends. His tone of voice was singularly musical and attractive, and his whole manner had a character of frankness and good humour which he well knew to be irresistible. When he attempted a bow to persons of rank his air was bashful and awkward, but when at ease with his friends, his motions and attitudes were in an uncommon degree graceful.[35]

The Move to Piccadilly

Prosperity called for larger premises. Ramsden's last payment of rates on the Haymarket house was in 1771. He increased his Sun Insurance cover to £1,000.[36] From 1773 he paid rates on 199 Piccadilly, adjacent to St James's churchyard.[37] St James's parish had been carved out of the ancient and extensive parish of St Martin-in-the-Fields, and was destined to serve the estate then being developed by Henry Jermyn, Earl of St Albans. The church was built in 1676–84 to a design of Sir Christopher Wren, with its entrance from Jermyn Street, at that time a more important thoroughfare than Piccadilly, and one lit by Heming's patent oil lights since 1688. The churchyard was enlarged in 1693 and again in 1748.[38] East of the church and bounded by Haymarket and Jermyn Street, lay a market, which was swept away when Waterloo Place and Regent Street were formed between 1813 and 1820.[39] Ramsden's house had a frontage of 19.3 feet, extending back for some 73 feet with a small central yard behind which stood a four-story wooden structure[40] (see Figure 1.5).

Sarah parted company with Jesse before 1786, by which time she had returned to the Dollond house at No. 35 Haymarket, from where she assisted her brothers with both their practical work and their accounts.[41]

Notes

1   Watson (1973).
2   Hargreaves (1991).
3   HLS (St James, Halifax, Parish Register (hereafter PR), baptisms).
4   HLS (St James, Halifax, PR, marriages); TNA, MS RG 4 (Northowram non-conformist chapel, marriages).
5   HLS (St James, Halifax, PR, baptisms). This parish register is in a very bad condition and it was not possible to find the baptisms of Ramsden's siblings, nor the dates of his parents' burials.
6   Cox (1879), 84.
7   Lalande (1790), 5–15.
8   Anon (1803).
9   Craik (1881), 486–505.
10   Walker, *Halifax Guardian*, 10 January 1857; Anon, *Courier and Guardian*, 24 March 1934.
11   Porritt (1970), 15–27.
12   I am obliged to Dr Ruth Wallis for this suggestion.
13   TNA, MS IR/20, 162.
14   WA, Gardner Coll. 62/2E.
15   Dawes (1988).
16   NLS MS 2866.
17   Noorthouck (1773), 414–15.
18   WA, St Martin Rate Book (hereafter RB), MS F546, F548.
19   LGL, Christ's Hospital Registers, MS 12876/5.
20   Aikin (1813), 451.
21   Clifton (2004).
22   Patent 721, 19 April 1758. Such patents were granted for 14 years.

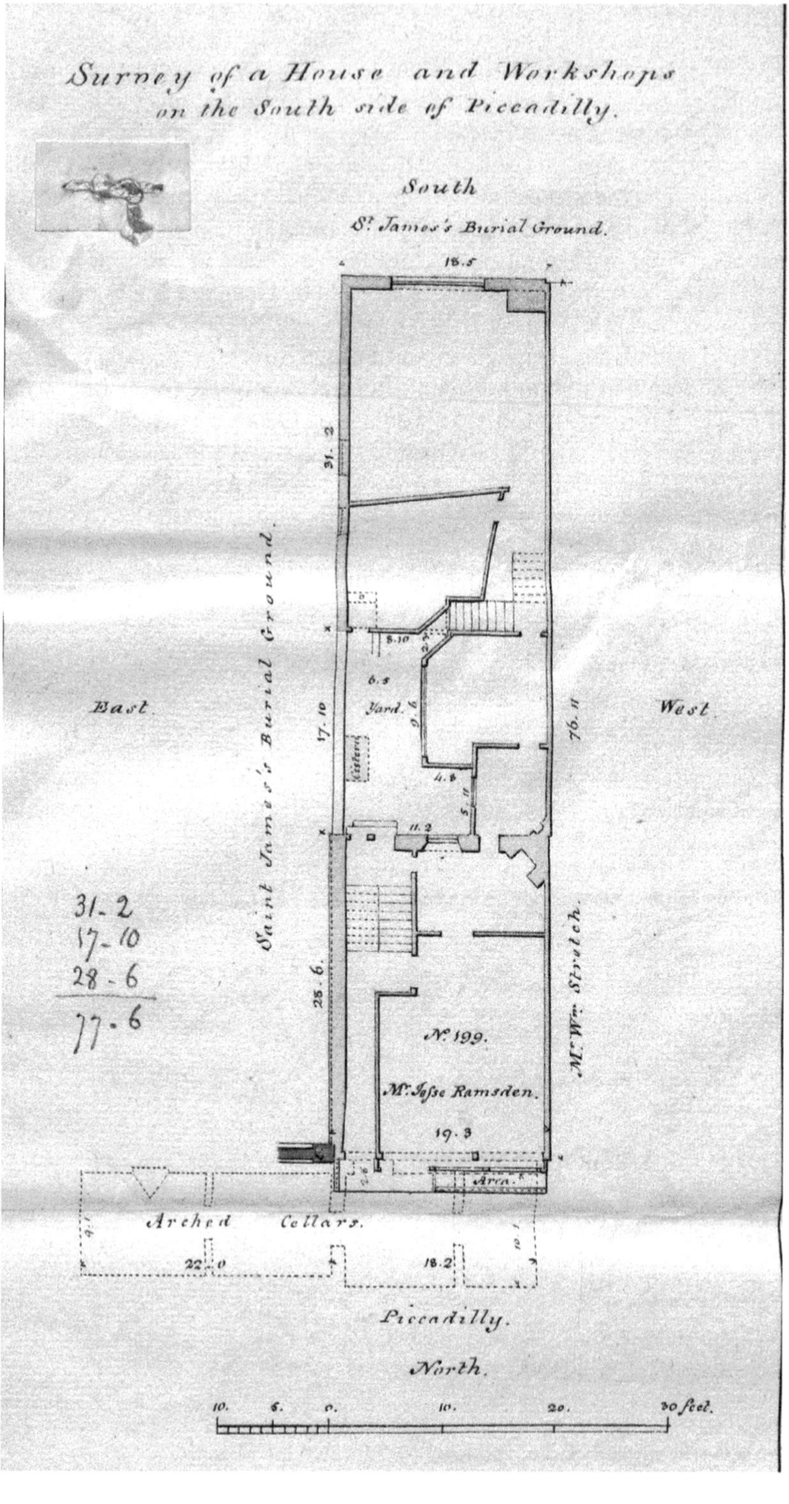

Figure 1.5   Ground plan of 199 Piccadilly, surveyed by Jonathan Marquand in 1799 for HM Land Revenue. TNA, CRES 39/139.

23 TNA, Plan of Haymarket, MS MR1452.
24 WA, St Martin, Exchange Ward, Poor Rate, MS F548.
25 TNA, Register of Crown Estates southward of Piccadilly, MS CRES 39/65; Westminster Archives, Str Martin RB, MSS F550, D6018, F2901.
26 LGL, MS 11936/170. Sun Insurance Policy 236541.
27 MHS, MS Blundell 10.
28 Anon (1803) 254.
29 WA, MSS St Martin PR baptisms and deaths.
30 BPU, Magellan to Mallet, 13 October 1775, 'cette machine dans laquelle travailloient les élèves et meme quelque fois la femme de Ramsden' MS Supp. 1654.
31 Royal Society of Arts, membership card index.
32 Bernoulli (1771), Lettre 5ème, 20 June 1769, 68–9.
33 Magellan to Bernoulli, 'Il est vrai que vous vous avez trompé sur quelques points dans vos Lettres Astr. relatifs à Dollond & à Ramsden. Je suis bien faché que vous n'aiez pas faite une intime connoissance avec ce dernier: & aussi que vous ayez touché si fortement sur [l']ignorance theoretique du premier. Lui (& moi aussi) a été bien faché, mais pour moi je connois que vous ne l'avez pas fait à dessein de le degrader. En verité il faut etre très delicat, en tout ce qu'on publie sur les qualités d'un autre; & plutot ne dire pas ce qu'on sait de disadvantageux, même lorsqu'on en pourroit donner toutes les preuves possibles'. Unpublished letter. I am obliged to Rod Home for a transcript of this letter.
34 Bugge (1997), 155.
35 Aikin (1813), 454b–455a.
36 WA, St Martin RB, MS F2901. LGL, 11936/210, Sun Insurance Policy 304380.
37 WA, St James RB, MS D116.
38 In 1793, when the burying-ground was full, a new cemetery and chapel was opened in Hampstead, north of London, which eventually became the separate parish of St James's Hampstead.
39 Anon (1938); Sheppard (1960), 31–5, 251–61.
40 TNA, MS CRES 39/139. On 15 May 1770 John Marquand, the Crown surveyor had reported on this site: 'Surveyed a piece or parcel of ground in Piccadilly in the parish of St James's Westminster belonging to Mr Pasmore abutting east on St James's Church Yard, Westminster on Mr Barnet, north on Piccadilly, south on Churchyard, on which stand a brick messuage, not in substantial condition, with a timber built workshop in yard, now occupied by Mr Thomas Thomas and let for fifty pounds per annum, may be valued at 36 pounds per annum, to make up the present term of fifty years.'
41 BPL, Sarah Ramsden to Matthew Boulton, Haymarket 7 September 1786. MS B&W Coll., Letterbook R1.

Chapter 2

Entering Trade

He was by nature endowed with uncommonly strong reasoning powers, and a most accurate and retentive memory ... with a quickness of penetration which enabled him ... to view in every light the subject on which he thought, and adopt at once the most advantageous mode of considering it.

J. Aikin, 'Ramsden', *General Biog.* v. 8, 454

This man also sells telescopes with a good reputation.

J. Bernoulli, *Lettres astronomiques* (1771), 69

This chapter continues the chronological story, leaving Ramsden's inventions to be examined in more detail in subsequent chapters.

During his journeyman years Ramsden learnt to shape brass and graduate all sorts of instruments and by handling old instruments brought in for repair he recognized design defects which could be corrected. Clearly, Ramsden was able to look at a quickly-drawn sketch and visualize it in three dimensions, a skill possessed by the most competent engineers and mechanics. Though well known for his great astronomical instruments and the development of the dividing engine, some of his minor innovations have not received the attention they deserve.

There was a considerable demand for small handheld achromatic telescopes, but while card and vellum tubes were light and inexpensive, they tended to bend at the joints between the tubes when fully extended. Ramsden replaced the card with laminated wooden tubes and with a master stroke of innovation he put a brass collar with a thread at each end of each tube so that when closed the telescope tubes would all screw into each other. But also when extended, the other ends of each tube would screw with reverse thread into the outer tube, making the extended telescope rigid and therefore optically more accurate. Actually this was not as difficult to make as it sounds because the reverse thread is the same as the screwing up thread, it is merely being approached from the other direction.[1]

A fine draughtsman himself, his 'JR' monogram identifies the illustrations in his published accounts of the sextant and the barometer. The science of metallurgy was still in its infancy, but once he moved into the range of heavy apparatus Ramsden had to learn about the properties and the sources of the various steels and alloys which he needed for tools and bearings. Here his early association with the forward-thinking industrialists among the members of the Lunar Society and especially his contacts in Birmingham – that hothouse of metal-working – would have kept him informed of new technology.[2] It is just possible that Ramsden met the future engineer James Watt (1736–1819) at this time; the two were of a similar age and Watt was in London in 1755–56 as a mature apprentice to John Morgan

(1712?–1758). Watt later knew Ramsden sufficiently well to have his letters directed to Ramsden when he was staying in London in 1767.[3]

In the second half of the eighteenth century, as British ships ventured into uncharted regions of the great oceans, there was a great push to improve navigation among officers of the Royal Navy, the mercantile ships of the Honourable East India Company (HEIC), and the other merchant adventurer companies trading to the Levant, Africa and the West Indies. The HEIC also needed a cadastral survey of those vast tracts of the Indian sub-continent now coming under its control. Elsewhere, military campaigns in Scotland – known then as North Britain – on the European mainland, and at sea, were becoming more reliant on telescopes, gunnery instruments and small cartographic aids. The theory and practical use of such devices was an important part of the curriculum at military and naval colleges.

Instruments from the Haymarket Workshop

Ramsden's first publication, his 12 page pamphlet, *Directions for using the new invented Electrical Machine, as made and sold by J. Ramsden*, is undated, but bears his address 'Near the Little Theatre, in St James's, Haymarket, and so precedes his move in 1773 to Piccadilly. The apparatus is further described in Chapter 7 and illustrated in Figure 7.2. The pamphlet concludes with the advertisement that Ramsden 'Makes and sells Refracting Telescopes of various Sorts, particularly those of a new Construction, for which Mr Dollond has *His Majesty's Royal Letters Patent*, in which the different refrangibility of the Rays of Light, also the Errors arising from the spherical Surfaces of the Glasses, are perfectly corrected'. A detailed list follows, ending 'and all other *Optical, Mathematical* and *Philosophical Instruments* according to the latest improvements. N.B. Spectacles of all the different kinds'.

Ramsden seldom expanded his signature beyond 'J. Ramsden London' but a sliding telescope with vellum tube, now in a private collection, bears a label with the Haymarket address, as does a box for a small gold-coin balance now in the Whipple Museum, Cambridge (see Figure 7.11). The Board of Longitude also records that in January 1768 an instrument devised by Samuel Dunn, a teacher of mathematics, for measuring large angles (but not further described), was to be made up by Ramsden under Dunn's supervision.[4] In November the Board resolved that the instrument, 'which is now constructing by Mr Ramsden' should be first tested ashore by the Astronomer Royal and then sent on any expedition going to Spitzbergen or the North Cape to observe the forthcoming Transit of Venus.[5]

Moving to Piccadilly – Equatorial Telescopes

One of the first instruments which Ramsden improved was the portable equatorial telescope, advertised in a small quarto pamphlet: *Description of a new Universal equatoreal, made by Mr J. Ramsden, with the method of adjusting it for observation.* (January 1773), simultaneously with a French equivalent: *Nouvelle instrument appelée Equatorial Universel,* The equatorial (the spelling was variable) mounting was not a new conception. Lalande reports that the first example which he had seen had been made by Philippe Vayringe (1684–1746), court clockmaker and mechanic to the Duke of Lorraine, at Lunéville, around 1735; he himself owned one of 7 or 8 inches diameter signed by Vayringe.[6] The instrument, often referred to as the portable observatory in recognition of its multiple uses, comprised an azimuth, or horizontal circle, representing the horizon of the locality and moving on a long vertical axis; an equatorial hour circle representing the equator, placed at right angles, and moving on a polar axis which represents the axis of the earth (see Figure 5.6). Thus the telescope is parallel to the Earth's polar axis, and once the star is brought in line with the telescope objective, it can be kept in view over a long period of time by means of a single motion. The instrument could also be used to measure zenith and altitude.[7]

The clockmaker Henry Hindley (1701–1771) of York had been working on an equatorially-mounted telescope when John Smeaton (1724–1792), instrument maker and later a famous civil engineer, visited him in 1741. It was finished in 1748 and taken to London where Smeaton tried to interest his mechanical and philosophical friends, among whom was James Short (1710–1768), maker of reflecting telescopes.[8] Hindley's instrument remained unsold, but the following year Short published a similar design for 'a portable observatory' explaining that his arrangement was nothing new, but was the first to hold a large telescope.[9] Many years elapsed before there was a renewed interest in such apparatus. In 1771 Edward Nairne (1726–1806) explained that his instrument was used in the same manner as that of Short but differed essentially in construction.[10] At about the same time, John and Peter Dollond's undated pamphlet *Description and uses of the new invented universal equatorial or portable observatory. With the divided object-glass micrometer,* was followed in 1779 by Peter Dollond's modification to the equatorial to correct errors caused by refraction.[11] Henry King states that 'Ramsden first turned his attention to the improvements of Short's equatorial mounting' – we may ask if someone brought such an instrument to his shop for repair – and that he 'provided [it] for the first time with circles and counterpoise of the correct size and weight'.[12]

According to the mathematician and astronomer Sir George Shuckburgh (1751–1804), for whom Ramsden made the monstrous instrument described in Chapter 6, Ramsden had made three or four small equatorials as early as 1770 or 1773; Sir Joseph Banks purchased one (see Figure 2.1) to take on James Cook's second voyage, departing in 1772, but Cook had refused to allow him to embark a wholly impractical quantity of goods and a brigade of servants. Banks went instead to Iceland, where a watercolour depicts the scientifically-minded physician James Lind (1736–1812) observing with this equatorial.[13] Other early models went to John Stuart (1713–1792), third Earl of Bute, one to Bute's brother, James Stuart Mackenzie (1719–1800) who, according to Piazzi, wrote the *Description of a new Universal Equatoreal* for the first edition of 1773[14] and one to Shuckburgh, which he took to France and Italy in 1774–75 and employed for astronomy and surveying.

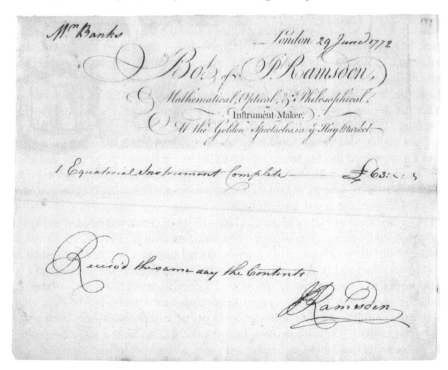

Figure 2.1   1 Ramsden's invoice to Sir Joseph Banks, 1772, for an equatorial telescope. State Library NSW, Banks CY3003-174.

The Dutch-American surveyor John William Gerard De Brahm (1717–1796) bought a Ramsden equatorial while he was in England in the 1770s. In dire need of money in later years, he was obliged to dispose of it and eventually, in 1792, Thomas Jefferson (1743–1826), Virginian patriot and later third president of the United States, bought it for the bargain price of $102.67.[15] Jefferson said of his instrument that it 'appears to be the only equatorial in the country'.[16]

Shortly after his move to Piccadilly, Ramsden took out a patent for his 'Astronomical Aequatorial Instrument'.[17] The German physicist Georg Lichtenberg (1742–1799), visiting London, wrote in November 1774 to his compatriot, the instrument maker Georg Frederich Brander (1713–1783), 'Ramsden is highly esteemed here, he is still a young man and has many ideas. He has considerably improved the portable observatory, and had a book published on it, which is however very rare and now available only when buying the instrument. He makes it for over £100'.[18] Lichtenberg is referring here to the *Description of a new Universal equatoreal*; Ramsden revised the text with each modification and improvement to the instrument. It was reissued in 1774, 1779 and 1791. Between the first known edition of 1773 and the last of 1791, it grew in length from seven to 28 pages to take account of later modifications and improvements, and included worked examples.

Interlude – Introducing J.H. Magellan

João Jacinto de Magalhãens (1722–1790), is generally known outside Portugal as John (or Jean) Hyacinth Magellan.[19] With a wide circle of correspondents and acquaintances in Britain and across Europe, Magellan flits in and out of Ramsden's life, both helping and interfering in his business.

Magellan was born in Aveiro, northern Portugal, was educated by the Augustinians at Coimbra, and entered their order in 1743. During his ten years as a priest, he absorbed a wide range of current science from the excellent convent library. By 1754 he had become dissatisfied with the cloistered life and he successfully petitioned Pope Benedict XIV for a letter of secularization, intending to travel abroad in order to study scientific and industrial progress. He was in Lisbon during the great earthquake of 1755, but then departed on a tour of Europe before settling in Paris, where the Portuguese physician António Nunes Ribeiro Sanches introduced him to the scientific community. Sanches later supported Magellan for many years, and this, together with an income from his family property in Aveiro, gave Magellan his financial independence. Although occasionally outspoken in political and religious matters, Magellan succeeded in keeping out of trouble, and opinions differed as to whether he privately retained his Roman Catholic faith. Otherwise his correspondence shows that he enjoyed his position at the centre of an extensive network of active men of science, to whom he wrote in Latin, French or somewhat shaky English.

In 1763 Magellan moved permanently to London, from where he kept the French minister of finances, Jean-Charles-Philibert Trudaigne de Montigny (1733–1777), informed of industrial progress in Britain and endeavoured, sometimes successfully, to smuggle machinery to France at a time when this was prohibited by Act of Parliament. The French lawyer Jacques-Pierre Brissot met him in London and drew a succinct pen-portrait, noting that Magellan was familiar with all the machines produced in England and made it his business to spread them across Europe. In this fashion, wrote Brissot, by corresponding with all the men of learning, he achieved his wish for an honorable independence.[20]

A member of the Society of Arts, Manufactures and Commerce, and from 1774 a Fellow of the Royal Society, Magellan introduced many foreign visitors into both societies. Sometimes referred to by modern-day historians as 'the Portuguese agent', Magellan was always willing to receive commissions from foreign individuals and establishments to procure clocks and scientific instruments from British craftsmen. He supplied entire collections of physics apparatus to colleges in Portugal and to the Italian physicist Alessandro Volta (1745–1827), then teaching at Pavia, and he sent nautical instruments to Cadiz for the major Spanish circumnavigation of 1789–94. Magellan also claims, in his own books and articles, to have assisted Ramsden with improvements to his sextants, his equatorial telescope, and his barometer. It is uncertain how far these claims are justified, but as later chapters indicate, Magellan was responsible for getting Ramsden's first dividing engine over to Paris, and he was certainly in and out of Ramsden's shop, delivering instruments belonging to out-of-town owners for repair and making other purchases on their behalf.

Equatorial Telescopes Continued

The peripatetic Swiss scientist Marc-Auguste Pictet (1752–1825), wrote from Paris to his compatriot J.A. Mallet on 10 May 1775, '... yesterday I saw Magellan ... who showed me a portable apparatus very nice, made by Ramsden and augmented by Magellan; truly a portable observatory. It cost 12 guineas'.[21] From London a few months later he informed Mallet 'Aubert introduced me to Ramsden and I shall dine with him on Sunday to see his shop and his instruments, which may cost me a few guineas'. On 21 November he reported:

> The equatorial of which you speak is not unknown to me – I saw and admired one at Ramsden's where I lived, so to speak, for several days – this instrument is astonishing in its precision – Ramsden claims to be able to read it to 2 seconds of arc. Ramsden has made me a little 3 inch pocket sextant, it is the third of its kind in Europe and Ramsden claims that it is the best of the three ... Ramsden gave me as a gift an artificial horizon.[22]

The Nairne, Dollond and Ramsden equatorials were broadly similar in construction, Ramsden alone of the three deciding to venture the expense of obtaining a patent. His instrument was described in Vince's *Treatise on Practical Astronomy* (1790) p. 152. Shuckburgh preferred Ramsden's instrument to the Dollond and Nairne versions, for its swinging level, the superb accuracy of its divisions, and its great portability. Ramsden had removed the endless screw which, bearing on the centre, destroyed its precision, placed the centre of gravity at the centre of the base, and constructed it so that all parts could be adjusted, and its movements operated in all directions. According to Piazzi, Ramsden added a small device to measure and correct for refraction before Dollond's publication.[23]

The Ramsden design was both steadier and more compact than its rivals, and its enormous popularity at home and abroad is attested by the reissue, in English and in French, of the pamphlet which described in the most careful detail the arrangements for setting it up, and its various uses. Set on its three feet, with all parts horizontal, the equatorial stood 29 inches high, the circles five and one-tenth inches in diameter. Its achromatic telescope had a triple object glass slightly exceeding 2 inches, its focal length being 17 inches. Its six eye-tubes gave a range of magnification from 44 to 168. Ramsden explained:

> By a new contrivance in this equatoreal, the telescope may be brought parallel to the polar axis, so as to point on the pole star in whatever part of its apparent diurnal revolution it be, and with this telescope it has been seen near 12 o'clock at noon, and the sun shining very bright.

Its two principal uses were, first, to find the meridian by one observation only, and second, to be able to point the telescope on a star, though not on the meridian, in broad daylight. He continues:

> After having stated these two examples of the uses *peculiar to this instrument* it is unnecessary to add, that it is applicable to all the purposes to which the principal astronomical instruments (viz, a transit, a quadrant, and an equal altitude instrument) are applied.

In his textbook *Principles of Natural Philosophy* (1784) George Atwood (1746–1807) included a section on the method of adjusting equatorial instruments, which explained how to correct errors in the position of the meridian and in the time as deduced from observation, adding 'This is one among many other improvements which have been applied to the construction of the equatorial instrument by Mr Ramsden.[24] We shall meet George Atwood again in Chapters 7 and 13.

The instrument could be made in different sizes. Rizzi Zannoni's equatorial (see Chapter 5 and Figure 5.6) had circles of 10½ inches diameter and a telescope of 14½ inches. An equatorial with 5-foot circles, to be driven by clockwork, was ordered by Ussher in 1774 for Dunsink Observatory but never constructed (see Chapter 6), and the 'Great Equatorial' with circles four feet in diameter, which Ramsden commenced in 1781 at the behest of Sir George Shuckburgh, was not entirely satisfactory (see Chapter 6). Ramsden, like Edward Troughton in later years, had been persuaded by a wealthy man to attempt a construction at a scale which his mechanical skill ought to have led him to decline.

Jean-Etienne Montucla (1725–1799), historian of mathematics, reports that King George III had an equatorial of 2½ feet diameter and this may have been the instrument that gave rise to the much-repeated story of Ramsden's delivery at the palace on the exact day promised but one year late.[25] The early model made for Lord Bute was probably not that sold after his death as 'Lot 200, A very complete and exquisitely finished new universal equatorial and apparatus by Ramsden containing rackwork, motions, adjustments ... in mahogany case.[26] This was bought by W. & S. Jones for £55, 12s, probably on behalf of Hans Moritz von Brühl (1736–1809), the diplomat and passionate amateur astronomer (see Plate 4). On learning of this acquisition, his friend Duke Ernst of Saxe-Gotha (1745–1804) sent his congratulations:

> My compliments, dear friend, on your acquisition of a Ramsden equatorial at the sale of Lord Bute's effects; if it is the equal of that described by his brother Mr Mackenzie it must be perfect and will render you excellent service.[27]

We may assume that the equatorial gave full satisfaction, for in 1794 Brühl penned a tribute to Ramsden in a treatise on astronomical observations made with that instrument. The original draft survives among his daughter's papers, she having annotated it down the left-hand margin: 'This letter from my father Count de Brühl to Mr Ramsden the Optician shews the very high estimation in which he was with Persons of Rank and Science for his great improvement in Telescopes & other Scientific Instruments – Harriet Polwarth 1841.'

Brühl's manuscript reads:

> To Mr Ramsden
> Sir
>   In prefixing your name to this small tract, I, in conformity with the universally established practice of authors, dedicate my insignificant publication to a great man for my private benefit.
>   I openly avow that my design in putting it under your protection is the accomplishment of that wish which I express at its end which was the only motive

that impelled me to publish it & which the approbation you Sir have bestowed upon the method of investigation therein explained cannot fail to promote.

I am, Sir, your constant admirer

June 14th 1794   C de B[28]

Duke Ernst's chance to acquire his own Ramsden equatorial came shortly afterwards. Franz Xaver von Zach (1754–1832), referring to 'Ramsden's superb equatorials', mentions that his patron Duke Ernst of Saxe-Gotha had bought the example belonging to the Parisian politician and collector Bochart de Saron and confiscated at the time of his execution in 1794.[29] Ramsden did not number these instruments and most of those now in museums are undated.

Sextants

Germane to the development of the sextant was the passage of the Longitude Act. Passed in 1714, the terms of the Act offered the huge sum of up to £20,000 to reward anyone who could discover a practical and reliable method of finding the longitude at sea, with smaller sums promised to inventions of lesser importance to navigation.[30] A Board of Longitude was appointed to assess the claims of the instruments and methods submitted.[31] Its ex officio members, the Astronomer Royal, the professors of astronomy at Oxford and Cambridge, delegates from Trinity House and the Royal Society, were joined by other worthies. Joseph Banks joined the Board in 1772 and became the second most important figure after Nevil Maskelyne, the Astronomer Royal. The Board ceased to function in 1828.

When Ramsden was starting in business, two methods of finding longitude were under active consideration. The chronometer method called for a clock set to the time at a reference meridian such as Greenwich, which would keep that time while carried on board ship and which could then be compared with local time as measured by the sun's noon passage. The earth rotates around its axis (360° of longitude) in 24 hours, or 15° each hour. Thus a difference of 4 minutes in time is equivalent to 1° of longitude. It had proved difficult to construct a clock which would remain accurate on board a moving ship and in all temperatures but the clockmaker John Harrison (1693–1776) had already received small sums for work on his chronometer.[32]

The lunar distance method required an observation of the moon's position against the background of stars and then finding in an almanac the Greenwich time at which it was predicted to be in that position. This method was still uncertain; the moon's position could not be measured to the required accuracy and the predictive tables were also unsatisfactory, but in 1757 the German astronomer [Johann] Tobias Mayer (1723–1762) sent to the Board a set of lunar tables which he claimed could be used to predict the moon's place with sufficient accuracy for the lunar distance method. He also sent a wooden model of his portable repeating circle for measuring the angle between the moon and a particular star.[33] Captain John Campbell (1720–1790) made trials of Mayer's tables and an instrument according to his model over the next two years, but having found the circle too heavy and awkward, consulted with John Bird. The wooden octant devised by John

Hadley in 1730, its arc extending over 45°, could by double reflection take angles up to 90° (being therefore often referred to as 'Hadley's quadrant').[34] The number of observing days in each lunation were limited to those when the moon was less than 90° from the observed star. Bird designed and constructed the first sextant, framed of flat brass plates.[35] By double reflection over its 60° arc, angles of up to 120° were possible, increasing to 15 or 16 the possible days each month (weather permitting) when observations could be made. These hand-held instruments are often referred to as 'Hadley' or 'reflecting' sextants, to distinguish them from the heavier non-reflecting 'astronomical' sextants on stands.

Aikin's biography tells us:

> Until this time, Hadley's sextant, though so much employed in the British navy, and so useful an instrument, was in a very defective state. The essential parts were not of sufficient strength; the centre was subject to too much friction; the index could be moved several minutes without any change being produced in the position of the mirror; the divisions in general were very coarse; and Mr Ramsden found that the Abbé de la Caille was right, when he estimated at five minutes the error which might take place in the observed distances of the moon and stars, and which might occasion in the longitude an error of fifty nautical leagues. Mr Ramsden therefore changed the construction in regard to the centre, and made these instruments so correct as never to give more than half a minute of uncertainty. His sextants of fifteen inches he warranted to be correct to within six seconds. From that size he made them to an inch and a half radius, and in the latter the minutes can be clearly distinguished; but he recommended for general use those of ten inches, as being more easily managed, and susceptible of the same exactness.[36]

In his treatise, *Description and Method of adjusting the Improved Hadley's Sextant, made and sold by J. Ramsden, Mathematical, Optical and Philosophical Instrument Maker, next St James's Church, in Piccadilly, London* (1775), Ramsden declares that many eminent astronomers and mechanics: 'in seeking to improve its construction and to correct the frequent observational inconsistencies, had in their ignorance of the true causes, complicated the instrument and made it subject to more errors, even during observation'. Ramsden rejected all adjustments save that for setting the index glass parallel to the horizon glass. He also remarks that some craftsmen, seeking to reduce the instrument's weight, made the frame so light that it flexed under its own weight. Ramsden's frames were of the strongest construction, with the least quantity of materials; his engraving (see Figure 2.2) shows a sextant frame of 'T' or edge-bar, section. This weight-saving design had been adopted by April 1770, when he received an order: 'for a Hadley's sextant in brass ... with edge bars £12, 12s, delivered to the Tower for use of the engineers going to Dominica'.[37] Another was taken on Cook's first voyage in the same year.[38] There is no comment on dividing. The pamphlet ends with 'Directions how to adjust the instrument'. Nothing was said about the invisible centre-work, but elsewhere we learn that sextant axes were formed of a polished steel cone, turning in bellmetal.[39]

Like his contemporaries, Ramsden employed a variety of frame patterns, depending on the size and presumably on the customer's instructions. Pearson

Figure 2.2 **Engraving of a sextant, 1775, from** *Description and Method of adjusting the Improved Hadley's Sextant, made and sold by J. Ramsden …*

notes that Ramsden's sextant frames were cast in a single piece, to which the edge-bars and limb were then screwed.[40] The lightweight T-bar was made in the basic 'A' and in a lattice form, elsewhere he used a flat plate construction. They ranged in size from the 'pocket sextant' of 3 inches radius to the usual navigation instruments of 12 to 15 inches radius. The 'bridge' pattern (see Figure 2.3) which aided stability but must have been difficult to construct and adjust, was followed by his successors Berge and Worthington. Ramsden also altered the mechanism

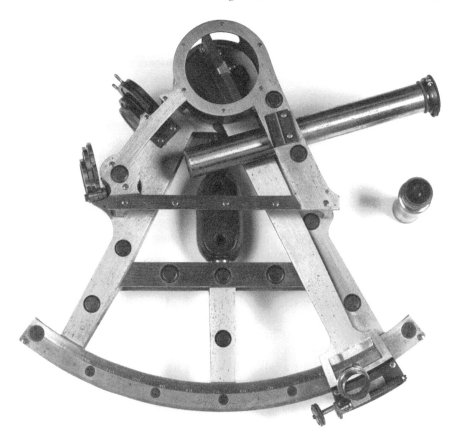

Figure 2.3 **Pillar and bridge sextant, radius 8 inches, numbered 1485, signed**
Berge London. The 'pillar' frame was patented by Edward Troughton
in 1788. National Museums of Scotland. Inv. T.1967.86.

to render the parts stronger and less liable to alter their positions. The drum
micrometer fitted to a Ramsden sextant in the National Maritime Museum is
now known to be a later addition.

Dividing Instrument Arcs

Wooden octants were widely available at a price seamen could afford but an
accurately divided brass sextant was a far more costly instrument, obtainable
only from the most skilled artisans. To graduate the arc of such an instrument,
it was centred on a firm circular table of 14 to 30 inches in diameter, its border
divided into degrees and quarters. An index of tempered steel, usually made
from a saw blade, was fitted with a brass plate which located the index on the
central arbor. Before cutting the divisions, circular lines were drawn with a beam
compass to limit the length of the strokes. The graduations were cut by a steel

knife with a beechwood handle, its rounded blade of the same thickness as the intended divisions. The blade, held at an angle of about 45° to the work piece, was drawn towards the workman, producing a furrow, displacing metal to either side. This burr was then scraped off and finally polished; if graduated on brass, with wet bluestone; if on inlaid bands of gold or silver, with willow charcoal and water.[41]

The large arcs for mural quadrants and other observatory apparatus were always hand-divided. The procedure devised by George Graham to divide a 90° arc into 96 equal parts allowed him to proceed by continuous bisection. His mural quadrants had two graduated arches – the inner one of 90°, the outer of 96 divisions. A conversion table converted the 96 graduation to that of normal degrees. Graham also turned the beam compass used for bisection into an accurate instrument used directly for engraving. Two slides ran on a mahogany beam, each slide with a steel point and fitted with a magnifying glass for greater accuracy in positioning the points on the brass plate. John Bird and Jeremiah Sisson continued and improved Graham's technique, and Bird's account of his procedure, *The Method of dividing Astronomical Instruments* was published by the Board of Longitude in 1767. The following year the Board published his treatise on *The Method of constructing Mural Quadrants* . He received awards totalling £650 for these publications, seen as benefiting the instruments trade; they were equally welcomed across the Channel. Bird and his successors were well aware that local heating of the arch being graduated resulted in uneven division. No heat-emitting illumination was allowed to come near the brass plate. The tools and the workpiece were kept at the same temperature and the work was restricted to certain times of the year – usually early mornings during spring and autumn – when the temperature of the workshop was fairly constant.[42]

Ramsden kept silent on his own technique for graduating arcs of large radius, confiding his method to a chosen few. He may have learnt the practicalities from Bird and Sisson; certainly he could have read Bird's text, although its is entirely possible that Bird omitted some small but crucial aspect of his process in order to retain his trade advantage. The author of the article 'Graduation' in the *Edinburgh Encyclopaedia* described Ramsden's method as uniting those of John Bird and the mechanically-minded French nobleman, the Duc de Chaulnes (1714–1769).[43]

During this period Ramsden was busy developing his dividing engine, the second and more perfect being rewarded by the Board of Longitude, and made public, as described in the next chapter. The labour and cost of this development suggests that he was not short of money. Day to day business was being handled by his workmen, among them Matthew Berge, Georg Dreschler and the apprentice Joseph Simpson. With his move to Piccadilly, this engine contributed to the increase in turnover needed to pay his increased workforce, purchase machine tools and materials, and pay the various rates and taxes on his premises (see Chapter 4).

Notes

1 I am obliged to Howard Dawes for this information.
2 The 'Lunar Society' – so-called because their meetings took place on evenings when a full moon would light their way home – consisted of an informal group of manufacturers, mainly based in the Midlands, including Wedgewood, Boulton, Watt and Keir, who discussed progress in the chemical, metallurgical and ceramics trades. Ramsden joined them on the occasions of their meeting at a London coffee house. Uglow (2002).
3 Hills (2002–05), 1. 58, 87.
4 RGO MS 14/2, 170.
5 RGO MS 14/5, 173.
6 Montucla (1799–1802), 4. 342.
7 This 'portable observatory' should not be confused with the portable observatories made from wood and canvas which served to house instruments taken on field surveys and expeditions.
8 Law (1971), 205–21.
9 Short (1749–50).
10 Nairne (1771).
11 Dollond (1779).
12 King (1955), 162.
13 The watercolour is BL MS Add.15511 f. 13.
14 Piazzi (1778), 246.
15 Bedini (2001), 509–11.
16 Bedini (2001), 651. This equatorial telescope remains at Monticello, Jefferson's home, now a museum.
17 UK Patent 1112 of 1775.
18 Joost and Shöne (1983), 483. Lichtenberg to Brander, 4 November 1774.
19 The best biographical account in English is by Rod Home in *OxDNB*; Professor Home is currently editing a collection of Magellan's correspondence for the press.
20 Brissot (1911), 1.363.
21 BPU MS Supp 1654, Pictet to Mallet, f. 60.
22 BPU MS Supp 1654, Pictet to Mallet, ff. 64, 66.
23 Piazzi (1788), 749.
24 Atwood (1784), 264.
25 Edgeworth (1969), 1.191–20. The story is amplified in Chapter 14.
26 Turner, G.L'E (1967).
27 'Je vous fait bien mes compliments, mon cher et digne ami, sur l'acquisition que vous aves fait pour votre observatoire d'un equatorial de Ramsden dans la vente des effets de mylord Bute, s'il est pareil de celui decrit par son frère M. M'Kenzie il doit être parfait, et vous rendre d'excellens services.' NAS, MS GD157/3379/30, Ernst to Brühl, 1 May 1793.
28 NAS, MS GD 157/2000. The dedication was printed on pp. 3–4 of the anonymous treatise *On the investigation of astronomical circles*, published in London in 1794.
29 Zach (1818) 452–3.
30 An Act for providing a Publick Reward for such Person or Persons as shall discover the Longitude at Sea, Anne 12 c.15.
31 Johnson (1989), 63–9.
32 Stimson (1996).

33   Forbes (1980), 191. (Mayer's reward from the Board of Longitude reached his widow after his death).
34   Stimson (1996); Clifton (2003), pp24–32.
35   NMM ref. NAV. 1177; Stimson (1985), 94–8.
36   Aikin (1813), 451a.
37   TNA, MS WO51/245, p. 282.
38   Beaglehole (1969), 721.
39   Magellan (1775), 13.
40   Pearson (1824–29), 2. 572–6.
41   Anon (1816), *EE*, vol. 10, art. 'Graduation'.
42   Anon (1816), *EE*, vol. 10, 366 (quoting Bird), 369, and elsewhere.
43   Anon (1816), *EE*, vol. 10, 455.

Chapter 3

The Dividing Engines

Cette machine, dont M. Ramsden s'est occupé pendant dix ans, a été regardé comme un trésor pour la marine angloise …

Preface to J.J. de Lalande, *Description d'une machine pour diviser …* (1790)

The fine Ramsden machine in France was the first model … Ramsden took care to spoil it before it was sold to de Saron.

J. de Mendoza y Rios, 15 February 1803, to Andréossy. CNAM, MS C.32

In this chapter I turn aside from a general chronological narrative to focus on the story of the design and construction of Ramsden's dividing engines, achievements for which he was best known to a wide readership in Britain and Europe. I am greatly indebted to three instrument historians: Randall Brooks[1] for the detailed information in his thesis, John Brooks,[2] both for his own published history and for his later personal help and guidance through the vocabulary and the technicalities of eighteenth-century mechanical engineering, and Michael Wright for helpful remarks on the subject. Remaining errors are my own.

With the increasing use of surveying and navigation instruments which measured degrees on graduated arcs or circles, the tedious business of setting-out and scribing the graduations was greatly eased by the invention of dividing engines. Ramsden mechanized the rotation of the workpiece whose scale was being graduated, through exact angular steps. The endless screw was fitted with a ratchet mechanism, worked by a treadle or lever, which caused the screw to make the same number of turns each time it was depressed. Until 1843, when William Simms built his self-acting dividing engine, the actual scribing continued to be done by hand, guided by a cutting-frame to ensure that the lines were truly radial.[3]

All circular dividing engines rely on a large horizontal wheel with teeth cut around the edge, and engaging with the 'endless screw' (a worm engaged with a wheel and so called because it never arrives at an end) that rotates the wheel through precisely controlled angles. These machines had evolved in the clockmakers' workshops, replacing by stages the original hand-filing of toothed wheels and pinions, so that during the seventeenth and early eighteenth centuries they grew from simple revolving platforms to incorporate a guide frame to direct the file, then to control the cut as required.[4] The cutting engine which the English emigrant clockmaker Henry Sully (1680–1728) brought from France in 1716 had a simple dividing plate with many concentric circles of different radii, each circle being divided with holes; an index dropping into these holes controlled the rotation of the plate and so of the wheel being cut. The Parisian clockmaker Fardoil constructed the first toothed wheel and endless screw, both being illustrated and described by Antoine Thiout (c.1694–1761) in his encyclopedic *Traité d'horologerie* ... (2 vols, 1741, repr. 1972).[5]

Henry Hindley of York

Thiout's treatise was among several works on clockmaking which helped to spread the knowledge and improvement of cutting engines. In England, Henry Hindley, clockmaker of York, had before that same date of 1741 made an engine with a toothed wheel which indented the edge of a circular workpiece, such that a screw with 15 threads acting at once would, by means of a micrometer, read off any number of divisions and thus serve to divide the circle, although it was still best suited to cutting toothed wheels for clockwork. Much of what we know of Hindley and his machine tools comes from an account written by his friend John Smeaton who, before he became famous as a civil engineer, had earned his living as a maker of philosophical instruments.[6] Hindley possessed a screw-cutting lathe with change wheels and a 'great lathe' capable of accurately turning a wooden block with a circumference of 8 feet (that is, about 30 inches in diameter). Besides his invention of a novel method of precise circular division, Hindley used the method in cutting the teeth of the wheel of an engine, but Ramsden followed a different procedure to achieve the same end.

Hindley demonstrated to Smeaton how he first cut a cylindrical worm on his lathe, matching its pitch to the teeth of the wheel, and from this intermediate form cut his 'hour-glass' worms (still known by some engineers as 'Hindley worms') on the engine.[7] The whole 15 threads of this worm could engage the circular plate. The engine, being 'for more nice and curious purposes, furnished with a wheel of about thirteen inches diameter, very stout and strong, and cut into 360 teeth, to which was applied an endless screw, adapted thereto', enabled Hindley to divide arcs for surveying and astronomical instruments. It seems that Hindley also contrived the cutter frame which was still in general use in 1809 when Edward Troughton (1753–1836) acknowledged him as its designer.[8]

Developments continued in France, most famously under the direction of Michel-Ferdinand d'Albert d'Ailly, duc de Chaulnes, who in his address to the Académie royale des sciences, published in 1765, mentioned the difficulties of perfecting the screw, on whose accuracy the operation depended.[9] In contrast to his notoriously unpleasant behaviour on the military field, de Chaulnes's serious interest in natural philosophy and astronomy led in 1743 to honorary membership of the Académie royale des sciences.[10] Between 1765 and 1768 de Chaulnes devised and had technicians build for him examples of both circular and linear dividing engines. Randall Brooks, on examining de Chaulnes's dividing engine of c.1762 at the Museo della scienza in Florence, concluded that 'the skill of de Chaulnes surpassed all previous screw-makers, including Ramsden'.[11] De Chaulnes was already familiar with the microscope as a tool and he extended his ideas to combine microscope and micrometer during the setting-out of his engines. The report on de Chaulnes's means of improving astronomical instruments[12] and that on a new method of dividing mathematical and astronomical instruments were both published in the *Mémoires de l'Académie des sciences* in 1768 and so could have been known to Ramsden, who made good use of these optical aids to accurate setting-out.

Hindley's contemporaries remark on his willingness to share what might have been his profitable trade secrets, which came to be known by the London craftsmen. The first link in this chain of transmission was from Hindley to John Stancliffe (w.1777–1812), according to Smeaton 'for some years a workman of Hindley's', adding that Ramsden learnt much from him in the construction of his second engine, including the design of cutting frame. Stancliffe was said to have finished building his own dividing engine in about 1788[13] and this date agrees with the remark by the Danish student of instrument making Jesper Bidstrup (1763–1802) that in 1789 there were three dividing engines at work in London, belonging to Ramsden, Stancliffe and Troughton.[14]

While no paper trail has yet been found to confirm the link between Hindley, working in distant York, and Ramsden, whose first engine caught the attention of the London virtuosi and craftsmen alike, James Allen and Edward Troughton both claimed that John Stancliffe was apprenticed to and worked for some years with Hindley, while Troughton, writing anonymously in the *Edinburgh Encyclopedia* says that he was foreman to Ramsden.[15] That would have been in Ramsden's Haymarket days, for in the early 1770s Stancliffe was living in Marylebone, close to John Bird with whom he was engaged on the large astronomical instruments for the Radcliffe Observatory in Oxford.[16]

Although Ramsden may have been inspired by hearing about Hindley's engine, Smeaton and Piazzi both say that Ramsden, who was laboriously graduating by hand small instruments of various kinds, had been considering a dividing engine since 1760.[17] Lalande adds that his friends, especially John Dollond, advised him not to embark on this project which was beyond his ability and would ruin his growing reputation. Ramsden, aware of the substantial reward which the Board of Longitude bestowed on Bird for his method of hand-dividing of large instruments, might reasonably have anticipated that the cost of the substantial weight of bellmetal, fine brass and steel, plus time spent perfecting the dividing engine, would be similarly rewarded.

The First Circular Dividing Engine

And so it was – in due course. After three years work, Smeaton saw Ramsden's first engine in the spring of the year 1768.[18] Its wheel of 30 inches diameter was built from sections of brass, rotating round a central arbor by treadle action. The table, the headstock carrying the endless screw and the frame carrying the tracelet were set on a wooden base plate. It sufficed to divide theodolites, sextants and other small items. It was easy to operate – the cutter was moved by hand and Ramsden's apprentices and occasionally his wife, worked on it.[19] But the labour of its construction and use immediately suggested ways to improve it, and Smeaton recounts that at their meeting, Ramsden showed him a pattern for casting a 45 inch diameter wheel, which he proposed to make, on the same general plan but with considerable improvements. The second model was ready by 1774, by which time Ramsden had moved to Piccadilly.

The first engine found a ready buyer in Paris, and it is worth asking if Ramsden preferred to send it to France, a country with whom Britain was then at war, rather than risk selling it in London where it might be seen by someone who could improve it before his next machine was completed. At this time there were comprehensive Acts of Parliament forbidding the export of textile machinery, but they were less clear on the export of metal-working machinery, which was more tightly regulated in 1785–86 by the Acts 25Geo3.c67 and 26Geo3.c89.[20] The buyer was Jean-Baptiste Gaspard Bochart de Saron (1730–1794), President of the pre-revolutionary Parlement and a passionate collector of scientific instruments; the person who eased the negotiations was Magellan, who reported his actions in a letter to Mallet in Geneva on 13 October 1775:

> I had the task of sending to France an excellent dividing engine, made by Ramsden, and for which I was in enough trouble here, for there were people who wanted it at any price, but the artist would not let them have it. You will find in Rozier's journal, February 1775, page 147, a summary of this engine, though the writer is mistaken in saying that the cutter was pedal-driven, whereas it was driven by hand. I have at the present time a circle divided on this engine ...[21]

Bochart de Saron made the engine available to French artisans,[22] among them Pierre Megnié, who in 1776 had constructed a straight-line dividing engine, and who was subsequently commissioned to build a large quadrant for the Paris Observatory – a project which came to naught as he fled the kingdom after having received the initial payment. Prior to his hasty departure, however, when applying in 1781 for the prize offered by the Académie royale des sciences, Megnié stated that he had constructed a circular engine:

> which enabled me to divide my sector into 120 parts drawn on the bevel, so that each of these lines might correspond to those set out at each 10° degrees on the edge of the quadrant's limb. I then adjusted one parallel to the scriber affixed to the sector. This scriber resembles that which Ramsden put on the engine which was seen at de Saron's house. My confidence in Ramsden made me adopt it without the exhaustive examination which I have now made but more than 1200 lines which are on the arch of the circle have brought me close to completing the instrument.[23]

Bochart de Saron was one of those unfortunate aristocrats executed during the French Revolution. His collection was dispersed, the dividing engine finding its way into government hands, where it was made available to French artisans. Etienne Lenoir (1744–1832), the leading Parisian instrument maker, modified it to graduate according to the new metric scale, with 400 divisions to the circle, and he was ordered to instruct young craftsmen in its use.[24] Much altered, it now resides in the Conservatoire des arts et métiers in Paris.

The Second Circular Dividing Engine

Ramsden's second engine was more stable, having a tough bellmetal cast wheel plate 45 inches in diameter with a brass limb on which the teeth were cut. It was completed by June 1774, when he submitted a memorial to the Board of Longitude:

> representing that he had invented an instrument for dividing sextants etc., the construction of which is such as leaves no dependence on the workman, as a boy can use it with the same exactness as the most experienced hand; and in the most speedy manner. And expressing the wish that the said instrument which has cost him £250 and another which he is constructing to divide streight [*sic*] lines may be the property of the Board on their paying him for them and giving him such a priority to him as, in their opinion, he may deserve. And the said Mr Ramsden, having been called in and discoursed with upon the subject, the professors who were present were desired to go to his house (taking with them Mr Bird if he was inclinable to go) in order to see a quadrant divided by the above instrument, and to report their opinion of it to the Board.[25]

Bird examined a sextant divided on the engine and reported that the graduations over the whole 120° were accurate to one two-thousandth part of an inch – a remarkable accuracy in comparison with previous methods of scale division. In December the Board instructed Ramsden to bring to its next meeting an account of the principles on which his engine was constructed.[26] In Bird's time, a cut of one four-thousandth part of an inch was considered to be the smallest visible graduation, this being less than one second of arc on a radius of four feet. Fine strokes cut on brass appeared rough and uneven when examined with a lens, though this roughness could be avoided by cutting the strokes on an inlaid band of silver or gold.

In March 1775 Ramsden jogged the Board's collective memory by submitting a memorial 'mentioning the names of other persons who had attempted but in vain to bring such ingines to perfection' and 'praying for some gratuity' and he produced a certificate from Captain Constantine Phipps (1744–1792) and Lieutenant-Colonel William Roy (1726–1790), who having examined a quadrant divided on the engine, had found that the greatest error did not amount to half a minute and most were under one quarter of a minute.[27]

The Board now sought the opinions of the Astronomer Royal, Nevil Maskelyne (1732–1811) and the Cambridge astronomer Anthony Shepherd (1721–1796), who likewise reported favourably, and the Board awarded Ramsden the sum of £300 for his invention, £19,000, in present-day values.[28] The Board also purchased the engine for £315, permitting it to remain with Ramsden, subject to certain conditions. He was to write a full and complete explanation of the engine, with illustrations; he must agree to instruct 'mathematical instrument makers, not exceeding ten, as shall be appointed by the Commissioners' in its construction; and he was to divide all sextants and octants brought to him by other craftsmen, at the rate of 3 shillings for each octant and 6 shillings for each brass sextant, with nonius [*recte* vernier] divisions to a half-minute.[29] His text, with a preface by Maskelyne and engravings of the engine taken from Ramsden's own drawings, was published as *Description of an engine for dividing mathematical instruments* by the Board of Longitude in 1777 (see Figure 3.1). No named craftsmen are known to have been instructed by Ramsden so that they might build further examples and Ramsden may have avoided this obligation. (See Dutens's letter, p. 47.)

Figure 3.1   Engraving of a circular dividing engine, 1777, from *Description of an Engine for Dividing Mathematical Instruments by Mr J. Ramsden* ...

Lalande adds to his reprint of Piazzi's letter prefacing his translation of Ramsden's *Description of an engine* (1777) the remark that while Ramsden was developing his second engine young Dollond spent three years trying to make a similar one, but despite having seen Ramsden's engine and having recruited some of Ramsden's best workmen, he abandoned his efforts and it was not until Ramsden's description was published, and with the assistance that Ramsden had given to other workmen, that more engines were built. The Dollond link was mentioned by Edward Troughton, writing in the *Edinburgh Encyclopaedia* long after Ramsden's death:

> Previous to the account of Ramsden's being published, and before its construction was generally known, Messrs Dollond made an engine, differing materially we believe from the former; but as it was never used except in the graduation of instruments made by them, the only judgement we can form of its quality, arises out of the respectability of that well-known house.[30]

And speaking of his brother's engine:

> In the year 1778, the late Mr John Troughton completed a graduating engine, which at the full strength of his pecuniary means had occupied him for three years. In its general construction, this differs in no material respect from Ramsden's, though it is generally, we believe, thought to be superior in point of accuracy. The trade were so ill satisfied with Ramsden, on account of his keeping their work for an unreasonable length of time, as well as for the careless manner in which it was often divided by his assistants, that Troughton immediately, at augmented prices, found full employment for his; and he has been heard to say, that by the care and industry of himself and his younger brother, he soon found himself as well remunerated for making his engine, as Mr Ramsden had been by public rewards.[31]

The instrument historian John Brooks, casting an experienced engineer's eye over Ramsden's text and drawings, intersperses his comments with extracts from Smeaton:

> ... the principle of Hindley's method was to use the change-wheels of his lathe to match the pitch of the [interim] cylindrical screw to that of the teeth of the wheel. He was then able to use the wheel itself to support the tool which cut the concave screw. Finally, this screw was [as Smeaton recounts] so ground with the wheel that the screw interlocked with the teeth of the wheel ... without ... the least sensation of inequality. Ramsden adopted an even more refined technique, cutting his screw almost to match the dimensions of the wheel, but then turning down the edge of the wheel by a small amount to get an exact match. However, his great innovation was to use the screw itself to cut the teeth of the wheel thereby obtaining the precise match in both pitch and shape necessary to achieve and retain high accuracy.[32]

In the same treatise, Ramsden described and illustrated his screw-cutting lathe. John Brooks discounts Allan Chapman's claim that Ramsden 'strategically avoided describing the critical process whereby he laid off the 2160 circumference teeth on the dividing wheel, the accuracy of which made the machine so significant'[33] but comments that Ramsden seems 'to have been somewhat disingenuous' in his straightforward description of the ratching process; certainly Edward Troughton, in the piece inserted in the *Edinburgh Encyclopaedia* article, p. 355, admits that he and his brother had problems with the notched screw, which 'cut sharper at one edge than with the other', giving an uneven division.

In adopting Hindley's design for his cutter frame, Ramsden found that the assembly lacked stability, being inadequately supported near the centre, allowing the cutter assembly to twist. He corrected this defect prior to 1788 when Lalande illustrated it in his French translation, mentioned below. The additional piece was a tubular bridge, spanning the wheel at right angles to the cutter frame bars, to which it is attached. Its ends are supported by small rollers, running on the surface of the wheel. This bridge also appears in the undated portrait of Ramsden by Robert Home, now at the Royal Society (see Plate 1), and in the engraving after that painting, dated 1790.

Among the bequests in Ramsden's will was 'to ... Edward Pritchard my small Dividing Engine together with the double Barr'd hand lathe now in the garratt'. This small engine, with a wheel slightly over 35 inches in diameter is now in the Science Museum, having come to the Museum from the firm of W.T. Parsons in 1930.[34] According to Parsons's information, it had been left to 'an apprentice', Ned Pritchard; was sold by one of Pritchard's sons to Dollond, and bought from Dollond by Parsons. Pritchard was named an executor in the will. He does not figure among the known apprentices, but Clifton lists 'E. Pritchard & Son, trading 1816–22, Divider of mathematical instruments, engraver, at 8 Porter Street, Soho', which fits the bill.[35] John Brooks describes this engine as based on Ramsden's principles but with significant variation from his second engine. Its cast frame has a downward extension to provide a lower pivot for the endless screw support, in place of the screw support being mounted on a brass pillar pivoted close to the floor, and there are marked differences in the design and operation of the cutter. Some or all of these modifications may have been made long after it left Ramsden's workshop.

After the introduction of Ramsden's second dividing engine in 1774 hand division of octants and sextants quickly gave way to engine division. To show this superior quality, a proprietor's mark was stamped on the centre of the scale, identifying the engine which had graduated the instrument. Ramsden's mark was a foul anchor flanked by his initials, and it is found on instruments from other makers.[36] Ramsden made some 1,450 sextants between 1760 and 1800; averaging 43 each year between 1780 and 1800. His contemporary, John Troughton, after building himself a dividing engine, made similar numbers annually between 1790 and 1800. Ramsden did not stamp numbers on his sextants until 1780 though he kept a tally; the lowest number so far known being 724.[37] When the engine passed to Berge and later to Nathaniel Worthington (1790?–1853) the numbering sequence together with the foul anchor with Ramsden's initials continued, since the engine remained the property of the Board of Longitude.[38] For its ultimate history see Chapter 13.

Shortly after Ramsden's death a letter was sent to Joseph Banks asking that the engine should not be left with a 'shopkeeper' on the grounds that:

There are several dividing engines now in London, in the possession of different shopkeepers. But they will not divide any work but their own, and if any mathematical instrument maker, does prevail with them sometimes to divide an instrument they must wait a long time before it is done, and pay a very extravagant price for it.

The writer made the further interesting point that should any instrument maker send a new instrument to be divided he risked the proprietor of the dividing engine holding onto it while he made a copy which he then passed off as his own invention.[39]

Ramsden's former apprentice Thomas Jones (1775–1852) also sought to acquire the engine. Jones was by then working for the trade and some private customers, and Louis Dutens appealed to Joseph Banks on his behalf, stating that:

[Ramsden] employed him more particularly on the dividing instrument belonging to the Board of Longitude and which had been deposited with Mr Ramsden on condition of dividing for the trade, when required, which was never done for reasons which Mr Jones will explain in a few words, if you will allow him the honour to appear before you …[40]

This statement contradicts the known fact that instruments by other makers were indeed divided on Ramsden's Board of Longitude engine. I suggest that Dutens misunderstood Jones's remarks, and that it was the condition of instructing other craftsmen that was not complied with. Jones was unsuccessful in his appeal; the Board chose to leave both engines with Matthew Berge, and it was only on Berge's death that Jones was granted custody of the straight line dividing engine, as recounted in Chapter 13.

Ramsden was contemplating the construction of a circular dividing engine to take even larger workpieces when he wrote to the mechanical engineer Matthew Boulton (1728–1809) of Soho, near Birmingham, on 6 September 1786:

Finding myself under some difficulty in the making a dividing engine, I take this liberty to request your advice on the subject and, sensible of your good disposition towards forwarding whatever may tend to the advancement of manufacture, will forbear troubling you with any apology.

The prodigious exactness with which we divide mathem[atical] instruments since I constructed the engine for that purpose, makes me wish to make one of a much larger radius than those at present, but this means a very great difficulty in getting that part or edge of the circle, in which the teeth that the screw works in are made, to be perfectly sound. I fear also that the inertia from the weight of a circle of that size, when put in motion, may be rather unfavorable; from a little experiment just made I am inclined to think that very good cast iron would bear cutting with a screw, if so, the circle might be cast much lighter with the same strength, than could be done with brass, it would also expand much less; pray have the goodness to favor me with your opinion, and if you think it will succeed permit me to know where you think it could be cast in the best manner. I do not regard the expence. I could wish to have it 5 or 6 feet in diameter.[41]

The outcome of this request is unknown.

Writing to his patron on 6 March 1789, Jesper Bidstrup, the Danish trainee instrument maker, remarked (in translation):

> The division of instruments here is no longer freehand, unless their radius is 2 feet or more; everything is done on machines of which there are three, those of Ramsden, Stancliffe and Troughton, the last being here considered the best ... The owners of these machines will not permit anyone to see these machines, fearing lest others should get similar machines, by which they might lose their share of the advantage they have by dividing.[42]

The straight line dividing engine

Having completed his second circular dividing engine, Ramsden built an engine for dividing straight lines. His memorial to the Board of Longitude in June 1777 stated that:

> he hath with great labour and expense just finished an engine for dividing straight lines into any number of equal parts which with his former engine makes the article of dividing of mathematical instruments complete and that he believes it will, upon trial, be found that the errors occasioned by the drawing of the lines etc., and also the errors of the instrument, will never exceed the half of a thousandth part of an inch, and at the same time enclosing a certificate from Colonel Roy and Mr Calderwood, that they had examined a scale divided by the said engine and had not found the error in any part to exceed the above quantity.[43]

This machine also made use of an extremely delicate screw, but unlike the circular engine, where only a small portion of the screw engaged with the base plate, here the entire length of the screw engaged with the teeth of the base plate. The screw had a pitch of one-twentieth of an inch, and to cut it, Ramsden constructed a variant on his previous screw-cutting lathe, but where the lead screw engaged with a circular toothed plate which controlled the movement of the tool. The machine itself consisted of an elongated rectangular brass plate on which was secured the workpiece to be divided. The plate moved on two runners, and was guided by a slide bar on which rested two springs to ensure contact with the endless screw. Like the circular engine, it was driven by a treadle mechanism. The scriber was carried by a metal carriage which rocked between two points, and could receive a lateral movement to mark the divisions. Thanks to the vernier, this machine could operate displacements of the order of one-thousandth of an inch. On this engine, Ramsden could divide scales to foreign measure simply by inclining the scale to the screw by the appropriate amount.[44]

This device was also purchased by the Board of Longitude, which in 1779 published Ramsden's *Description of an engine for dividing straight lines on mathematical instruments.*[45] It was soon copied by other craftsmen in England and France. One such was François-Antoine Jecker, (see Chapter 4) who had spent five years in Ramsden's workshop before returning to Paris by 1794 where he established an instrument making dynasty. Jecker claimed that his machine was 'almost' that which he saw at Ramsden's, with various costly parts omitted, which as a young mechanic he could not afford. Regarding the screw, so crucial to the success of such apparatus, his was a simple direct construction, such as he had seen at Ramsden's workshop, having nothing in common with the screw-cutting machine published in the collection of Ramsden machines in 1790.[46] He insisted that the screw-cutting machine he saw at Ramsden's was not as figured in the published work.[47]

When he came to England in 1777, the Danish astronomer Thomas Bugge (1740–1815) was able to purchase a copy of Ramsden's treatise on the circular engine,[48] just before the stored copies were lost in a warehouse fire. This destruction obliged Lalande to borrow Professor Shepherd's copy in order to produce his French translation, with his own additions and comments. He remarks that the central steel arbor, with a truly vertical hole, centred in the bellmetal plate, was one of the most difficult parts to make and its most important feature. Together with the description of the straight line dividing engine which had been translated by the amateur astronomer abbé Jean-François Blachier (1737–1794) of the Royal Society of Nancy, and other related notices, Lalande's booklet was published at Paris in 1790 as *Description d'une machine pour diviser les instruments de mathématiques, par M. Ramsden, de la Société Royale de Londres; Publiée à Londres, en 1787* [*recte* 1777] *par ordre du Bureau des Longitudes; Traduite de l'anglois; Augmentée de la description d'une machine à diviser les lignes droites, et de la notice de divers ouvrages de M. Ramsden. Par M. de la Lande, de l'Académie royale des sciences, de la Société Royale de Londres, etc. Pour faire suite à la description des moyens employés pour mesurer la base de Hounslow-Heath.*

Notes

1 R.C Brooks (1989a). This thesis includes a section on the development of dividing engines.
2 J. Brooks (1992).
3 Simms (1839–1843).
4 Rees (1802–1820). *Encyclopedia*, 2. art. 'Cutting engines'.
5 Opinions differ as to whether Sully's engine had been made by Antoine Fardoil (Daumas (1972), 117–18), or by De La Fandrière (Watkins (1980), 731).
6 Smeaton (1776).
7 J. Brooks (1992); Setchell (1972), 43.
8 Troughton (1809), 128.
9 'Une plate-forme à diviser les instrumens [*sic*] de mathématique proposée par M. le Duc de Chaulnes', p. 140 in 'Machines ou inventions approuvées par l'Académie en 1765'. *Histoire de l'Académie royale des sciences*, A. 1765 [ie, for that year], [published in] 1768, pp. 133–43.
10 Provost, *DBF* art. 'Chaulnes'.
11 Brooks, R.C. (1989b), 8. But how much credit for this skill should be accorded to Chaulnes' technician?
12 'Sur quelques moyens de perfectionner les instrumens [*sic*] d'astronomie', *Histoire de l'Académie royale des sciences,* A. 1765 [ie, for that year], [published in] 1768, pp 65–75, and [same title] *Mémoires de l'Académie royale des sciences*, A. 1765, 1768, pp. 411–27.

13 *Edinburgh Encyclopedia* (1816) vol. 10, art. 'Graduation', 353.

14 CRL, NKS 287, ii. Bidstrup to his patrons 6 March 1789. Transcribed and translated by Claus Thykier, adapted by AMC. Stancliffe is an indistinct figure, his vital dates remain unknown. His nephew, who he trained, emigrated to the United States.

15 Anon (1816), 353.

16 WA, Marylebone RB, Stancliffe at Little Marybone Street from 1773.

17 Smeaton (1776), 17.

18 Smeaton (1776), 18.

19 BPU, MS Supp. 1654 f.42^v. Magellan to Mallet, 13 October 1775.

20 Jeremy (1977).

21 BPU, MS Supp. 1654 f.42^v. Magellan to Mallet, 13 October 1775. 'J'eu l'adresse de faire passer en France une excellente machine à diviser, faite par Ramsden, et pour laquelle on m'a blamé ici assez forte, car quelques uns la vouloient à tout prix, mais l'artiste n'a pas voulu le leur ceder. Vous trouverez dans le journal de Rozier à page 147 du mois de février 1775 un precis de cette machine, quoique l'auteur l'a trompé disant que le tailloir étoit mis en action avec le talon, au lieu que c'est la main qui ment le tailloir. j'ai actuellement une cercle divisée par cette machine ....' The reference is incorrect: Ramsden's publication on his engine was noticed in *Journal des savans* 1777, from where the younger Jean Bernoulli copied it to his *Nouvelles litteraires de divers pays ...* (Berlin, 1776–79), 4. 64–5.

22 Cassini (1810), 381.

23 Paris, Archives de l'Académie royale des sciences, 'Prix', MSS Carton I, 1774–1781, 'Instruments de mathématiques ...' No. 2. '[qui] m'a procuré la facilité de diviser mon secteur en 120 parties tracées sur un bizeau pour que chacun de ces traits puissant correspondre à ceux établi du 10° en 10° sur le bord du limbe du quart de cercle j'ai adjustai en suite un à pareille du tracelet fixé sur le secteur, ce tracelet resemble à celui que Ramsden a placé sur la machine à diviser qu'on vit chez M. le president de Saron ... La confiance que j'ai en Ramsden me la fait adopter sans l'examen vigorous que j'en ai fait aujourd'hui mais près de 1200 traits qui sont sur l'arc de ce quart de cercle m'ont mis à portée du projet sur cet instrument'.

24 AN, F^{17}1307^B Minutes of the Commission temporare des Arts, séance du 10 floreal II [29 April 1794]; Turner, A.J. (1989).

25 RGO, MS 14/5, 262–3.

26 RGO, MS 14/5, 268.

27 RGO, MS 14/5, 270–71, 277.

28 Howse (1998), 411.

29 RGO, MS 15/5, Board Minutes, 25 May 1775, 276.

30 Anon (1816), 353.

31 Anon (1816), 353.

32 Brooks (1992), 111.

33 Chapman (1993), 422.

34 Inv. 1930–366.

35 Clifton (1995), 224.

36 The old letter-form 'I' represents the initial 'J'.

37 Stimson (1976).

38 Proprietors of other dividing engines followed this custom – engine marks of Troughton, Spencer, Browning & Rust and others are known. Stimson (1985).

39 RS, MS 2/8/23, B. Bennitt to Banks, 2 December 1800.

40 RS, MS MM 8.24, Dutens to Banks, 4 December 1800.

41 BPL Boulton & Watt Collection, MS MB 251, 86.

42   CRL Bidstrup to Bugge.
43   RGO, MS 14/5, 318–19. See Chapter 9 for the association between Roy and Captain William Calderwood FRS (d.1787).
44   Brooks, R.C. (1989), 148–9.
45   Howse (1998), 411.
46   Lalande (1790), Préface.
47   CNAM, C.25g Report on [François Antoine] Jecker ; Turner , A.J. (1989), 20; Blémont, *DBF* art. 'Jecker'.
48   Bugge (1997), 203.

Chapter 4

'At the Sign of the Golden Spectacles': The Piccadilly Workshop

Ramsden, the famous Ramsden, to whom astronomy and the physical sciences are so beholden.

Fontana to the Duc de Chaulnes, c.1776,
Opuscoli scientifici di Felice Fontana (1783), 156–70

That every part of his instruments may be fabricated under his own inspection, Mr Ramsden has in his workshops men of every branch of trade necessary for completing them.

European Magazine 15 (1789), 96.

This chapter examines Ramsden's business life, his premises and his employees, and ventures into the great workshop in Piccadilly.

Expanding premises

Ramsden's transfer in 1773 from Haymarket to Piccadilly was to lift him out of the normal run of instrument manufacture and was an essential factor in enabling him to lead in the design and construction of major new observatory and surveying apparatus. The site occupied Crown land and is thus well documented.[1] The houses were first numbered in 1772, from Titchbourne Street on the north side, consecutively west to St James's Street, then back on the south side, to Haymarket.[2] By the time of Ramsden's arrival, several of the houses extending westward along Piccadilly were occupied by booksellers, a situation which lingered into the twentieth century. Nearly opposite Ramsden's premises stood Burlington House, one of the great Palladian residences along Piccadilly and Pall Mall belonging to the peerage.[3]

Ramsden moved into No. 199 with its rear wooden workshop as a tenant, paying £317, 10s for the lease in 1782.[4] At least one upstairs room in the house was used as a workshop.[5] Some of the Haymarket workmen accompanied him – it was only a short walk from one to the other – but he placed advertisements in the *Daily Advertiser* of 22 August and 27 October 1774, 'To the Mathematical Instrument makers. Wanted immediately several good hands, or of any other trade that can file and turn exceeding well, here they will have constant Employ and good Wages. None must apply but such as are very capable and whose Characters will bear the strictest enquiry. Enquire at Mr Ramsden's, Optician, opposite Sackville Street, Piccadilly'. On his bill-heads Ramsden continued to use the sign of the

'Golden Spectacles', for his Piccadilly premises. There was no monopoly on such signs, the Dollond business had been trading under 'The Golden Spectacles and Sea Quadrant' since at least 1752, and large pairs of golden spectacles advertised opticians' shops in various London streets.

Shortly after 1780, Ramsden acquired No. 196 Piccadilly, which had a wider frontage but less depth, as it backed onto a large and lofty wooden building formerly occupied by a coach-builder (see Figure 4.1). The broad exit to Piccadilly ran beneath the upper stories of No. 196. Both house and workshops were somewhat dilapidated, and in 1784 Ramsden began negotiations with the Crown surveyors, saying that he wished to rebuild the workshop and asking for a new lease to make this worthwhile. This was agreed, with Ramsden paying £80 for a new lease, to run for 29 years from 28 February 1806 (the expiry of the old lease), at an annual rent of £25, 5s. Ramsden took out an insurance policy dated 1787, covering against fire for stock and tools in a timber workshop, to a value not exceeding £300.[6]

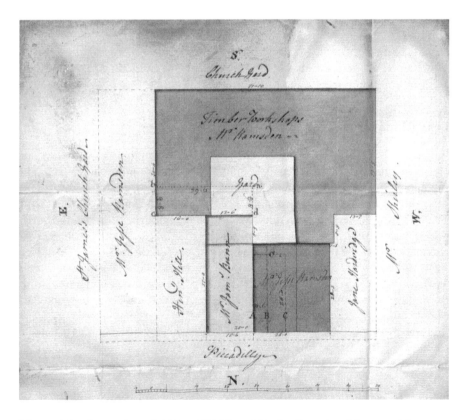

Figure 4.1   **Street plan of 196 to 199 Piccadilly, by Jonathan Marquand in c.1776 for HM Land Revenue. TNA, CRES 2/730.**

Rebuilding the Great Workshop

A view of the workshop from St James's Square probably taken before Ramsden's occupation, appears to show a flat-roofed building but this may be a misrepresentation. The old irregular structure was 25 feet wide at ground level, 21 feet 4 inches above. The passage ran back 37 feet to open up behind the houses fronting Piccadilly. This section was 71 feet 10 inches from west to east (where it abutted on No. 199) and extended south 40 feet 4 inches, backing onto the churchyard. The yearly rent was £8, 15s. Plans in the Crown Estates documents indicate a covered building on three sides of an open yard (see Figure 4.2). Fire would be a hazard in such a wooden structure and Ramsden would have sent his patterns for casting in brass to one of the many nearby brassfounders, and probably obtained his hard metal castings from a foundry outside London. Beneath No. 199, and extending both east and west, was an arched cellar. The sequence of events is complicated by the Crown Estates documentation not always giving the house number, and by the subsequent renumbering imposed with the building of Regent Street and Piccadilly Circus.

In December 1784 Ramsden's formal petition to the Lords Commissioners of His Majesty's Treasury (the necessary avenue for negotiations over Crown land), stated:

> Your Lordships petitioner lately purchased an underlease of which 26 years are as yet unexpired ... at present held of the Crown by Mr Bedard of Grays Inn Lane, Cowkeeper; on these premises your petitioner has for some time past constructed his larger Instruments, and purchased this Underlease, with intent to make the Workshops capable of receiving yet larger Instruments, which Your petitioner had just begun; but finding the whole of the premises in a very ruinous condition, and on enquiry, that the Grant now held by Mr Bedard expires in 21 years from February next, above 5 years before the Underlease, the shortness of the term will render the expense of rebuilding, and repairing such parts as are necessary, extremely heavy on him.
> Your Lordships petitioner therefore
> Most Humbly Prays
> That Your Lordships will be pleased to grant to Your petitioner, a lease of the premises he now occupies, for a term of 29 years, to commence from the year 1806, the period from which the present Lease expires, On Your petitioner paying such fine or Ground Rent, as Your Lordships Surveyor may be pleased to appoint.[7]

On 19 November 1785 Ramsden paid £80 for the renewal of the lease of the great workshop. The new lease included a clause prohibiting 'erection of any new edifices or buildings whatsoever on the premises therein mentioned'. As he intended to rebuild, Ramsden appealed and on 1 May 1786 the clause was duly removed.[8]

The petition engendered a flurry of official correspondence, and a lengthy report by George Augustus Selwyn, Surveyor of Crown Lands, rehearsing the history of all the leases and sub-leases relating to the house and workshop, together with three other houses fronting Piccadilly. Since the original grant of 1744, the property had been divided and sub-let. Ramsden had paid £450 to Bedard for

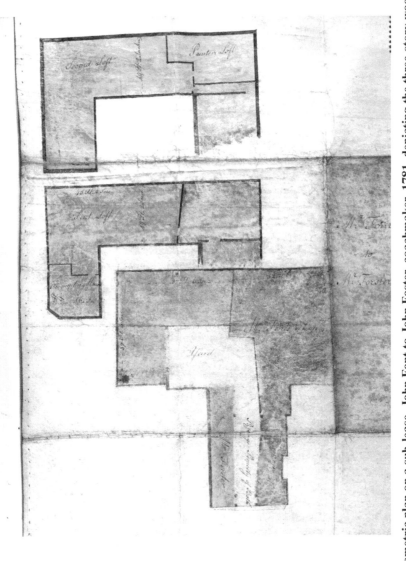

Figure 4.2 Isometric plan on a sub-lease, John Font to John Foster, coachmaker, 1781, depicting the three-story wooden workshop behind 196 Piccadilly, acquired by Ramsden. Bethlem Royal Hospital Archives.

the underlease, and informed Selwyn that his reason for agreeing to such a large sum was in order to fit up the house and workshop to make larger instruments, for which alterations and repairs would be needed but that he felt that Bedard's control was so great that he might have difficulties, and that this was his reason for applying for an extension of the lease. Selwyn discovered that in fact Bedard had no power to grant this underlease, which he had done without authority of the proprietor, and that the whole business had been 'a very irregular transaction'.[9] After several pages of legal argument, including the assessment of the recent agreement reached between Bedard, Ramsden and James Bunn, occupier of the adjoining house, the grant was made to Ramsden, who thus increased the size of his workshop area, in three dimensions.[10] Along the rear wall of the workshop he put a 60-foot 'great bench', on which the rods and bars for Roy's triangulation base could be accurately marked.

The declared intention to rebuild may have been a formality, for in 1795 another series of exchanges with the Surveyor's office sanctioned the reconstruction in brick of the workshops at 196 Piccadilly, and of the house fronting Piccadilly. In 1795 the party wall between Ramsden's house and that adjoining was in such bad condition that he began negotiations with the various authorities concerned, presumably with the aid and advice of a lawyer, for which he had to follow the course laid down under the 1774 Building Act, 14GeoIIIc78. This Act defined the required thickness of party walls between properties, and decreed that in cases where – as here – properties intersected and where disputes arose, the site of such walls was to be determined by juries at law. Two surveyors or 'able workmen' were to be appointed by each party to inspect prior to building, and declarations of expense of rebuilding were to be filed with the other party.

Having given notice in writing to John Fitch, co-owner of the party wall, on 2 November 1795 Ramsden submitted the required certificate to the Westminster Justices of the Peace. It was agreed that the structure was so decayed that it would not be acceptable as a party wall when either of the adjoining houses were to be rebuilt. Further documentation was needed, and on 22 May 1796 two Crown surveyors reported favourably. More papers must have been shuffled between officials, for the next surviving letter, dated 19 October 1796, refers to three Ramsden communications.[11] The first was a petition for an extension of his present lease in order to help defray the cost of rebuilding, the second was a request that the new lease comprise the whole of his premises, and the third, that the term be extended to 99 years, subject to an increased ground rent as the Crown thought fit.

The Crown's Surveyor-General, John Fordyce, noted that the house was currently uninhabited, that the surveyors estimated the cost of rebuilding to be £900, and as it was to the Crown's advantage to encourage him to proceed, he recommended that Ramsden should be granted a new lease for a reversionary term of 60 years and 36 days, to commence from 28 February 1835 and thus to extend the full term to 99 years from 5 April 1795, and that a new annual rent of £51, 11s would, all things considered, be reasonable. The Treasury Commissioners agreed with these proposals and on 5 December 1796 instructed Fordyce to deal with the necessary terms for the documentation.[12] In July 1796, while this reconstruction was in hand, Ramsden wrote to the Surveyor offering to take the other two houses fronting his workshops if the present tenant was agreeable. He was willing to pay the tenant £500, which was £200 over the recent valuation and would take the houses in their present condition. His bid was not successful.

Number 199 Piccadilly remained Ramsden's residence and his formal address, where visitors and customers could be dealt with by Ramsden or his shopman; it seems that only a favoured few were admitted to the workshop. Matthew Berge, and probably some other apprentices and long-term employees, occupied No. 196, though it is unclear where Berge resided during the rebuilding. Nor do we know whether Ramsden and Berge, one separated, the other unmarried, employed a resident housekeeper, or whether they were cared for by the wife or wives of other resident employees.

Impositions on Business

Finances

Ramsden's financial situation at this time remains unclear. His account with Messrs Coutts, the bank favoured by the Board of Longitude and by Maskelyne, was opened on 28 March 1772 with a cash payment of £178 but thereafter records mainly cash transactions until 1775–76 when a flurry of payments-out includes one to Messrs Dollonds, balanced by a cash payment-in of £540. The following year a payment of 5 guineas to Berge, and in 1778–79 two payments to a Mr Arnold – perhaps the watchmaker John Arnold (1736–1799) – is balanced by payments-in of £200 and £100. In 1783–74 the account was closed on a balance of £107, 14s, 6d. No other account has been located. In April 1781, when there were no movements in the Coutts account, Ramsden wrote to Matthew Boulton asking him to pay his bill for 40 guineas, outstanding from the previous year, 'My order for the Navy and some large instruments I have now making distresses me for money, or I would not have taken this liberty'.[13]

Ramsden also used Maskelyne as his petty cash box. From 1776 to 1791 Maskelyne frequently noted in his account book such entries as: '1778 August I lent Ramsden on his note of hand £7, 7s'. After other small handouts, he notes '1780 March 26 received this back', but again on '1780 May 27 lent Ramsden £20'.[14] And so it continues. He needed an advance to undertake an order for the Board of Longitude: '1791 September 7 received back from Mr Jesse Ramsden what I advanced him July 2 for the Board of Longitude order for a universal theodolite for Mr Wm Gooch £126.'[15] Yet when importuned by Cassini in January 1788 to make instruments for the Paris Observatory Ramsden declined a deposit: 'My workshop turnover provides the necessary money'[16] although, as we shall see in Chapter 13, the Paris Observatory did make progress payments.

In 1772, on the death of George Adams senior, appointed supplier to the Board of Ordnance, the Board transferred its custom to Jeremiah Sisson, whose financial difficulties landed him in the bankruptcy court in 1775. Several orders were then placed with Ramsden, who may have been in line for the next appointment.[17] His slow delivery brought a warning in July 1777 that no further orders would come his way unless he complied with those in hand – particularly some drawing instruments ordered six months previously. The Board gradually transferred its custom elsewhere, including George Adams junior. Ramsden continued to dally; of 24 sets of drawing instruments ordered in January 1777, by April 1780 only 12 had been delivered and they were now in need of repair.[18] This was not entirely the parting of the ways for Ramsden's involvement soon afterwards with General Roy and the triangulation survey brought a good deal of work, commissioned and paid for through the Board, for which, see Chapter 8.

Properties in St James's parish were assessed on their 'rateable value' (for houses this was equivalent to the rent). Land Tax varied from year to year until 1772 when it settled at 4s in the pound. It was collected quarterly, along with the Paving Rate, Watch Rate and Poor Rate. Number 199 Piccadilly had a rateable value of £40 and in 1780 Ramsden paid 16s 8d for Watch Rate, £2 + £1 for Paving Rate + Lights and £4, 13s, 4d for Poor Rate and Cleansing. In 1799 the charges were Paving rate: £2, Watch Rate: 1s, Poor Rate: £7. Window Tax was sometimes exempted on shops and business premises but no records survive for Piccadilly at this period. Ramsden probably avoided the short-lived Shop Tax, and he died just as Pitt was introducing Income Tax. Insurance against fire, damage or theft, was not obligatory, but most prudent businessmen insured their buildings and contents, in Ramsden's case with the Sun Insurance Company.

Weather and Political Events

The steady expansion of Ramsden's business must have suffered from natural and political events. Both temperature and the quality of natural light controlled the times when large instruments could be hand-divided. The entire workpiece had to be kept at an even temperature, and illuminated without the use of oil lamps which could cause one section of the brass arc to heat up and expand. Extremes of weather affected the progress of work, while floods, droughts and frozen rivers delayed deliveries of supplies and the export of finished items.[19] The winter of 1775–76 was extremely cold in Britain and across much of western Europe.[20] In London, snow fell throughout January and the Thames was frozen.

In 1783 the sun was obscured during the summer months by 'The Great Dry Fog'. The darkened sky was first apparent in London on 23 June and extended across much of Europe. Pictet referred to it in a letter to De Luc, 28 September 1783: 'as for the fog which has covered Europe ... it first appeared on 17 June ... interrupted by several rainy days ... began again on 23 June and lasted almost without interruption for the remainder of that month and all July. The sun gave scarcely more light than the moon ...'.[21] It brought a reddened sun, increased the sulphurous quality of London's already smoky air, and was thought to have been responsible for lowered harvest yields. Its cause, unknown until news reached Copenhagen in late autumn, was a series of eruptions of the Laki basalt fissure in Iceland which commenced on 8 June and continued intermittently for several months thereafter.[22] It was followed by a severe winter, beginning in late

December, with the Thames frozen in February. In 1796 the temperature fell to –6°F (–14°C) over Christmas, with the Thames again frozen.

During the 1780s the fierce political struggle between the parliamentarians William Pitt the younger (1759–1806) and Charles James Fox (1749–1806) led to several riots, including those connected with the Westminster elections of 1784 and 1788, regarded by many contemporaries as notorious even by eighteenth-century standards. Mobs were hired, and in return for money, alcohol or both, surged along the streets in pursuit of their declared enemies, attacking their carriages and, on election day, surrounding the hustings always held under the church porch of St Paul, Covent Garden. The 1788 election was the more violent; the government employed gangs of sailors while the opposition paid Irish sedan-chairmen five shillings a day to participate. Crowds of 200 or more clashed with the Westminster constables on several occasions, leading to several deaths. Both sides tried to use their gangs to prevent opposing voters from reaching the poll, while the hustings were guarded by young men who interrogated voters in violent terms.[23] Ramsden, not being a freeholder, was not entitled to vote, and so did not have to brave the belligerent crowds surrounding the hustings, but on those occasions when the mob surged along Piccadilly he must have feared for the safety of his windows.

A Specialized Workforce

Some 35 men have been identified as having at some time been employed by Ramsden. His workforce comprised apprentices, journeymen, workmen, foremen and shopmen. There were probably two or three apprentices in the shop at any one time, going through their seven-year training. Journeymen had completed their training with Ramsden or elsewhere and were expanding their knowledge and experience. Both apprentices and journeymen operated the dividing engines and learnt to make some of Ramsden's speciality items for which there was a good demand. Workmen with specialized trade skills would have included carpenters, joiners, lens-polishers, tool-makers, and metal-workers, able to cut, shape, file and turn, make screws, make wooden patterns for casting, lacquer-finished instruments, and to engrave. The foremen supervized the workforce and on occasion went out to deal with repairs or installations. The shopmen dealt with customers calling at the shop at 199 Piccadilly. Somebody, perhaps Ramsden himself, must have kept the accounts.

British Apprentices

Eleven British apprentices are known, mostly from registrations in the Inland Revenue Apprenticeship Tax registers. On 28 February 1763, when Ramsden was still at the Strand, James Ware was 'admitted from St Andrew's Holborn, to serve five years, £5 cloathes included'.[24] He was indentured on 4 March 1763 for £16 but turned over in June to a shoemaker.[25] Simon Spicer was indentured 10 March 1768, for 7 years, for £30.[26] Joseph Simpson, was indentured on 15 June 1769 for £30.[27] Thomas Porter was indentured on 24 June 1778, for 7 years for

£30.[28] John Rooker was said by a descendant to have been 'an indoor apprentice' to Ramsden from about 1784. Jonathan Mac[k]neil was indentured on 1 February 1786 for 7 years for £25.[29] Edward Dixey was apprenticed on 4 September 1787 for 7 years for £21.[30] Thomas Jones (1775–1852) was indentured on 14 March 1789 for 7 years for £50 guineas.[31] Richard Adams was indentured on 11 June 1796 for 7 years for £21;[32] Charles Swift, not indentured, 7 July 1796, 7 years, £30[33]; Joseph Green, not indentured, 8 December 1798, 5 years, £100.[34]

The most important of Ramsden's workmen was undoubtedly Matthew Berge. Born in or about 1753, Matthew was said to have been apprenticed to Ramsden, probably in about 1767, but his name does not appear in the apprenticeship registers. He was a younger brother of John Berge (1751–?1808), who had been apprenticed to Peter Dollond. To Matthew, Ramsden confided his method of hand-dividing great circles, and the accuracy of Matthew's graduation of the circles on the Shuckburgh Equatorial (see Chapter 6) was lauded by its owner. After Ramsden's death Matthew Berge ran the Piccadilly workshop until his own death in 1819.

Former apprentices enhanced their later reputations by advertising their training. William Cary's bill-heads and his trade card proclaimed him to be a former Ramsden apprentice.[35] His trade card announced him as 'Apprentice to Ramsden'. Dixey was probably the partner in Willson & Dixey, whose card proclaimed them as 'working opticians, 9 Wardrobe Place, Doctors Commons. Apprentice [sic] to the late J. Ramsden'.[36] Mackneil engraved on a navigation night glass 'Mackneil London Apprentice of the late Mr Ramsden'.[37]

Other successful former apprentices included Porter, who traded at 18 Aylesbury Street, Clerkenwell, as an optical, philosophical and mathematical instrument maker. In 1777 Spicer, then at Three Falcon Court, Fleet Street, insured his household goods and clothes for £130, his tools and stock for £70.[38] His prosperity is indicated by a second policy, for premises at Charing Cross, on 26 March 1784, where his utensils, stock and goods in trust are covered for £450, his household goods, clothes and books for £340.[39] Rooker stayed until 1800 as he witnessed Ramsden's signature on his will. His descendants continued to trade as instrument dividers for several generations.[40]

Foreign Apprentices

Various foreign workmen joined Ramsden, as apprentices or as journeymen. Two Portuguese youths were apprenticed but not indentured: João Maria Pedroso on 22 November 1798, for 7 years for £157;[41] Gaspar José Marquez (1775–1843) on 22 December 1798, also 7 years for £157.[42] The high fees (probably stipulated as 150 guineas) were paid by the Portuguese Sociedade Real Marítima which, among its other interests, saw the need to set up workshops in Portugal, to manufacture ships' compasses and other navigation instruments. The Sociedade consulted the eminent Portuguese naturalist José Francisco Correia da Serra (1750–1823), then a political refugee in London, as to the best master craftsmen and he presumably suggested Ramsden.[43] In 1800 Pedroso sent home a sextant of his own making, now in the Museu de Marinha, Lisbon.[44] He returned from

London in 1807 and worked as an instrument maker in the Arsenal at Rio de Janeiro, later in Lisbon. Marques moved from Piccadilly to join Boulton & Watt at Birmingham where he familiarised himself with Watt's steam engines, one of which was bought in 1811 for the Mint at Rio. He left England that year, was in Brazil in 1826, then in Lisbon, where he had a workshop in the Rua do Tesouro Velho.[45]He died in 1843.[46]

One of the Haas brothers is said to have been a pupil of Ramsden; either the well-known Jacob Bernard Haas (1735–1828), alleged maker of an artificial horizon – now lost – supplied with Ramsden's sextant to Palermo[47] or, according to the naval historian, Antonio Estacio dos Reis, his younger brother, Carl Friedrich Haas.[48] Both men were born in Biberach, in Swabia, and came to work in London as instrument makers before going to Lisbon where they settled.[49] In 1788 Jesper Bidstrup reported to his patron in Copenhagen that the King of Sweden had paid Ramsden £200 to take a Swedish youth, Apelquist, as his apprentice, but there is no record of such a binding. Bidstrup added that Apelquist had to leave the workshop prematurely, because of a dispute with Ramsden.[50] It is known that Carl Apelquist (c.1749–1824) spent two years with Ramsden from about 1775, probably funded by one of the Swedish royal enterprises. He later came back to London and while working for the instrument maker William Fraser, wrote to Pehr Wilhelm Wargentin (1717–1783), secretary of the Swedish Academy of Science, to remark on the many interesting new inventions he had seen, including Ramsden's dividing engine. He returned to Sweden in 1779.[51]

British Journeymen

Some trained workmen are known from earlier employment, others from later years. Unfortunately much of the information relating to them comes from unreliable or inexact sources, and in some cases only a surname is known. Three men came from Matthew Boulton's Soho works in 1777: Peter Kelly and Webb in January, William Baddely in October. Kelly appears to have worked for Ramsden on an earlier occasion, to have then spent time at Soho but wished to find employment again in London. Boulton had written on his behalf to Ramsden, who replied:

> I am exceedingly obliged to you for favoring me with yours respecting Kelly, he is a man who is very usefull to me and should be glad he would return. His mother who I find is a very bad woman and a great torment to him set out from hence to Lancashire the latter end of last week. If you could in any way prevail on him to return, it will add to the many favors I receive from you.[52]

Two years later Kelly was accused of theft of certain workshop tools and instrument parts (see below). Webb was also known, as Ramsden wrote to Boulton:

> … you mention Webb. If you should have no further occasion for him I would gladly employ him. Though a rough hand he was usefull enough but he is always in some mischief or other.[53]

Baddely was equally welcome, and Ramsden wrote to Boulton to inform him:

> A workman whose name is William Baddely and says he formerly worked at Soho is just come to work with me and says he intends to settle here in London. As he works tolerably well [he] has prevailed on me to give him some assistance to get his wife and family up here. If therefore she should wait on you at Soho, [I] will esteem [it] a very great favor if you will be so kind [as] to give her 4 guineas on my Acc^t, and [I] will immediately return it when you shall be in London.[54]

European Journeymen

Four men had trained in Europe before coming to London as journeymen. François-Antoine Jecker (1765–1834) was born in Alsace into a family of Swiss origin. In 1784 he was apprenticed to a clockmaker at Besançon; he came to London in 1786 and was there for six years.[55] By 1794 he was in Paris, where a post-Revolution report speaks well of his attainments.

> He worked five years with Ramsden and it is the works of this artist that he seeks to bring to fruit in France ... the machine for straight line dividing built in France is almost that which he saw at Ramsden's, having left out various costly parts which could be sacrificed, he being a young mechanic in difficult times ... Regarding the screw, so crucial to the success of such apparatus, his was a simple direct construction, such as he had seen at Ramsden's although having nothing in common with the screw cutting machine published in the collection of Ramsden machines as translated by de Lalande ...[56]

Later, Jecker affirms that the screw-cutting machine he saw at Ramsden's was not as figured in the published work. Jecker's certificates show that he was back in Paris by Germinal, An II, [1794]. The Jecker brothers and their descendants established a successful manufactory in Paris.[57]

Edmund Nicholas Gabory was born in Strasburg. He trained in London under Ramsden and John Dollond, and established his own workshop in Holborn in 1790. He married at St Andrew Holborn in 1795, taking his wife and baby daughter to Hamburg where in 1796 he founded his own workshop with a retail shop attached. Gabory's daughter married into the Krüss family, their name being perpetuated in the optical business which continued into the late twentieth century. Gabory died at Hamburg in 1813.[58]

Under the Hanoverian kings, Britain and the Electorate were parts of George III's kingdom, with freedom of movement between the two. Little is known of Georg Drechsler, beyond the fact that he came from Hanover, joined Ramsden in 1770 at the Haymarket shop, accompanied him to Piccadilly, and went to Hamburg in 1775. His younger brother worked for Edward Troughton and chose to accompany Piazzi's circle to Palermo (see Chapter 6), where he remained thereafter.[59] Wilhelm Gottlob Benjamin Baumann (1772–1849) was born in Tumlingen, in the German Duchy of Württemberg, and died in Stuttgart. He was already a trained engraver when he joined Ramsden, probably in 1795, and was back in Stuttgart by 1797.[60] German authors spoke well of his skills; Repsold comments that 'Baumann made sextants of 4 inches radius to the same precision as Ramsden'.[61]

British workmen of unknown origin

Several workmen were beneficiaries in Ramsden's will (see Chapter 11). James Allen junior had by then been Ramsden's shopman for three years and he continued in that post under Berge, being thus identified in his father's will in 1816.[62] It is unclear whether James Allen senior (1739?–1816) had also been employed by Ramsden; however, by the time of Ramsden's death he had been in business on his own account for 12 years.[63] Of John Hill and one Curtis, (even Ramsden did not know his first name and left a space in his will) nothing more is known. Edward Pritchard was both executor and legatee, receiving a small dividing engine and a lathe. He was probably the senior partner in E. Pritchard & Son, divider of mathematical instruments, at 8 Porter Street, Soho from 1816–22. Peter Lealand, or Layland, mentioned in relation to the Blenheim instruments (see Chapter 5), also testified at the trial of Peter Kelly. (It has not been possible to trace a relationship to Peter and Richard Lealand, later partners in the microscope makers Powell & Lealand). George Pope, described by Sir George Shuckburgh as Ramsden's foreman (see Chapter 6), died on 18 February 1805. Samuel Pierce stayed on after Ramsden's death, being described by William Kitchiner as 'telescope maker at Mr Berge's'.[64]

We are left with four men whose relationships with Ramsden are the least documented, and some other names which emerge from evidence at two Old Bailey trials. Leonard Bennet is an uncertain candidate; William Harrison, instrument maker to the amateur scientist the Hon. Henry Cavendish (1731–1810) is said by Cavendish's biographers to have worked under Ramsden.[65] John Stancliffe, who died between 1812 and 1816, was said to have been apprenticed to Henry Hindley and had then come to London and been foreman to Ramsden, this connection having played a part in Ramsden's development of his dividing engine (see Chapter 3). Stancliffe later worked with John Bird on the Radcliffe Observatory instruments.

One Higgins is described by Jesper Bidstrup (see below) at some length.

> … Higgins … was Ramsden's first apprentice and worked afterwards for a long time with him, not long ago he was in a kind of Comp. [company?] with R-n [Ramsden], lived in his house and managed his workshops, but Ramsden's indifferent character would not allow them to be together more than 8 months, he was with R-n and participated in the work on the exceptionally great Theodolite, which General Roy used for the observations at Hounslow Heath. Now again he is engaged by Nairne & Blunt …

In his letter of 1790 Bidstrup gave Higgins' address as 32 Manor Road, Walworth.[66] In an undated note, Maskelyne mentions a Higgins of Clerkenwell, who may be the same.[67] A Higgins, again not further identified, was a testator at Peter Kelly's trial (see below). This trial throws up two more names which, from their statements, are men from Ramsden's workshop: Henry Pye, who identified a pair of cutting pliers as belonging to Ramsden, and John Harnes, who worked at Ramsden's for five years. The Old Bailey trial of 1801,[68] concerning theft from Matthew Berge's house, supplies the name of Philip Coleman, who worked briefly with Ramsden in his final year of life.

Lastly, the Hungarian politician and engineer Miklòs Vay, Baron de Vaja (1756–1824), during his visit to London in 1787–88, assisted Ramsden with the mathematical calculations needed for making Piazzi's circle.[69]

Life on the shop floor

The remarkable size of Ramsden's workshops and the number of craftsmen busy under his roofs, seems to have been unique in London's instrument trade. Other major instrument makers put work out to sub-contractors known as 'chamber masters' or 'garret masters' working in their own homes, or as small groups of ten or 12 men. This system, while it was efficient for businesses such as those run by the Adams family, delivering quantities of identical small items for the Board of Ordnance, or for the Dollonds, whose telescopes were standardized, or for Edward Nairne, another quality supplier of mostly small standard items, even for the barometer makers who supplied such reputable 'names', was unsuited to construction of large instruments or the development of new designs which needed constant supervision. John Bird, meeting orders for big mural quadrants, had been obliged to move to what was then the northern margin of London, where a large shed could be built behind his house, and by the early nineteenth century the Troughtons, whose Peterborough Court premises had only a narrow access to Fleet Street, had the larger castings for their big astronomical telescopes made at the Bermondsey engineering works of Bryan Donkin (1768–1855) and took the small precision parts from Peterborough Court to Bermondsey for assembly.[70]

The practice of bringing together a large workforce, where each man was assigned to a single task, but no central power source was involved, had been long-established in shipyards, printing houses, glasshouses, and tapestry works. Outside London, machinery, replacing craftsmen with unskilled machine-minders, was first introduced in the water-powered textile mills of northern England, and soon reached the manufacturing region around Birmingham. Large factories, powered machinery, and division of labour began to feature in the metal trades; one of the earliest of such establishments was the Soho Works of Boulton & Watt, outside Birmingham. With the completion of the Birmingham Canal and its junction in 1772 with the Grand Trunk – later the Grand Union – Canal more heavy engineering works appeared around Birmingham. The crowded streets of London north of the river were not suited to heavy industry; the engineering works of Henry Maudslay, Brian Donkin and others were established on the Surrey bank, where river boats delivered their iron and coal and carried away their heavy manufactured goods. Upstream of London Bridge, which presented a barrier to large vessels, numerous brass battery, paperworks and other small mills were powered by the Thames tributaries flowing from the South Downs.

Ramsden was a close acquaintance of both James Watt and Matthew Boulton but it is not known if he ever visited Boulton's manufactory at Soho, near Birmingham. Boulton was one of the social group composed of forward-thinking industrialists, known as 'The Lunar Men'[71] who met regularly in the Midlands and from time to time in London coffee houses, where Smeaton and Ramsden joined them.[72] Boulton's Soho factory was powered by Watts' steam engines, and had adopted a division of labour, with each man allocated to a particular stage of the manufacturing process. The Boulton letter-books reveal that Ramsden and Boulton traded with one another, sub-contracted work in both directions, and on occasion exchanged workmen.[73] If Boulton's practice inspired Ramsden, it was for him an ideal system. Ramsden needed to supervize the large observatory instruments constructed to his designs, so that when a problem arose, he could modify his plans. These instruments could then be set up and tested in the workshop's central well. It is clear that no two were exactly the same, for if repeat orders came in, Ramsden generally made improvements, based on his own experience or that of the astronomers working with the previous instrument. The finished apparatus was then dismounted and packed into sturdy crates for transport to the river and embarkation. The large arcs and circles could not be taken apart, and loading their crates into a ship's hold could present difficulties – notably that of the mural quadrant destined for Brera (see Chapter 6) – as did manhandling the heavy crates and contents onto the wheeled transport of the time. Edward Troughton's crated mural circle fell from its waggon on the long overland journey from Hamburg to Leipzig in 1803, and in 1829 Thomas Jones's circle was damaged when its crate was dropped onto the dockside at the Cape of Good Hope.[74]

This varied demand, and the trust that was confided in Ramsden to improve existing instruments, or to devise something new, must surely have come about because of the high reputation his sextants and other mathematical instruments had gained among men who moved in high government or social circles. He was respected by peers and politicians with a taste for astronomy and scientific endeavour. Those with well-lined pockets built themselves private observatories and travelled round Britain and Europe carrying in their baggage Ramsden's portable observatories, his barometers and telescopes. His contact with the Board of Longitude opened the way to a wider world, for the Board was composed of the professors of astronomy at Oxford and Cambridge, the Astronomer Royal, and other astronomers and scientific teachers and practitioners of high repute, and they in turn were in regular communication with their opposite numbers in the universities and observatories of Europe, India and North America. Having tried out, and admired, the latest work of art from Piccadilly, they would negotiate for something similar – and usually grander – and this instrument in turn, would be shown off by its foreign owner, and would generate further orders.

Rules and regulations

Jöns Mathias Ljungberg (1748–1812), former professor of mathematics at Kiel, travelled three times to England to gather information on various technical processes. On his visit to London in 1787–88, he called in at Ramsden's establishment and noted the workshop rules:

The ordinary workmen, who file, turn and plane &c, receive from 18 to 21 shillings per week of six days. They work for 12 hours daily, from 6 in the morning to 8.30 in the evening. They have one hour from 1 to 2 for lunch, ½ hour in the morning for breakfast, and after twelve hours he gives them another ½ hour to drink tea.

A slate tablet hangs in the workshop where each one writes his time; when he arrives 2 minutes after 6, he must write 6.15, and is paid accordingly. If he signs in 15 minutes late he is docked 7d from his pay, and for each hour late, 2s 4d. He [Ramsden] stays one hour in the shop and thereafter the workmen control the slate. When anyone kicks another, he is fined one shilling, when anyone hits another, 2s.6d. Anyone who shows up drunk is fined and to bring a stranger into the shop there is a fine of 2s.6d. The workmen are paid each evening [Christensen says every Saturday]; the hours that they have not worked, together with any fines, are deducted from their pay, and the workmen go to for a drink, spending part of their wages on ale or porter.

When they come to the workshop in soiled clothes or without having shaved their beards, they are fined. Elsewhere other rules apply. The rules are hung up in the workshop.[75]

The Danish historian Dan Christensen who deciphered Ljungberg's notebook explains that Ljungberg probably recorded these rules because he found them unusual; in Denmark the guilds made the rules and these were followed in every workshop.[76]

Ramsden's rates of pay are less than those cited by the Cambridge mathematician William Ludlam (1717–1788), who employed Bird and other London craftsmen to make his lathes, tools and instruments. His notebook of 1758 refers to '5/– a day for a good workman'.[77] Campbell's *London Tradesman* (1747) mentions a working week of six days, 6 am to 8 pm, with foremen earning 18 shillings per week, and assistants 15 shillings. Ramsden's men may have taken time off at the Christmas festival, as Ramsden explained on 27 December 1787 to Miklòs Vay. Vay was kept waiting for his instruments until he was about to sail from Dover:

Dear Sir – I am really sorry I have not been able to have your Instruments ready for this Evening but flatter myself being able to send them by tomorrow's Coach – these holidays are terribly ag[ain]st me.[78]

Another reporter was Jesper Bidstrup, the young Dane whose patrons had sent him to London to learn the craft of instrument making.[79] Bidstrup did not join Ramsden and perhaps learnt about the working conditions from employees whose tongues were loosened in the alehouse. In a letter home, he wrote (in abridged translation):

With Mr. White, with whom I work, I am very much satisfied, he is an honest man and has with the greatest readiness taught me all he knows, and even introduced me to artists, whose acquaintance I have turned to account and hope to benefit from in future. I am pleased with the decision I made, for I am now fully convinced that it was not Nairne & Blunt, Adams or Dollond or their peers from whom I would have learnt anything, as their most distinguished instruments are manufactured all around the city, and the men employed in their houses just repair instruments or execute some [illeg]. … Ramsden is the only one having the most important instruments manufactured in his house, he has often got 40 to 50 workers each of them manufacturing various parts of an instrument, some are planing or filing, others turning on the lathe, making screws and so on. Anyone designing or even manufacturing an instrument is not doing anything else, and he must already be known as a good and experienced workman. Had I entered his workshop as a hired man, which would have been the only way, this would have been of little advantage to me for I would, like anyone else, have been put to plane, file, turn on the lathe, and so on. I would have to do what was ordered, which would not have improved my knowledge of other branches of the art, not because Ramsden could not give the best instruction, but because no mortal would make him do that.[80]

Tools and Materials

The vast range of tools needed for the various tasks is illustrated in 50 plates in the first volume of Antoine Thiout's *Traité d'horologie* (1741) and it is likely that many of these would be found at the Piccadilly workshop. Of the machine tools, little has come to light. At Ramsden's death his lathes and other machine tools, and hand tools were valued at £628.[81] Two of the dividing engines were the property of the Board of Longitude, a smaller circular one belonged to Ramsden himself.

Thanks to Ljungberg's industrial espionage we also have details of a tube-drawing machine which Ramsden was said to have owned for ten years.[82] In June 1789 Ljungberg, who had not seen this device, questioned one of Ramsden's workmen, perhaps over a jug of ale. The workman explained that the machine, which had cost £200, produced brass tubes for telescopes, and that Ramsden and a Mr Wright had recently obtained patents for plated telescope tubes. On reading his notes, Ljungberg realised that he had insufficient detail for the machine to be copied and he asked the workman to write down the operating procedure and make a measured drawing (see Figure 4.3). The originals are in Ljungberg's notebook and the text is translated in full by Christensen.[83] The hand-operated process consisted in drawing the brass tube through a series of mandrels, lengthening and thinning the metal until it was the required dimension.

Although Ramsden was not named in any patent, his friend the instrument maker Joshua Lover Martin had patented such a machine for plated tubes, in 1782,[84] and Ramsden was certainly involved with the outcome, as he explained to Matthew Boulton, when enquiring about a supply of plated copper:

> A friend of mine, with whom I am concerned is taking out a patent for drawing tubes for optical purposes plated with silver or gold. This patent I expect will be out in about 14 days or a little longer and shall have the whole of the tube drawing in this metal. By this you will judge if the patent succeeds of the quantity [of plated copper] that will be wanted and I am determined to make it a ready money article in buying as well as selling.[85]

A series of letters and invoices over the following years shows that Boulton was regularly supplying Ramsden with sheet and tubes of plated copper. In July 1782 Boulton also sent some solder, 'which I hope will prove to your wishes. I have to

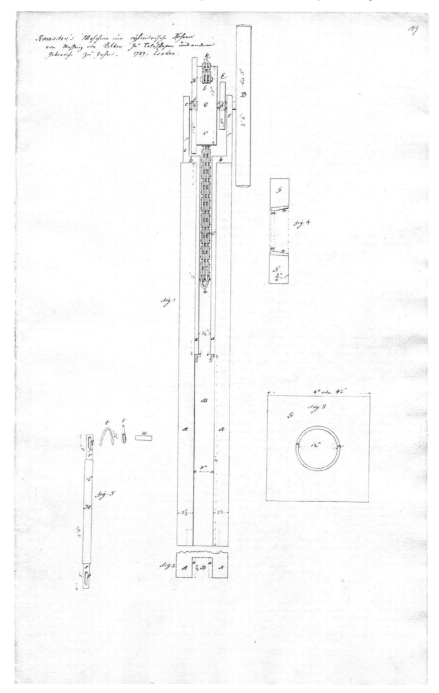

Figure 4.3   Ramsden's tube-drawing machine, drawn by a Ramsden workman
for Jöns Mathias Ljungberg. Stockholm, Kung. Myntverkets arkiv,
Varia Id., Fig. 1.

observe that a great deal depends on the workman knowing his business to solder well the tubes etc'.[86] Silver-plated tubes were used for opera-glasses and small telescopes (see Figure 4.4). In 1784 Boulton had been promised a tube-drawing machine of his own, and its non-appearance triggered a letter to Ramsden asking if it had been sent.[87]

Figure 4.4   Opera-glass in plated tube, signed *Ramsden*. Enamelled in red and yellow stripes. Length closed 3½ inches. National Museums of Scotland. Inv. T.1927.20.

A sidelight on Ramsden's provision of hand tools emerged from the proceedings of a trial held at London's Old Bailey court on 13 January 1779. Peter Kelly was charged with having stolen 'two quadrant glasses value 1/–, three steel arbors value 2/–, 3 steel broaches value 1/–, a steel countersink value 6d, 2 steel files value 2d, a brass and steel centre, value 6d, a steel chamfering tool value 6d, and a pair of steel dies value 6d', allegedly all Ramsden's property. In his testimony, Ramsden declared that each of his many employees had his private drawer where he kept his tools locked up. The tools being frequently lost, and Kelly coming under suspicion, Ramsden and Peter Layland had opened Kelly's drawer, marked two arbors, and replaced them in the drawer. As Kelly was about to leave his employ, Ramsden obtained a search warrant, and in a drawer in Kelly's lodging he found the items mentioned in the indictment. He asserted that he provided tools for the workmen, and was not aware that they brought their own tools. Peter Layland and Matthew Berge asserted that items found at Kelly's home were Ramsden's property.

In his evidence, Kelly stated that, having few tools when he had joined Ramsden, he had asked that they be provided, but Ramsden being 'affronted' by this request, Kelly had at various times bought his own tools. He asserted that he was a toolmaker by trade, and as he was about to go away, he thought he had a right to take them. John Harnes, who had worked five years with Ramsden, testified however that Ramsden's shop was very poorly supplied with tools, and that men commonly brought their own. Simon Spicer, who had been an apprentice when Kelly arrived, was questioned as to whether the workmen usually brought their own tools. His response was 'At times it may happen so', but that he could not say if Kelly had brought his own tools. Kelly was found 'part guilty', since there was no proof that the unmarked items had belonged to Ramsden.[88]

Ramsden's bank accounts do not record payments to named suppliers of steel, brass, glass and other components. There are, however, records of his purchases of glass from the Whitefriars Glasshouse between 1784 and 1793, showing regular purchases both large and small. His account was settled annually, by cash and by credit for cullet returned. Whitefriars produced fine glass for domestic and semi-industrial purposes. The nature of the glass is not indicated, apart from two purchases in April and November 1786, when the ledger is annotated 'plate'.[89] The glassmaker Apsley Pellatt, in his *Curiosities of Glassmaking* (1849), states that 'for many years subsequent to the time of Dollond English flint glass was almost the only heavy glass used for telescopes at home and on the continent … it was sold to opticians in the form of annealed plates, 14 × 10 inches × ½ inch thick'.[90] There was a general shortage of optical-quality glass, brought about by the British government's excise duty imposed on all glass, including any which had to be rejected and went back into the melting-pot. Faced by this imposition, glassmakers were reluctant to venture on a type of glass which was, in comparison with domestic or decorative glassware, needed only in small quantities.[91]

In 1786 the scarcity of good quality brass of large dimension led Ramsden to seek Boulton's help, as he wrote:

… if you think it possible to have brass made in England sufficiently long for a quadrant of nine feet radius. The Bristol Company with whom I have long dealt and who on many occasions have done everything in their power to oblige me, tell me that it cannot be done here, that the brassmakers have no spirit to put themselves out of the way to improve their manufacture ... I would willingly give the price of Dutch or more if it can be had good and as this size might be out of the common size would pay £50 or £100 in the beginning to undertake it and the remainder when it is ready to be delivered.[92]

Ramsden's Election to the Royal Society

The Royal Society, as its second charter of 1663 declared, was founded 'for the aim of improving natural knowledge' and 'for experimental learning'. There was no barrier of religion or nationality, the justification for election as a Fellow being principally a manifest interest in the advancement of physical or natural sciences, preferably backed by ability to pay the dues and donate specimens or books to the Society. Outside the summer months, the Society held regular evening meetings at its London house, when papers were read by one of the two Secretaries or experiments were demonstrated. The substance of the papers, and any discussions, were minuted in the Society's manuscript journal and the Society's council selected those judged most worthy for printing in the *Philosophical Transactions.* The Society was financially independent of Crown or Parliament, but government desires were sometimes conveyed through the Society to appropriate members. In this way the Society was drawn into planning for astronomical and geographical expeditions and, in the 1780s, into the measurement of the specific gravity of water and alcohol mixtures, for HM Excise.[93]

Ramsden certainly moved within a circle of active Fellows including the Astronomer Royal, naval officers, members of Parliament, diplomatists, medical men, chemists, naturalists and assorted cognoscenti, and many were his customers. Industrialists seen as contributing to the improvement of their trade were elected, and in his own craft, Ramsden was preceded by horologists John Ellicott and George Graham, and instrument makers James Short, John Dollond and Edward Nairne.

From January 1773 Ramsden was a regular visitor to the Royal Society's evening discourses. Non-Fellows – 'strangers' in the Society's terminology – had to be introduced by a Fellow and formally given leave to be present. The names of both men were entered in the Society's journal. Among those who introduced Ramsden were Alexander Aubert, Charles Francis Greville, Jean-André De Luc, William Roy, John Hunter, William Philip Perrin, Samuel Horsley, Nevil Maskelyne, Edward Poore, Nathaniel Pigott, George Atwood and William Wales. On 12 May 1785 Ramsden – allegedly without his knowledge – was proposed for election. At subsequent meetings, his numerous supporters, half of whom figure in this history, added their names to the proposal: Thomas Anguish, George Shuckburgh, Alexander Dalrymple, Henry Englefield, John Lloyd, William Roy, James Watson, George Atwood, Anthony George Eckhardt, Charles Layard, Nevil Maskelyne, Matthew Raper, John Smeaton, William Wales, William Calderwood, J. M. Nooth, Alexander Aubert, Charles Blagden, Edward Nairne, Patrick Crafurd, Edward Grey, John Paradise, Richard Kirwan. The proposal was balloted on 12 January 1786. Ramsden was duly elected, and admitted the following week. Thereafter his attendance is not recorded in the journal, except on the few occasions that he himself introduced 'strangers'.

Notes

1   TNA, MSS CRES 2/729 and 730 (unfoliated long bundles) document Ramsden's applications for leases and renewals, with the Crown Estates Surveyors' responses and plans.
2   The building of Regent Street and Piccadilly Circus caused a renumbering, with 199 becoming 196, and 196 becoming 193. In 1830 the freehold of the site of the present numbers 181 to 196, including Ramsden's great workshop, was granted to the Governors of Bethlem Hospital in exchange for the hospital's property at Charing Cross, where the government intended to make improvements.
3   Sheppard (1960).
4   Beckenham, Bethlem Royal Hospital archives. Indenture dated 6 February 1782 recording Ramsden's purchase of the lease from George Pasmore, gent., for £317.10.0.
5   Berge testified in an Old Bailey trial for theft of his household property in 1801 where one of the accused had previously worked in that room and knew the security arrangements. www.oldbaileyonline.org. Ref: t18010520–58.
6   LGL. MS 11936/342, Sun Insurance policy no. 524057 of 1787.
7   TNA, MS CRES 2/729. Among these 'larger instruments' now under construction was a large equatorial telescope for Sir George Shuckburgh. In Ramsden's head, on paper, or perhaps beginning to take shape, was a transit telescope for Gotha and, commissioned by General Roy for the Ordnance Survey, the pyrometer for testing the steel surveying chains, and the three-feet geodetic theodolite.
8   TNA, MS E 367/6004.
9   TNA, MS CRES 2/729.
10 TNA, MS CRES 2/729 and CRES 2/730. Indentures for leases and sub-leases are in the archives of Bethlem Royal Hospital, to whom the property was transferred in the early nineteenth century.
11 TNA, MS CRES 2/729. Fordyce to HM Treasury.
12 TNA, MS CRES 2/279. Lords Commissioners to Surveyor General, 5 December 1796.
13 BPL Boulton & Watt Coll. MB 251, 83.
14 RGO, MS 35/AC I.
15 RGO, MS 35/AC II.
16 PAO, MS D5-37. 'Le courant de mon atelier me fournit suffisament l'argent necessaire'.
17 Millburn (2000), 176, citing TNA WO 47/89 pp. 216 and 598.
18 Millburn (2000), 177, citing TNA WO 47/91 p. 69 and WO 47/95 p. 434.
19 Brazell (1968); Andrews (1887).
20 Lavoisier (1862–93), 3.349–420.
21 Bickerton and Sigrist (2000), *Pictet corr.*, 3.235. '… le brouillard qui a couru l'Europe … il a paru pour le premier fois le 17 juin … interrompu par series de jours pluvieux …recommença le 23 juin et dura presque sans interruption le reste du mois et tout celui de juillet. Le soleil n'avait guère plus d'éclat que la lune ….'
22 Stothers (1966).
23 Stevenson (1992), 206–7; McAdams (1972), 25–53.

24   LGL. MS 12876/5, Christ's Hospital registers.
25   TNA, MS IR 1/23. Not all bindings attracted tax and the registers are not a complete record.
26   TNA, MS IR 1/25.
27   TNA, MS IR/1/26.
28   TNA, MS IR 1/29.
29   TNA, MS IR 1/33.
30   Wallis (1985), 911.
31   TNA, MS IR 1/34.
32   TNA, MS IR 1/37.
33   TNA, MS IR 1/37.
34   TNA, MS IR 1/37.
35   British Museum, Heal Coll. H105.15 and 105.18.
36   Gloria Clifton, personal communication 2005.
37   Offered for sale in *Tesseract Catalogue* 47 (Winter 1994/5) item 20.
38   LGL, MS 11936/260, policy no. 389160 of 1777.
39   LGL, MS 11936/321, policy no. 490375 of 1784.
40   CRO, MS D PEN/223, Letter from Mansell Swift to Colonel Ramsden, 14 March 1928.
41   TNA, MS IR 1/37.
42   TNA, MS IR 1/37.
43   Estacio dos Reis (1991), 86–7.
44   Estacio dos Reis (2006), 33–5 and Figure 5.
45   Estacio dos Reis (2006) is the story of Marques and his life as a steam engineer.
46   Estacio dos Reis (1991), 86–7.
47   Foderà Serio and Chinnici (1997), 63.
48   Estacio dos Reis (1991), 92 n. 21A, citing Arquivo Geral da Marinha, caixas da Cordoaria.
49   Estacio dos Reis (1991), 10–12, 91–2, n. 14.
50   Christensen (1994), 212.
51   Bergman (1952), 130–31.
52   BPL, MS MB 251, 80, Ramsden to Boulton 22 January 1777.
53   BPL, MS MB 251, 80, Ramsden to Boulton, 22 January 1777.
54   BPL, MS MB 251, 81, Ramsden to Boulton, 30 October 1777.
55   *DBF* art. 'Jecker'.
56   CNAM, MS C.25g Report on [François Antoine] Jecker.
57   Turner, A.J. (1989), 54. The Jecker family were the first to set up a large manufactory in Paris, emulating Ramsden's methods.
58   Krüss (1966), 7 and 9.
59   Ragona (1857), cols. 262–3, n. 2.
60   I am indebted to Wolfgang Schaller of Stuttgart, for his summary of Baumann's career including that gleaned from Stadarchiv Stuttgart. Baumann's work was praised in *Monatliche Correspondenz* 6, 450. See also H. Minow, *Historische Vermessungsinstrumente* (1982).
61   Repsold (1908), 93.
62   TNA, MS PROB 11/1578. Will of James Allen senior.
63   Allen (1811), 107–19.
64   Kitchiner (1818), 85.
65   Jungnickel and McCormmach (1988), 386.

66 CRL, MS NKS 287 ii. Bidstrup Letter 2, 1789, and Letter 3, 1790. Transcribed and translated by Claus Thykier, adapted by AMC.

67 RGO, MS 356/MBIII, 8.

68 www.oldbaileyonline.org. Ref. t18010520–58. Proceedings of the Old Bailey, 20 May 1801.

69 Ferenc Kazinczy (1960). *Pályám emlékezete* 1. (Diary of my work), 197. Kazinczy states that when two Hungarians met Ramsden in 1799 and asked if he knew Vay, Ramsden told them how Vay had helped him with the calculations for a new instrument. I am much obliged to Magda Vargha for providing a translation of the excerpt.

70 McConnell (1994).

71 Uglow (2002). While some London streets were illuminated, the informal title derived from their evening meetings elsewhere being timed to coincide with the full moon.

72 Edgeworth (1969), 188.

73 See also Robinson (1958).

74 On Troughton's circle, McConnell (1992), 14; on Jones' circle, Warner (1979), 8.

75 McKay (1979–1980), 222 (in German, with some abbreviations); Christensen (2001), 55–6. The variations between the two reports may arise from the difficulty of reading the original, and from Christensen having translated the original into Danish then into English.

76 Christensen (2001), 56.

77 MHS, MS Museum 221.

78 Sàrospatak, Hungary, Protestant College, Archive of the Vay family Kii IV/1I am indebted to Magda Vargha for a copy of this letter.

79 Christensen (1994).

80 CRL, NKS ii,. Bidstrup to his patrons, 6 March 1789. Transcribed and translated by Claus Thykier, adapted by AMC.

81 TNA, MS PROB 31/929/337.

82 I am indebted here to the English text of Dan Christensen (Christensen, 2001), who deciphered the old-style German of Ljungberg's notebook, now in the Riksarkiv, Stockholm.

83 Christensen (2001), 51–4.

84 Patent 1316 of 14 January 1782. The process of silver-plating copper originated in Sheffield in the 1840s but was not patented. The son of the well-known Benjamin Martin, J.L. Martin went bankrupt in 1782 and departed to Naples.

85 BPL, Ramsden to Boulton, 29 December 1781. MB 251.

86 BPL, Boulton Letterbook M, p. 5, Boulton to Ramsden, 2 July 1782.

87 BPL, Boulton Letterbook N, p. 22, Hodges *pp* Boulton to Ramsden, 7 June 1784.

88 www.oldbaileyonline.org. Ref. t17790113–115, Proceedings of the Old Bailey, 13 January 1779.

89 MOL, MS Whitefriars Glasshouse archive.

90 Pellatt (1849), 41.

91 Turner, G.L'E. (2000b), 404–8.

92 BPL, MS MB 251, 86. Ramsden to Boulton, 6 September 1786.

93 Hall (1991).

Chapter 5

Observatory Instruments and Expeditions, before 1786

Is there no news from the sieur Ramsden? I believe I must go and beat up his quarters myself.

Hornsby to the Duke of Marlborough nd.?1784.
Royal Astronomical Society, Radcliffe A2. 89

A man full of constructive ideas ... a worthy man, but sluggish in the making of instruments.

Jan Komarzewski to King Stanislas of Poland,
cited in SML Court MSS folder 'T'

This chapter deals in turn with each of the observatories which Ramsden supplied, and a few of the expeditions which took his instruments.

Observatories and Their Instruments

Two major 'Royal' observatories were founded in the seventeenth century. The French Observatoire Royale was established in Paris in 1667. Its director, invited from Bologna in 1669, was Giovanni Domenico (Jean-Dominique) Cassini (1625–1712), first in the dynasty of four members of the Cassini family who in turn directed this great institution until 1793. Across the Channel, the Royal Observatory at Greenwich, completed in 1676, occupied a hill overlooking the Thames downstream from London. John Flamsteed (1646–1719) was the first holder of the title of Astronomer Royal. Both observatories were intended for positional astronomy, in an age when only simple telescopes of very long focal length were available, and star positions were measured with portable quadrants. Increasing the quadrant's radius expanded each division on the graduated arc, but such enlargement increased precision at the cost of accuracy. Also, as their size increased, the quadrants ceased to be portable, and to maintain rigidity, were secured to the observatory wall. The first mural quadrants at Greenwich had iron frames and brass graduated arches, but differing co-expansions of the two metals tended to deform the arch and in 1749 John Bird was commissioned to construct an all-brass mural quadrant.[1] This change in the style of instruments in turn changed the architecture of observatories. Accurate clocks were needed to time events such as the starting and ending of solar or lunar eclipses, the transits of Venus or Mercury across the sun's disc, or the meridian passage of certain stars. Considerable technical effort went into the design and manufacture of clocks, both for astronomical observatories and for navigators, in the form of marine chronometers which could function reliably at sea. The craft of clockmaking has its own history and historians and will not be considered here.[2]

At the University 'degli studi' of Bologna, an ancient secular foundation although located within the Papal States, an observatory tower was erected by 1725 as part of the Istituto delle scienze, established within the university. In 1739 it was decided to acquire the latest types of instruments, and with the assistance of the Royal Society (which had elected several Bolognese scientists) and Thomas Derham, British ambassador in Rome, orders were placed with Jonathan Sisson (1690–1747), pupil of George Graham (1673–1751), the renowned maker of clocks and instruments, for a transit with a 3-foot telescope, a 3-foot mural quadrant, and a 2-foot portable quadrant. These arrived by sea at Leghorn and were set up in the observatory in 1741. This mural quadrant was the first of its kind to have both frame and arch made of brass, preceding those from Bird's workshop, which some historians have hitherto considered to be the earliest. Sisson constructed his frame from a latticework of plates, a method followed by Bird and Ramsden. Thus newly supplied, there was no call for Ramsden instruments.[3]

Several observatories purchased mobile quadrants and other apparatus from the Paris shop of Jacques Canivet (d.1774).[4] The fine reflecting telescopes of James Short were much sought after and Short, who in 1738 moved his workshop from Edinburgh to London, was represented in many public and private observatories.[5] As the number of observatories increased, many being founded with the aim of supplying an accurate grid of latitudes and longitudes on which to peg a local cartographic survey or to support a school of navigation, Bird's reputation often made him the first choice as supplier. By 1770, with Jeremiah Sisson in financial difficulties, and with Bird preoccupied with instruments for Oxford's Radcliffe Observatory, buyers turned to Ramsden. By the time of Bird's death in 1776 Ramsden had risen to prominence as the leading designer and builder of mural quadrants and transits.

Seen from the front, the mural quadrant appears to be a fairly straightforward construction, but the rear face, shown in Bird's publication of 1768 and repeated in 14 large plates by Pierre-Charles Le Monnier (1715–1799), in 1774, shows how complex these great instruments were.[6] Each joint was reinforced by flat or angled plates, fastened with more than a thousand binding screws. The arcs were built up from short lengths of plate, held together by dove-tail joints, a construction also seen on portable quadrants. Le Monnier illustrates the steel bearing plates for the axis, the telescope and its counterpoise, and how the arc carries the two divisions, into 90 and 96 parts. The skill and labour involved is reflected in the time needed to construct these mural quadrants and their considerable cost.

The first regular journal offering prompt news of astronomical matters was Franz Xaver von Zach's monthly *Allgemeine Geographische Ephemeriden* which began in 1798.[7] Prior to that time, pamphlets and correspondence were the only rapid means of conveying news and the all-important claims of priority, but the printed word was no match for the practical comparison of instruments. Ramsden's fame was broadcast across western Europe by the astronomers' pleasant custom of travelling round the best public and private observatories in order to meet old friends and exchange information on new discoveries and instrument development. Back home, they set about persuading their paymasters to commission the best and newest instruments; by the third quarter of the eighteenth century, that generally meant applying to the London craftsmen, principally the Dollonds for achromatic telescopes, and Ramsden and later the brothers John and Edward Troughton, for other large apparatus. Contracts were signed by the astronomer responsible if he was in London, otherwise by diplomatic representatives of his home state; payment went through mercantile banking houses where it was converted to sterling. Nevil Maskelyne, as Astronomer Royal, was often called on to inspect the finished product and declare it fit and proper for its intended use, before the apparatus was partly dismantled and carefully packed for shipment, on the first stage of its journey across Europe.

Throughout his working life, Ramsden supplied instruments to private and public observatories in Britain, France, and the various kingdoms, dukedoms and small republics which then made up much of central Europe.[8] It was to satisfy this demand for large apparatus, much of it novel in design, which led Ramsden into the more spacious premises in Piccadilly, where he was able to supervize their day-to-day construction. He also supplied both standard and new types of instruments to the Admiralty for naval expeditions.

Vilnius Observatory

In 1570 a small college, known as Accademia Vilnensis, was founded by the Jesuits in the city of Wilno, now Vilnius in Lithuania; this became a university in 1579. Mathematics and astronomy were taught, Galileo's telescope and its observations of Jupiter's satellites were described by J. Rudamina in his *Illustriora theoremata et problemata mathematica* (1633) and, possibly under Oswald Kruger (1598–1665), a telescope was acquired. The college expanded in 1639, with the support of Wladislaw IV Vasa, Duke of Lithuania and King of Poland, the year when Alberto Dyblinski published *Centuria astronomica*, reviewing the work of eminent astronomers of the time.

The seventeenth century closed with a series of wars bringing economic disaster and famine to the region, and the university began to revive only in the mid-eighteenth century, as Enlightenment ideas stimulated science and education. In 1752 Thomas Zebrowski (1714–1758), architect, mathematician and astronomer, returned to Vilnius from studies in Vienna and Prague. Although Vilnius enjoyed only about 100 clear nights per year, an observatory occupying additional floors and three observational towers was established in the college in 1753, funded by the Polish Countess Elzbieta Oginska-Puzynina (1700–1768), who had a great interest in astronomy. The first telescope, a 13.5cm reflector, was donated by Michaelis Radziwill (1702–1762), Commander-in-Chief of the Lithuanian army; a 10cm reflector was donated by Josephus Sapiega (1708–1754), bishop of Vilnius.[9]

Zebrowski, as the observatory's first astronomer, was succeeded in 1758 by Jokubas Nakcijonavicius (1725–1777), then in 1764 by the Jesuit Marcin [Martin] Odlanicki-Poczobutt (1728–1810). Between 1761 and 1763 Poczobutt[10] studied mathematics, physics and astronomy in Germany, Italy and France, where two years at Marseille Observatory under Esprit Pézenas, inspired Poczobutt with a lifelong passion for astronomy. He returned to Vilnius in 1764, then went to visit the observatories of Greenwich and Paris.[11] He attended meetings at the Royal Society in December and January 1768–69 and was elected FRS on 30 May 1771. In 1769 he observed the transit of Venus from Talinn.[12]

Language was one of several problems confronting travelling astronomers. Poczobutt was staying 'at Mr Barrets, Suffolk Street, Strand' in 1768 when he was invited by the clockmaker John Holmes, who also lived in the Strand:

> Mr Holmes presents his compliments to Mr Pozobut [*sic*]. He has engaged Mr Smeaton, Mr Michell and Mr Ludlam to dine with him next Sunday at three oclock and will be very much obliged to Mr Pozobut if he will favour him with his company at the same time.
>
> He thinks none of the gentlemen speak French but believes they do Latin but if Mr Pozobut pleases to bring his interpreter with him Mr Holmes will be glad of his company. Strand, Fryday [*sic*] nine oclock.[13]

The Jesuits were expelled from Poland in 1773 but Countess Puzynina continued to fund the observatory; under King Stanislas Augustus Poniatowski (r. 1764–1795) it became the Royal Observatory, with Poczobutt as King's Astronomer.[14] (see Figure 5.1a) During these years he sought to acquire modern instruments, commissioning from Ramsden a 4-foot transit, probably with Dollond optics, built while Ramsden was still at his Strand address (see Figure 5.2). It was delivered in 1765 and was followed in 1770 by a 3½ foot achromatic telescope, again with triple lens by Dollond (see Figure 5.1b); an 8-foot mural quadrant in 1777 (see Figure 5.3a) and in 1788 a meridian circle with a 4-inch telescope objective. The mural quadrant arch, which is now partially de-zincified, shows how it was built up from short lengths of brass, with dovetail joints (see Figure 5.3b).

A slip of paper in Polish gives the expense of shipping the mural quadrant to Danzig in unspecified currency, together with the costs of 'correspondence with the Revds Lind, Buhaty, Ramsden and Aubert'.[15] Historian Romualdas Sviedrys suggests that the quadrant's likely route from Danzig was up the Vistula river to Warsaw and from there overland to Vilnius.[16] In preparation for its reception, the observatory was extended to the south, with the construction of a classical structure with two towers for observations and a substantial wall in the plane of the meridian.

Jean Bernoulli, writing in 1779, remarked on 'Le magnifique observatoire de Vilna en Lithuanie' and finding it the more remarkable, given the climate there:

> It is known in England and France that M. Poczobutt, during his lengthy journey some ten years ago to obtain excellent instruments from the best craftsmen made under his own eyes, let nothing stand in the way of acquiring a complete understanding of their use. His pupil, and now his worthy colleague, M. Strzlecki, who found it easy to progress under such a master ... has returned from a long journey in France and especially in England, and brought from these countries, towards the end of last year, new information and good instruments.[17]

Figure 5.1a Vilnius, the 1770s Observatory building today.

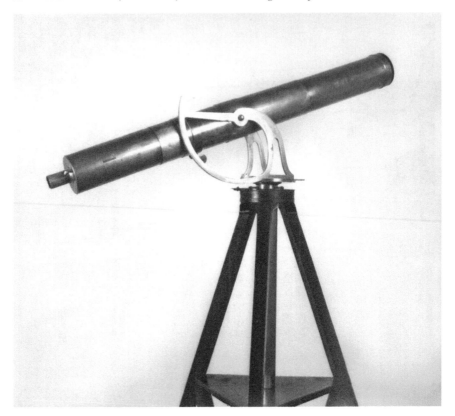

Figure 5.1b Vilnius Observatory, 3½-foot achromatic telescope, signed Ramsden.

Poczobutt was the first to determine the latitude and longitude of Vilnius. He contributed to the *Mémoires de l'Académie royale des sciences* in Paris and some of his observations were published in *Cahiers des observations astronomiques faites à l'observatoire royale de Vilna* (1773).[18]

Ramsden's transit, the mural quadrant, and an achromatic telescope, survived the university's closure between 1832 and 1919, the years under periods of Lithuanian, Polish, Russian and German control, the hazards of World War II and the predations of scrap-merchants. With Lithuania now independent, the instruments can once more be seen in their original setting in Poczobutt's refurbished observatory.

Cartography in Poland–Lithuania

Stanislas Augustus was the first ruler of the vast region comprising Poland and the Duchy of Lithuania to introduce ideas from the European enlightenment into his kingdom. In his effort to revive Polish cartography, he employed a German cartographer, Herman Karol [Charles] de Perthées (1740–1815), who borrowed telescopes, theodolites and timekeepers from Vilnius observatory.[19] Jan Sniadecki

Figure 5.2   Vilnius Observatory, 4-foot transit telescope, signed *Ramsden London.*

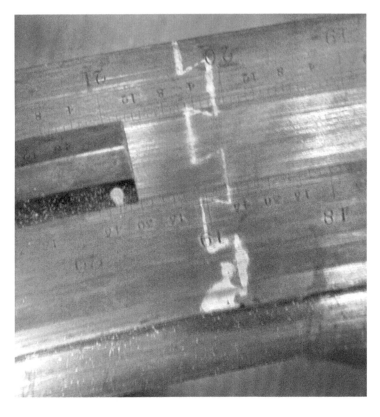

Figure 5.3b Detail, showing how the quadrant arch was built up from small sections, fastened by dove-tail joints. Photographs: Arminas Stuopis.

Figure 5.3a Vilnius Observatory, 8-foot mural quadrant signed *Ramsden London*. An access platform half-way up the instrument prevents a full view.

(1756–1830) a distinguished Polish mathematician and astronomer who had visited the astronomer William Herschel in 1787, was head of the observatory from 1807–24.[20] Sniadecki and his military colleague, Jan Chrzciciel Komarzewski (1748–1809) were sent to London to acquire instruments from Ramsden. A large scale survey was to begin as soon as instruments arrived, but the project was never started, being interrupted by the first partition of Poland in 1772 and the expulsion of the Jesuits. A catalogue of the King's mathematical instruments, compiled in 1782, includes two astronomical quadrants of 14 Paris inches radius and with fixed and moveable telescopes, in mahogany boxes; a pocket telescope in shagreen; and a theodolite, diameter 11 Paris inches, all by Ramsden. There is an inserted note that one of the Ramsden quadrants was given in 1796 to the King's astronomer (Poczobutt), along with the second of two Shelton clocks.[21]

Komarzewski, fluent in English and French, spent much of his time with William Herschel (1738–1822), and stood godfather to his son, John Frederick William Herschel (1792–1871). In a letter to the Polish King, Komarzewski tells of his many visits to Ramsden, 'A man full of constructive ideas … a worthy man, but sluggish in the making of instruments' who repeatedly assured Komarzewski that he would not have to leave London without his theodolite.[22] He was elected to the Royal Society in 1792. The King then sent Poczobutt's assistant, Andrew Strzlecki, to study in Paris and to go to London where on behalf of the King he ordered instruments from Ramsden and Dollond. Strzlecki travelled to London, bearing letters of introduction written by astronomer Edme-Sébastien Jeurat (1724–1803), from the Paris observatory, dated 22 January 1778:

> Jeurat has the honour of sending to the Revd Strecki [*sic*] a letter for M. Maskelyne, protestant minister who will willingly show him round the superb Greenwich Observatory, and also a letter for M. Magellan, catholic priest, who will arrange for M. Strecki to be admitted to the Royal Society and to the Society of Arts.[23]

He obtained the prices of instruments from Maskelyne, whose note to Strzlecki 'at Montforts, Comb maker in Davis Street Berkly Square' was dated Greenwich 16 April 1778:

> Sir, The expense of instruments which you require is as follows. This I send in answer to a note from Mr Narvoys to me. I am – your very humble ser[t].
>
> | | |
> |---|---:|
> | Mural Quadt in the common way | Guineas 400 |
> | Package | 7 |
> | Internal micrometer | 15 |
> | Mr Aubert's contrivance for taking off the weight off the telescope from the centre | <u>5</u> |
> | | 427 |
> | If screws be added to go thro' the flat bars into the edge bars the additional expence will be | <u>50</u> |
> | | 477 |
> | An apparatus of a plumb line and a level on a board for examining the total arc of the mural quadrant | <u>25</u> |
> | | 502 |

A 2 feet moveable quadrant, with 3 telescopes, two having a horizontal motion, & one of them a cross axis to produce a vertical motion; with two levels to set the cross axis of the telescope and the plane of the quadrant horizontal, will come to 105 Guineas.

A transit instrument like Mr Hornsby's with a counterballance [sic] 110 Guineas.

A plum [*sic*] line apparatus similar to that lately made for the Royal Observatory at Greenwich 15 Guineas, in all 125 Guineas.

P.S. I wish when you write to Pere Hell you will mention that there are other instrument makers as capable of executing these instruments as Mr Sisson; & that I desire to be at liberty to employ one of them, or Mr Sisson, as I shall see to be most proper & convenient. I speak this, particularly, because I know Mr Sisson expects an order to be sent for another mural quadrant, & other instruments, & it may not be proper that he should have two such considerable pieces of work in hand at one time.

P.S. I desire you will not send [paper torn] me by the stage but by the [paper torn] post.[24]

'Pere Hell' was Maximilian Hell (1720–1792), the Jesuit astronomer at Vienna. 'Mr Aubert's contrivance' refers to the amateur astronomer Alexander Aubert (1730–1805), who possessed a 4-foot mural quadrant by Bird. Strzlecki may have visited his observatory, then at Loampit Hill in Kent.

Ramsden was able to write in French to Strzecki:

Ramsden fait bien ses compl[es] a Mons Struki [*sic*] et comme il va envoyer son commission a la manufacture de plaque de cuivre aujourdhui. Pour son quart de cercle si Mons Struki peut lui obliger avec £50 sur son accompt Ramsden lui remercierai infiniment.

  Piccadilly 6 Aug[t] 1778.[25]

The report on the survey, which was delivered to the Commssion of National Education, was lost in the subsequent turmoil of war and the second partition in 1793. With the third partition, in 1795, Poland and Lithuania disappeared from the map and Vilnius passed into Russian hands.[26]

The Duke of Marlborough's observatory at Blenheim, Oxfordshire

The observatory at Blenheim Palace, in Oxfordshire, was set up in about 1780 by George Spencer (1739–1817), fourth Duke of Marlborough.[27] There are contrary appraisals of the Duke's character; the anonymous author of *Jockey Club, or a sketch of the manners of the age*, a journal which aimed 'to hold the mirror up to nature, to show vice its own image'[28] depicted him as manifesting 'a uniform hauteur of deportment ... sullen and overbearing in general demeanour',[29] whereas in the *Complete Peerage of 1775* he is 'a noble man of great worth, generous, humane, and hospitable, no abject courtier'.[30] If the first description related to his life in politics and high society, the second is more true of the Duke happily at play in his observatory. His passion for astronomy led the Duke to spend many cold clear nights star-gazing and long daytime hours with pencil and paper on his mathematical calculations, while dispensing enormous amounts of money on his hobby. Certainly, in his dealings with Ramsden, the Duke, who was elected FRS in 1786, showed a humorous affection, if tinged with despair at Ramsden's occasional unreliability and lack of timekeeping, and this pleasant aspect of the Duke's character was remarked by the diarist Joseph Farington, who noted on 6 September 1800 'though nervous in the extreme and reserved, he possesses a great deal of humour'.[31]

The Duke worked closely with Thomas Hornsby (1733–1810), Savilian Professor, and astronomer at the Radcliffe Observatory on the outskirts of Oxford. This observatory was built and equipped by the Radcliffe Trust, on land provided by the Duke, on whom Oxford University had recently conferred an honorary doctorate. Under Hornsby's tuition, the Duke learnt his observing technique on the fine suite of instruments delivered by John Bird, and was taught the necessary mathematics.[32] When he came to create his own observatory at Blenheim Palace, Bird had died and the Duke turned to Ramsden as the leading London maker. It was the start of a long and close relationship, with Ramsden not simply delivering the finished instruments but being frequently summoned to make the long coach journey to Blenheim in order to effect adjustments or repairs.

The Duke's 6-foot transit and a 6-foot mural quadrant of unusual design, were supplied by Ramsden in the early 1780s, and although the first dated astronomical observations from Blenheim were made in 1781, it seems that the quadrant had to be shipped back for adjustments and re-division. Unusually, it was supported on pillars rather than fixed to a wall. It could be easily rotated to face north or south, since it was carried one foot out from four columns, spaced 20 inches apart, counterpoised to the centre of gravity, and with two other counterweights, one at the centre of the apparatus, the other at the end of the horizontal radius. The four columns were connected above and below by two large pivots around which the apparatus turned.[33] This was not the first example of a mobile mural quadrant. At the Collège des Capuchins in Paris Le Monnier had a 7½ (Paris) foot quadrant made by Bird in 1753 which was secured to a mobile wall which turned on a polished cannonball as its bearing, allowing the quadrant to be swung to face north or south.[34] In 1783 Charles Blagden inspected it and reported to Joseph Banks that it did seem to have a real advantage, but the design was not taken up elsewhere.[35]

Confronted by a series of delays in the delivery and modifications to his quadrant, caused mostly by Ramsden's lack of care, the Duke, who had doubtless paid out a huge sum of money, had simply to grit his teeth and wait for his new treasure to be installed in working order. A series of letters from Hornsby, relating events in London and Oxford, brought him cheer and despair alternately:

14 September 1781 – Your Grace is expecting Ramsden every day. I am impatient to hear, and indeed wonder that I have not heard from Admiral Campbell. Dr Maskelyne in a letter last night tells me that Ramsden has invented a new system of eyeglasses in which the wires are in the common focus of the two glasses which are on the same side of the wires; in my instrument the wires are between the two eye glasses. I hope

this contrivance is applied to your Grace's instrument. Mr Bird made one for me 20 years ago but it did not quite answer.[36]

21 December 1782 – If Ramsden really should come tomorrow ... let me beg your Grace to give him a good box of the ear for the sake of α arietis, and every other observation where the coincidence is very near the beginning of the nonius.[37]

24 December 1782 – [Re the new quadrant etc] I hope Ramsden proposed to alter the reading glass to bring the first division of the nonius to the centre ... but this cannot be done without sending the whole to London, or at least the spring tube.[38]

30 July 1783 – What a shatter-pated fellow that Ramsden is! To send the first man down without the smallest wire.[39]

26 August 1784 – I saw Ramsden on Monday evening. He was at Hounslow on Saturday during the time that the King attended, when it rained heavily as before. I asked Ramsden how the Quadrant went. He said all the men were at work upon it. I asked when it would be finished. His answer was, in a month – or so.[40]

30 September 1784 – I hope with Mr Ramsden that the instrument or at least the axis will be at Blenheim in the course of next week ...[41]

Is there no news from the sieur Ramsden. I believe I must go and beat up his quarters myself.[42]

15 May 1788 – I am sorry that Ramsden is indisposed ...[43]

Favourable Reports

Ramsden's instruments were considered the finest of their kind and much admired by the many visiting foreign astronomers who included Blenheim Observatory on their 'must see' list. In 1784 Professor Patrick Copland (1748–1822) of Marischal College, Aberdeen, was in London overseeing the construction of some of the apparatus he was planning to install in what may have been Britain's first observatory accessible to the public. John Stuart, third Earl of Bute and Chancellor of Marischal College, had donated a Ramsden 4-foot transit, and an equatorial made by Sisson but now to be redivided and refitted by Ramsden. In a letter to the Revd Alexander Cock of Cruden, dated 16 July that year, Copland wrote:

> Ramsden had not touched the great equatorial which [had] been with him many months but the Prof'r has acquired so great influence over him that he has prevailed with him to lay aside instruments for his favourite the Duke of Marlborough and the apparatus for the operations that interest the philosophers so much at present for connecting the Observatories of London and Paris by actual mensuration and sett a dozen men at work on his equatorial and transit instruments.[44]

Barnaba Oriani (1752–1832), the astronomer from Brera Observatory in Milan, was similarly impressed. He had come to England in 1786 to look round the best-equipped observatories and to order a mural quadrant from Ramsden for Brera. He judged Thomas Hornsby's Radcliffe Observatory at Oxford, with its collection of large and excellent instruments, to be the most magnificent. On 26 July, after a swift tour of Oxford's Bodleian Library, he proceeded to nearby Blenheim Palace. His diary records that, after viewing some of the Duke's art treasures, he climbed up to the observatory where he saw Ramsden's admirable transit and quadrant.

Back in London on 5 August Oriani recorded his thoughts. Although the Duke had only two instruments in his observatory, these were unarguably the most splendid instruments Oriani had seen in England. The quadrant he is ordering from Ramsden will be exactly like that at Blenheim, which is the most beautiful and the most accurate example he has ever seen. The only difference will be in the suspension: the Duke's quadrant can rotate on an internal axis and in less than a minute can be turned to face the north. At Brera, for want of space, the quadrant will have to be fixed to the wall. The manner of illuminating the wires on Ramsden's transit was like that at Mannheim, which could be increased or decreased by means of a coloured glass prism, more convenient than any others to be found in England, and infinitely better than that at Brera. (He may be referring here to the transit instrument which had been built by Giuseppe Megele of Milan).[45]

In 1787 three men from the Paris Observatoire: Jean-Dominique Cassini (1748–1845), Pierre-François-André Méchain (1744–1804) and the mathematician Adrien-Marie Legendre (1752–1833), were in England for the geodetic survey to establish the longitude between the observatories of Greenwich and Paris, as described in Chapter 8. Cassini had been instructed to call on the leading London instrument makers and to inspect other important observatories. Learning that Méchain and Cassini intended to go to Blenheim, Ramsden arranged their visit with the Duke and himself introduced them to the observatory. Cassini was overwhelmed by the splendour of the observatory instruments. He greatly admired the design and workmanship of the Duke's rotatable quadrant, immediately coveting a similar apparatus for the Observatoire.[46]

Another Italian astronomer, Giuseppe Piazzi (1746–1826) of Palermo, was also in London in 1788, to order a new instrument from Ramsden; he too expressed admiration for the Blenheim quadrant:

> The six-foot mural belonging to the Duke of Marlborough at Blenheim is another instrument which you, like myself, admired; there is an arrangement of four columns which turn on two pivots so that it may be turned to face north or south in a minute. This instrument is both beautiful and perfect, but no-one is more worthy of owning it than the Duke, no professional astronomers have more zeal, perseverance or accuracy.

But Piazzi noted another feature of the quadrant, the means of ensuring a true vertical, and an error-free 90-degree arc.

> It was for this beautiful instrument that M. Ramsden conceived a means to check the arc of 90 degrees where a competent astronomer had found difficulties, but with

a horizontal wire and a plumb-line making a sort of cross which did not touch the circle, he showed him that there was not so much as a second of error in the 90 degree and that the difference arose from a mural made by Bird, where the 90 degree arc contained several seconds in excess, and which had never been verified by a method as precise as that of M. Ramsden.[47]

This useful adjunct was supplied along with the with the circle sent to Palermo in 1789 and also with the mural quadrant sent to Brera in 1790 (see Chapter 6).

Suspicions of the Quadrant's Accuracy

Despite paeans of praise from eminent visitors – they could admire the instruments at Blenheim but were apparently not invited to put eye to telescope – in the 1790s the quadrant was suspected of error, since the Duke's observations of certain stellar events differed from those of other highly reputable astronomers resident nearby. In 1795 Ramsden was again summoned to Blenheim, a visit which led to some hilarious moments which fortunately the Duke took in good part. Hornsby is our informant:

> 4 April 1795 – If your Grace has not heard from the Gentleman of Piccadilly I think I can venture to say that you will see him tomorrow or Monday. One of his men who is to be at Blenheim with his master was with Powell this morning, and is to be with him again in the evening. And his master is to be here tomorrow morning. ... I suppose Ramsden will not return before Tuesday or Wednesday: and if your Grace will recollect I ventured to prophesy as is the fashion at this time, that he would not come before Easter.[48]

> April 1795 – When Ramsden got up on Sunday morning at the Angel and enquired after a working man, they told him that no such person was there, though Pope was even then in the house and in bed. He immediately posted away to me between 7 and 8, and I told him that Pope was at the Angel. When Pope got up and enquired after his master Mr Ramsden, they told him they knew of no such person. He then went to Powell and Powell did not then know that Ramsden had come. I had planned it beforehand with Pope that they should be with me at breakfast, because Powell was to go to prayers, and that they should be at Balliol after breakfast when I proposed to go to church. I sent my servant down to Powell to inform him that Ramsden was come and to learn what they were doing. He announced the gentlemen's arrival but said no more: Pope came up to see me and was here above half an hour after Ramsden had left me. However they could not produce that flood of light in which we had failed the night before, nor have they succeeded yet, tho' the lamp is now placed in the direction of the axis. Ramsden promised to return on Monday and not on Sunday as was intended. He came on Sunday on horseback and without a surtout for he had left it in London and Pope had his only clean shirt at Blenheim. He forgot something for your Grace and Powell's counterpoise, tho' they were both packed up together, but he did bring my meridian mark, tho' he forgot the adjusting screw. I told him I wondered he had not left his head behind. He went away on Tuesday about 12 oclock, Pope having left his coat for him, and he walked to New College etc in one of mine.[49]
> I believe I did not mention to your Grace that when Ramsden was to set out on Monday morning he was sadly distressed because Pope had his only clean shirt with

him at Blenheim. He said he was not a little hurt that day when he had the honour to see the Duchess. The clean shirt however made its appearance on Tuesday morning; and he packed up the dirty one and directed it, meaning to leave it at the Angel to be carried up by one of the coaches. I told him I might perhaps be in Town the week following and would bring it up for him but he said no: and upon my observing that it was not too bulky for his own pocket, he replied that he did not care to be embarrassed with anything when on horseback. He laid the parcel down and went out leaving it behind him, as I expected. And after he was out of the room I immediately called him back, and reminded him of it.[50]

In 1796–97 the Duke of Marlborough, Ramsden, and Hans Moritz, Count von Brühl (1736–1809), whose observatory was at Harefield House, north-west of London, debated as to why their observations differed.[51] Brühl, mindful of the Duke's status, suggested that some of the Duke's observations made before the redivision of the quadrant agreed better with his own, and again queried the accuracy of the graduation. The Duke consulted Ramsden, who was critical of Brühl's reliance on a small sextant with artificial horizon, and also of Maskelyne, whose mural quadrant he suspected of zenith error. He had offered to check the divisions on both instruments free of charge but was not allowed to do so. The argument ran for some months, the surviving correspondence ending without resolution.

Professor Samuel Vince (1749–1821), astronomer at Cambridge, wrote that after Ramsden had finished this great quadrant, the Duke wished to have the total arc examined, when 'Ramsden, concerned lest the weight laid on the telescope by Bradley's method should have led to some error, invented a piece of apparatus for testing this'.[52] This confirms an anonymous and undated manuscript 'A method of examining the total arc of a quadrant' now preserved at Palermo Observatory.[53] We may take Ramsden to be the author since he is assisted by 'Leyland one of my workmen'. The method employed by Bradley and John Bird on the 8-foot quadrant made by Bird for Greenwich Observatory is categorized as:

> too objectionable to give that satisfaction which is requisite in these nice experiments. … No instrument yet made I have occasion to believe being able to support such pressure without considerably altering its figure and in all probability the length of the arc that we are attempting to measure. … [the tests should not impose additional weight on the quadrant] … Nor is it sufficient to measure the length of the arc contained between zero and 90 degrees – it is also necessary to determine the angular motion of the line of collimation the telescope makes between the index being set to the 0 and 90th degrees on the quadrantal arch.

Having set out the desiderata, Ramsden explains:

> His Grace the Duke of Marlborough whose ideas of accuracy and whose zeal, generosity and encouragement for the practical art of astronomy does honor to his high rank, he order'd me to make him an apparatus constructed on the above principles wherewith to determine the length of the arc of his six feet pillar quadrant at Blenheim. In the measure of that arc the apparatus had not the least connection with the quadrant during the time of the experiment. The examining of the measure of the arc was by

His Grace himself which he did to the exactness of the fraction of a second thereby obtaining that satisfaction which an accurate hand cannot possess by a reliance on the observation of another. I shall now endeavour to convey an idea of the above properties by a description of His Grace's apparatus and his experiment.

The manuscript describes the apparatus and the exact manner of its application. In summary, it consisted of a 7-foot mahogany frame, its weight carried from the observatory ceiling. A plumbline and the various brass parts, cross-wires, micrometers and so on, were attached to the frame and did not touch the quadrant itself. There were divisions at each end for each of the two arches. The space between the divisions for the 90° arch subtending 28° 50′ and that between the divisions for the 96-divided arch having 31 primary and 12 secondary divisions.[54]

The experiment, made by the Duke, Ramsden and Peter Leyland showed the accuracy of the graduations along the 90° arch, there being less than 1 second of error. 'But', Ramsden continues:

On a comparison with some observations made by His Grace with similar ones made by the Revd Dr Hornsby at Oxford a doubt arose concerning the exactness of the experiment, which had been only once made and not repeated. From the above comparison it was conjectured the quadrantal area of His Grace's instrument erred 8 or 9 seconds. An inaccuracy of this sort not at all fitting with His Grace's ideas and precision, he desired me to come down to Blenheim to repeat these experiments which was done in the presence of Dr Hornsby and with which he declared himself perfectly satisfied. The results were much as before namely that there was not an error in the measurement of the quadrantal arc amounting to half a second.

Bird's quadrant in Hornsby's observatory at Oxford had several seconds of error as it had not been verified by such an exact method.

Hornsby sought the Duke's approval for Ramsden and himself to examine the gregorian telescope by Short which the Duke kept at Marlborough House, his town house in Westminster, before it was dispatched to Blenheim.

11 April 1795 – If I find myself tolerably well I think of going to London on Tuesday … We talked about Mr Short's telescope. Will your Grace be pleased to permit Ramsden and myself to survey it as far as its present state will permit. I have always thought that it should be examined before it be sent to this place.[55]

This telescope is now in the Museum of the History of Science, Oxford. The quadrant and transit have not survived and there is no known illustration of either instrument.

Ramsden's Trial Surveys at Sion Hill and Kew

The Duke of Marlborough also had an observatory at his house at Sion Hill, near Kew, where he had a 3-foot mural quadrant and an equatorial telescope, both by Ramsden.[56] In the summer of 1796 Ramsden was making some trial surveys between the Pagoda in Kew Gardens, the King's Observatory at Kew and the Duke's house. On 17 July he wrote to the Duke:

Having made a 25 feet measuring chain one Friday evening along with Mr Demainbray, Mr Staunton etc I measured a Base of 1050 feet and by taking angles at the extremity of the Base and thence computing the distance between the Pagoda and the Observatory it came out to be 4401.5, by computing with the greatest care from the triangles taken by General Roy it appears to be 4403 feet. There may be something of chance, my Lord, in the nearness of these results but such is the case of it and should afford your Grace any amusement. I will have the honor to transmit to you the particulars which I thought to have done but that I am continually interrupted when I get down to work.

I now think my Lord of ascertaining the latt^e of some place not far from the King's Ob^sy by a circle.

In the last letter of this series, he wrote on 24 July, that he was 'under no necessity but amusement … and by practising the use of instruments to see what improvements may be made in their construction', revealing that even at this late stage in his life, Ramsden was still looking for ways to improve his apparatus.

Padua Observatory

The north Italian city of Padua (properly Padova) had been home to the secular university, 'degli studi', of the Venetian Republic since the thirteenth century. Giovanni Alberto Colombo (1708–1777), professor of astronomy from 1746–64, repeatedly petitioned the Venetian Senate to establish an observatory and in 1761 a decree was issued to this effect, as part of a general project to reform teaching at Padua. This decree might have remained on paper, had it not been for Giuseppe Toaldo (1715–1797) who succeeded Colombo as professor; he in turn appointed Domenico Cerato (1715–1792) as architect. The tower of the university was unsuitable, being located in the city centre without clear views to the horizon, and Cerato converted the ancient tower of an old castle, known as Torlonga, built in 1242. This structure was adjacent to a small waterway, along which building material, and later the large instruments, could be transported from Venice (see Figure 5.4a). The necessary architectural work, involving the modification of the internal structure, allowing a pendulum to be suspended through several floors and the addition of an upper structure, was carried out between 1767 and 1777 under Toaldo's supervision, but the acquisition of small instruments enabled him to commence giving lessons in astronomy before building commenced.[57]

Toaldo had been educated at a seminary in Padua where he became an enthusiastic disciple of Galileo's applied mathematics. He took his doctorate in theology, and while undertaking parish duties, studied astronomy, meteorology and the geology and natural history of the region. He had been hoping for a chance to apply for the university's chair of natural philosophy, but in 1764, following a realignment of the teaching duties, he was appointed professor of astronomy and meteorology, with the additional requirement to teach geography. Needing to learn more about recent scientific ideas, he visited the observatories at Bologna and Pisa, and later consulted astronomers Joseph Liesganig (1719–1799) at Vienna and Ruggiero Boscovich (1711–1787), who had worked at Brera (on Boscovich, see below).[58]

In a letter of 1766, addressed to his paymasters, the 'Riformatori' of the Venetian Senate, Toaldo listed the apparatus needed to teach astronomy, geography and navigation, principally telescopes, armillary spheres, and a clock. He also compiled a list of major instruments for the observatory, as found in the best-equipped Italian observatories, Pisa, Bologna and Milan, and in those north of the Alps, such as Vienna. Although the Venetian instrument maker Lorenzo Selva (1710–1800) could make simple refracting telescopes, he did not have access to good-quality, dense flint glass needed for achromatic lenses, which would therefore have to be procured from the Dollond workshop. The local craftsman Antonio Fabris could make the small brass quadrants, but the major instruments could only be obtained from London. Toaldo had Lalande's London prices; he had also consulted with Nevil Maskelyne, but Maskelyne's letter with prices did not reach him until the summer of 1768. Toaldo – one of the few from this period who did not go to London – was elected FRS in June 1776, following letters of support signed by his friends in Paris, London and Venice.

The assortment of cited values – French livres, English pounds, and guineas, had to be converted into Venetian ducats or lire. According to Maskelyne, an 8-foot mural quadrant from Bird cost £370; in Lalande's list this was given as 8,000 livres – by Toaldo's calculation, 2,035 ducats. But if an achromatic telescope was included, this would raise the price to 2,750 ducats.[59]

On 28 November 1770 Toaldo submitted two lists, one of essential apparatus, with an estimated cost of 4,662 ducats:

Two English astronomical clocks
An English achromatic telescope with micrometer
12 ft achromatic telescope by Dollond
Mural quadrant, 8 ft radius, by Bird
Portable quadrant, 3 ft radius, by Bird or Ramsden
Two globes by Senex
Equatorial, with telescope made in Italy or Paris
Transit and stand, from Italy or Paris
English 2 ft level
Graduated toise of the Paris Academy
Declination needle
Barometers, thermometers and other small instruments.[60]

John Bird probably declined the order for the quadrant; he was more than fully occupied with similar instruments for Oxford. Maskelyne's notebook carries the entry '20 October 1775. Mr Ramsden received of Mr Pisani the Venetian Resident 200 gns being moiety of an 8 foot mural quadrant to be finished in 12 months' and this corresponds to a letter dated 4 September 1775 from the Riformatori for a similar amount to be sent to the Venetian Resident in London.[61] Maskelyne examined the quadrant when it was finished and found that no error exceeded 2½ seconds. He added to his original notebook entry 'and finished Novr 1778 in 3 yrs'[62] (see Figure 5.4b). Unlike the later 8-foot quadrant for Brera, there seems to have been no difficulty in finding a suitable vessel to transport it to Venice where

Figure 5.4a Padua, the old Observatory building today.

Figure 5.4b Padua Observatory, 8-foot mural quadrant, signed *Ramsden*.

it arrived in 1779, Toaldo being paid in April to deal with its shipment along the canal to the Observatory.

The next problem was to get it erected and set up within the tower, and Toaldo was permitted to visit Brera, where Boscovich had previously supervised the setting-up of a large mural quadrant by Canivet. Back in Padua, Toaldo addressed one of the Riformatori about the matter of erecting the Ramsden quadrant. Ramsden had insisted on sending one of his young men to set up the quadrant but this demand had been refused because of the exorbitant cost. Toaldo had then considered inviting Giuseppe Megele, the craftsman responsible for erecting the quadrant at Brera, but as Megele was on a salary Toaldo could not negotiate privately with him, and the Riformatori would need to arrange this with the governor of Milan. If Megele did come, not only could he instruct Toaldo's mechanics, but he could at the same time set up the two small quadrants brought from the Jesuit college at Venice. This plan was likewise rejected; Toaldo had to make do with Antonio Fabris, who was paid for twelve days work during the autumn months to mount the mural quadrant. The cost was considerable, involving as it did assorted craftsmen – masons, carpenters, painters and Fabris himself. Included in the account is an entry of 18 lire paid as an encouragement to the 12 men who raised up the quadrant. The university paid for an assistant to observe with the quadrant, to which post Toaldo's nephew, Vicenzo Chiminello (1741–1815), later his successor, was appointed.[63]

The astronomer Giovanni Santini (1787–1877), who arrived at Padua Observatory in 1806, described the principal instruments there in his *Elementi di astronomia,*[64] categorizing the mural quadrant as the 'opera perfettissima del celebre artefice Ramsden'. The quadrant, now in fine condition, is displayed with other instruments within the old Observatory in Padua.

Introducing the Cartographer Gianntonio Rizzi Zannoni

Here I re-introduce the cartographer Gianntonio (Giovanni Antonio) Rizzi Zanonni (1736–1814), who was born in Padua and returned there many years later, to assist Toaldo in his observatory.[65] While earlier historians thought that his colourful accounts of his life and activities in Poland, Germany, France, Britain and North America, were largely invented, recent studies confirm the high quality of his contributions to French, Polish and Italian cartography. From Paris, where he was employed in the Depot de la Marine, collaborating with the royal cartographer Nicolas Delisle and other map-making notables, Rizzi Zannoni wrote to the map publishers Jefferys & Faden, asking for the price of a Ramsden astronomical quadrant of 2-foot radius, and if one could be bought from stock.[66] He reports that he had such a quadrant made in London (see Figure 5.5).[67]Rizzi Zannoni returned to Padua in 1776, where he instructed Toaldo's nephew Chiminello in the use of the observatory instruments. In 1780 he issued his *Gran Carta del Padovano*, and the following year moved to Naples, where he set up the Topographical Department. He remained thereafter in the Kingdom of Naples.

On the unification of Italy, Rizzi Zannoni's quadrant, and a Ramsden equatorial telescope with circles of 10½ inches diameter, for which we do not have a date of purchase, passed to the Istituto Geografico Militare in Florence, where they remain (see Figure 5.6).[68]

Two Failed Commissions

An Equatorial Sector for Greenwich Observatory

In about 1735 Graham had supplied Greenwich with an equatorial sector of 2½ feet radius, the first of its kind. In 1770 Maskelyne complained to the visitors that Graham's venerable equatorial sector was inadequate, its old telescope unable to see faint comets – the principal task for this instrument being the measurement of the angular distance between a comet relative to a star whose position was known. In 1773 two sectors were purchased from Jeremiah Sisson. In 1775 the

Figure 5.5   Astronomical quadrant, radius 24 inches, signed *Ramsden*, used by Rizzi Zannoni. Florence, Istituto Geografico Militare.

Figure 5.6   Equatorial telescope, circles of 10 ½ inches diameter, signed
Ramsden London Inv et fecit; used by Rizzi Zannoni. Florence,
Istituto Geografico Militare.

visitors, with Bird and Ramsden in attendance, examined Sisson's sectors and found them useless. Bird declared himself too busy to give more than advice and the visitors decided to consult Ramsden on the cost of the necessary repairs.[69] He duly reported:

> In obedience to your commands I have remov'd the Equatorial Sector from the Royal Observatory to my house. I have diligently inspected and examined it and find the state of it to be as follows.
>
> 1st The tube of the telescope is strong and very good, the finder also will do very well
>
> 2. The large brass dovetail pieces which support the two ends of the Polar axis will do again with some little alterations
>
> 3. The Polar axis is very weak and will require considerable strengthening
>
> 4. I am of opinion that the meridian and Equatorial circles are much too small and others should be substituted in their stead, as large as can conveniently be made
>
> 5. A long axis is absolutely necessary to be applied to the declination circle
>
> As far as I can judge at present the expense of these alterations in order to make it a complete instrument will not exceed 150£.[70]

The old sector was not worth repairing and in July 1778 he was ordered to make a new instrument with a 5-foot telescope and 5-foot circles, at a cost of £500. The Board of Ordnance advanced Ramsden £200 and on 11 August 1781, the visitors having called at his Piccadilly workshop, 'the equatorial sector now making by Mr Ramsden was found to be in great forwardness'.[71]

But by January 1783 the visitors were most unhappy about Ramsden's lack of progress. On 16 January:

> Mr Ramsden, who had been summoned at the request of the Astronomer Royal, attended, and being questioned concerning the state of the equatorial instrument which he had, five years ago, received orders from the Board of Ordnance to make for the Royal Observatory and what was the reason for the long delay in sending it in. He answered that the delay in great measure arose from the tardiness of the payments of the Board of Ordnance which rendered it necessary for him to attend more particularly to such orders as he knew he should be paid for immediately.

This was a good excuse, cash flow being always a problem where large instruments were several years in the making, meanwhile the workmen had to be paid daily or weekly.[72]

Ramsden was then questioned about the £200 advance, to which he replied reasonably enough that it had been spent on the acquisition of the materials. But worse was to come: the Council told Ramsden that they had heard that another would-be buyer had made him an offer, which he had not absolutely refused. Ramsden admitted that he had been offered 500 guineas (£25 more than he had agreed with Greenwich), with immediate payment on delivery. He admitted that this was a tempting offer but if he could be sure of prompt payment from the Board of Ordnance, the equatorial would definitely go to Greenwich. This did not go down well. 'Mr Holford, upon this acquainted Mr Ramsden that having given a receipt for an imprest, he might be liable to a prosecution if he did not fulfil the conditions for which he had received that imprest.'

The Council also assured him that if he would apply himself with more diligence in completing the instrument, they would endeavour to speed his payment, and with some guile, they asked for a delivery date in order that they might forewarn the Ordnance as to when payment would be needed. Ramsden obligingly promised to deliver in July that year.

After he had left the room, the visitors ordered that the Ordnance should be asked to keep the outstanding £300 in readiness, and to let the visitors know that this could be done, in order to press Ramsden. But having met their side of the bargain, the visitors were probably not surprised to learn, on 25 July, that Ramsden had failed to deliver. They copied the extract of their minutes to him, with a note that they would meet at his house on 2 pm of 14 August. There they saw the meridian circle ready for dividing, with basket bracers for receiving the telescope axis; the polar axis, complete with ten cones for supporting the equatorial circle; the equatorial circle and the telescope, without its object glass. Ramsden promised if possible to finish the instrument within six weeks. On that day, 25 September, Aubert, Cavendish, Maskelyne and Planta were shown the meridian circle, now divided; Ramsden promised that the equatorial circle would be divided within three weeks – by 16 October. The committee was summoned, but the visit was postponed at Ramsden's request.

The Board's minutes give no reason for this postponement but on 11 November 1784, more than six years after the placing of the order:

> The Astronomer Royal reported that finding that the equatorial instrument making by Mr Ramsden is not likely to be ready for some time to come, he was desirous that the old sector which stood in the new observatory at the west end of the terrace be put in order and replaced. The Visitors approved ...[73]

Possibly the Board recognized that Roy's urgent need for Ramsden's great theodolite for the Greenwich–Paris triangulation should take priority (for which, see Chapter 8).

An Astronomical Circle for the Florence Observatory

The observatory established in 1750 by Leonardo Ximenes (1716–1786), Jesuit, astronomer, mechanic and naturalist, was the first to be established in Florence and the first in Italy to be set up in a Jesuit college, where Ximenes taught physics.[74] Ximenes corresponded widely with mathematicians and astronomers elsewhere in Europe, but his masters obliged him to confine himself to 'useful' observations in astronomy, meteorology and seismology. He was involved with Rizzi Zannoni on the plan to map Tuscany and had listed the necessary instruments for the Tuscan civil authorities. Instruments were acquired from Florence, Rome and Vienna, and from three London opticians, a transit and a dioptric microscope by John Cuff (w.1731–1772), with prepared slides, a telescope by Dollond, and another – surely rather antiquated by this time – by John Yarwell (1646–1713). Ximenes set a craftsman to build a mural quadrant of 10 foot radius, but the divisions were never finished.[75] Gaetano del Ricco succeeded Ximenes as astronomer and new instruments were acquired, among them a 'macchina' from Ramsden, which turned out to be unsuited for the tasks envisioned for it.[76]

Ximenes continued in his observatory without coming into conflict with Duke Pietro Leopoldo of Tuscany, who arrived in Florence in 1765 and decided to establish his 'Imperiale e Regio Museo di Fisica e Storia Naturale' in his rebuilt Palazzo Torrigiani. Felice Fontana (1730–1805) known to the Duke as an experimental physicist, was appointed director, assisted by the naturalist Giovanni Fabbroni (1752–1822), a protégé of the Duke.[77]

The energetic Fontana set about the construction of physics apparatus for the Florentine Museum's cabinet, much of it made by craftsmen under his own direction, to his own designs or those published by the eminent physicists Willem Jacob 's Gravesande (1688–1742) and Jean-Antoine Nollet (1700–1770). We shall meet Fontana again in Chapter 7. He spent the years 1775–78 in Paris and London, buying physics apparatus, chemicals, and natural history specimens, but his ambitious plans included the establishment of observatories for meteorology and astronomy. He was aware that achromatic telescopes could not be obtained in Italy, due to the lack of good quality flint glass, neither was there in Italy a great mural quadrant which would serve as a pattern for his workmen to replicate. For these items he must place orders in London.

Fontana and Fabbroni arrived in London in January 1778, and at one period lodged at 13 Haymarket. Magellan immediately took them under his wing, accompanying them in and around London, introducing them at the Society of Arts and the Royal Society. They dined at the Globe, favourite haunt of the Fellows, in the company of the King's physician Sir John Pringle and Nevil Maskelyne. Fontana presented several communications to the Royal Society while searching for natural history specimens and apparatus. He made friends with Ramsden, who encouraged him to provision the observatory with the finest instruments. In this way, as he informed the Tuscan court, the Florence observatory would be a world leader.

One of Fontana's grandest dreams, never realized, was an astronomical great circle. He brought back from London Ramsden's 14-page list of 'The most perfect instruments for a complete astronomical observatory' headed by 'No. 1, two quadrants of eight feet radius' and 'No 2, a complete circle of twelve feet diameter, which will serve the purpose of two mural quadrants; …'.[78] Undated, but probably written in 1778/9, the list can be found in Appendix 2.

This is an early declaration of Ramsden's preference for astronomical circles. Jean-André de Luc heard of it by 1781, as he wrote to Marc-Auguste Pictet:

> Ramsden would have astronomers employ circles rather than quadrants, for mural instruments or for all instruments where delicate observations are made, and where a circle can be used. Thus, for example, in place of a quadrant of 4 feet radius he would have you employ a circle of 3 feet radius, and these are his reasons.
> 1. The framing of a circle is made up of concentric elements which by going from the centre to the circumference will maintain the form of the instrument whatever

the forces acting on the metal, which is not the case with a quadrant, however well its frame is designed.

2. A circle of 3 feet radius will take a 6-foot telescope, which becomes the diameter of the circle, moving about its centre.

3. The division of a circle is assured. Six times the radius gives six points reliably more certain than the best rules, divided in equal parts, can give a 90° angle, and with these six points, all others can be found by continual bisection.

4. The line of sight is infallible in such an instrument, whose telescope carries a nonius at each end.

5. All the changes which such an instrument may sustain, even all the errors of division, are revealed by this double nonius, and by always reading the observations on both sides, and studying the reasons for the differences, one is certain of avoiding all error.[79]

In the summer of 1779 Fontana sent Ramsden's list, with a translation, to the Tuscan finance minister. Silence ensued until Fontana wrote to Maskelyne in January 1782 to say that he wanted Ramsden to make the 12-foot circle and asking Maskelyne what the price would be.[80] His *Memoria del Direttore* of 1786 was still optimistic:

> The other project is an astronomical mural circle of twelve feet diameter which the famous English Ramsden lacked the boldness to make either for the English Admiralty or for your Honour's Museum, though he contracted to make such an instrument eight years ago, [here Fontana may have had Dunsink in mind] and by its enormous size and novel construction, and for its great utility, and the accuracy needed, it will be unique in Europe, and I hope to conclude the matter in two years, or a little more …[81]

Back in Florence, years passed; in 1782 Jeremiah Sisson delivered a 10-foot zenith sector and an 8-foot transit, the payments made through Antonio Songa.[82] But nothing arrived from Ramsden, and in 1792 the Grand Duke appointed an astronomer to oversee the construction of the circle at Florence. But there were so many practical difficulties: where it could be constructed, who could build such a large and novel instrument without models, and how, when finished, it could be carried into the observatory. Fontana had been reading Piazzi's description of his great circle at Palermo Observatory, and in April 1794 he allowed that 'the builder of such circles must know astronomy and other sciences, and this is exactly why they are found only in England, and all the most renowned observatories in Europe are supplied with instruments made in London …'.[83] A wooden model of the Palermo circle was constructed at Florence in 1796; there was talk of making wooden moulds for casting parts in brass, and more arguments and discussions, but the great circle was never built.

Scientific Expeditions

Its extent and remoteness from the Old World kept 'the great Southern Ocean' largely unknown to navigators from the nations bordering the North Atlantic. There had been famous circumnavigations, notably those of Drake in 1577–80 and

Cavendish in 1586–88 and again in 1591, and merchantmen plied a trade route passing diagonally from the tip of South America to the East Indies. But the very term 'Southern Ocean' reveals that the northern part was as yet unexplored and as the Spanish mariners on the 'Manila galleons', trading between Central America and the Philippines knew to their cost, crews on the longer return crossing were often decimated by scurvy.

The voyages of Edmund Halley (1656–1742) in 1698 and 1699–1700 in HMS *Paramour*, are generally taken to be the first made specifically for scientific purposes; in this case 'to improve the knowledge of the Longitude and variations of the Compasse'.[84] Accounts of the four circumnavigations of William Dampier between 1695 and 1711 aroused popular and scientific interest and were published in many editions and combinations.[85] But for the next century and a half the only voyages combining geographical discovery with commerce were that of the Dutchman Jacob Roggeveen (1659–1729) in 1721–23 and two Russian voyages commanded by the Dane Vitus Bering (1681–1741) in 1725–30 and 1733–42.[86]

A rare event which contributed to a rising interest in scientific voyages was the pair of transits of Venus due in 1761 and 1769. Pairs of transits, when Venus is seen to pass across the face of the sun, occur at intervals of 113 years. By stationing observers at various points of the globe to measure the exact times when the planet begins and ends its transit across the sun's disc, and thus the length of the chord of its transit, it should be possible to calculate a fundamental astronomical dimension, namely, the distance between the earth and the sun. A total of 120 observers (18 of them English) hoped to observe the transit of 1761. The mid-Atlantic island of St Helena, then one of the East India Company's possessions, was thought to offer an excellent viewpoint in southern latitudes. The expedition was judged to be sufficiently important for the government to fund the cost of instruments. The astronomer appointed for this task was Nevil Maskelyne, who benefited enormously from this early seafaring experience when he was later appointed Astronomer Royal. Sadly, however, as the time of the transit approached, clouds obscured the sun and no observations were possible. Elsewhere around the world, they were not sufficiently precise to yield the desired measurement.[87]

During the 1760s, changes in European power politics, and in Anglo-Spanish relations, favoured British and French peaceful ventures into the Pacific, making it possible to plan one transit observation from that region. A series of major French and English expeditions were dispatched, in the course of which the prevailing winds and currents led most vessels to circumnavigate the globe. Although their declared aim was science, many captains carried secret political instructions to look out for new lands which could be profitably colonized. The voyages of Louis-Antoine, Comte de Bougainville (1729–1811), in 1766–69, in *La Boudeuse* and *L'Etoile* and of James Cook (1728–1779) in 1768–71 in HMS *Endeavour*, were the first of many, commanded by naval officers, provided with detailed scientific instructions, and equipped with astronomical instruments, as well as means of investigating the depths of the sea, and magnetic variation. Pierre-Anton Véron, Bougainville's astronomer, was the first to use lunar distances to estimate the width of the Pacific Ocean. Within a few years the earlier optimistic sightings of the 'great southern continent' were resolved into inhospitable capes of Antarctica, temperate coasts of Australia and New Zealand, and sundry mid-ocean islands, and duly laid down on the charts.

Captain Cook's Expeditions, 1768–71, 1772–75, 1776–80

Cook's first voyage to the Pacific was primarily to observe the 1769 transit of Venus and for this he was provided with instruments from the leading craftsmen.[88] These scientific voyages, both French and English, gave an enormous boost to the London trade. The development of marine timekeepers, in Britain and France, and new instruments such as the sextant, led to the recruitment and training of more boys and men to serve as navigators.[89]

The Admiralty supplied Cook from its own reserve and on the first voyage Ramsden was represented only by a Hadley sextant, a barometer and a dipping needle. On the second voyage both astronomers, William Wales (1734–1798) and William Bayly, or Bailey (1737–1810), Maskelyne's former assistant, had Ramsden sextants.

During the third voyage Cook was killed during a fracas with the native population of the Sandwich Islands (now Hawaii) and Captain Charles Clerke, who took over the command, died in August, leading to various moves between the two vessels. Lieutenant James King (1750–1784), who had been astronomer on HMS *Resolution*, took command of HMS *Discovery* and Bayly transferred to the *Resolution*. Two sextants had been used on board: one by Dollond of 15 inches radius, the other by Ramsden, of 12 inches radius. When the ships returned, King reported to the Board of Longitude that he had found 'a consistent error in Ramsden's otherwise excellent sextant'.[90] In an undated letter to Banks headed 'Some account of a 15 inch sextant made by Mr Ramsden', Bayly explains that he had four sextants. Two belonged to the Board of Longitude, one by Dollond and one by Ramsden, both of 15 inches radius. King also had two of his own, one of 12 inches radius by Ramsden, and one of 5 inches, by Stancliffe. On the passage out to the Cape he employed all four instruments, but found that longitudes taken by Ramsden's 15-inch generally differed 'near a degree' from values given by the other three, whose readings were within half a degree of each other. On leaving the Cape he kept the Ramsden 15-inch solely for taking altitudes when measuring lunar distances, since any error would hardly affect the main computations. As a small sting in the tail, he added:

> If Mr Ramsden's 15 inch had measured angles ever so exact, very few distances of the Moon from stars could have been observed with it for want of proper glasses ....[91]

A subsequent letter, referred to but not identified, roused Bayly to speak again to Joseph Banks, confirming that the smoked glass and telescope belonging to Ramsden's sextant had been with it from the time it had been issued until the instrument was returned to Maskelyne after the voyage. Although the telescopes of this and other instruments had subsequently been jumbled together in the

Royal Observatory, Bayly had recognized the telescope belonging to his sextant by various markings on it.

Ramsden was never one to accept accusations of poor workmanship. Bayly reported:

> In answer to this Mr Ramsden says that two sextants were made and fitted alike, that they were packed each in its respective case in the same manner and thus delivered to the Board of Longitude for the use of the Discovery. That he apprehends the said two sextants are now lodged at the Royal Observatory with the said their packages. That if the two telescopes new appropriated [sic] the sextant which Mr Bayly had, fit their respective places in the packages as well and truly as those in the sextant used by Captain King he will acknowledge that the telescope in question is the identical one delivered at the commencement of the voyage but that if he finds that it will not fit he must remain convinced that it is not. Provided also that the telescope is furnished with the same powers and the clamp ring to secure the eye tubes in their places as the similar one in Captain King's package. Mr Ramsden I apprehend means that as Captain King's sextant and Mr Bayly's were fitted alike with similar packages, if the telescopes found in Mr Bayly's package now are similar to those in Captain King's and fit in as good a manner the places intended to hold them.[92]

The outcome of this dispute is unknown. Another sextant employed by the then midshipman George Vancouver (1757–1798) 'had been much complained of'[93] and again Ramsden must have taken issue. This time Wales wrote to Ramsden on 16 March 1781:

> In answer to your note I assure you that I examined both the divisions and total arc of Mr Vancouver's sextant, which was so much complained of with the utmost rigour, and am certain that the error nowhere amounts to more than ¼ of a minute – indeed I can scarcely say that it was anywhere sensible and you are welcome to make what use you think proper of this declaration.

As this is the original posted letter, Ramsden must have forwarded it to the Board, to be shown to the complainants. Wales's letter also bore a sting in the tail, for he adds a postscript:

> Mr Vancouver has sent me word that he is in great want of his telescope and begs that you will let his sister have it as well as a dark glass for his sextant and they will pay you for it immediately.

This would not be the last irritation that Vancouver had to suffer at Ramsden's hands; delays in delivery of instruments for his own later Pacific voyage are recounted in Chapter 6.

Phipps's 'Voyage towards the North Pole', 1773

The naval officer Constantine John Phipps (1744–1792), later the second Baron Mulgrave, commanded HMS *Racehorse* and HMS *Carcass* on a northern voyage at a period when it was believed that, once past the barrier of sea ice, clear water extended to the North Pole. In the Appendix to his *Voyage towards the North Pole,*

1773 (1774), Phipps describes two new instruments which Ramsden provided: a megameter and a manometer. The megameter, a split-lens sighting device which Phipps employed to survey the passing coastline, may be identified with Magellan's adaption of the common octant, – the first example being constructed on one of Ramsden's instruments, a later example being taken by Phipps on his northern voyage,[94] and was perhaps akin to the micrometer telescope which Ramsden made for Phipps in 1775.[95]

The manometer was a small glass instrument for measuring changes in pressure and hence the density of the atmosphere – important when correcting astronomical observations for the effects of refraction caused by a thin layer of extremely cold air overlaying sea ice. Phipps's log[96] recorded the daily manometer reading for a month, and thereafter irregularly, making no judgement on its value.[97] Ramsden may have sought to improve on his first model, for in September 1777 the Danish astronomer Thomas Bugge called on Ramsden who mentioned this instrument to find the state of the air, to calculate refraction. From Bugge's description, this was the manometer taken on Phipps's voyage.[98]

The Expedition of Lapérouse, 1785–88

The French expedition of Jean-François de Galaup de Lapérouse (1741–1788) was modelled on those of Cook. Instruments were sought in London and Ramsden supplied sextants of 9 and 12 inches radius, two 7-inch theodolites, four steering compasses, two night telescopes and four thermometers.[99]

During the voyage, astronomical observations were made simultaneously with Borda circles and Ramsden's sextants. Paul-Antoine-Marie Fleuriot de Langle, astronomer on this voyage, declared that the sextants were 'very mediocre'[100] Lapérouse, on the other hand, judged that they had a better finish than the Borda circles. The second astronomer, Joseph Lapaute Dagelet, believed that the errors on the sextants did not exceed 30 seconds, those on the circles, 15 seconds; divisions on both instruments were excellent but it was difficult to locate the zero on the sextants' scale. Where observations differed, the Borda circles were given preference.[101]

Notes

1   Brass and iron expand and contract at differing rates in response to a given change in temperature. The lower cost of an iron-framed quadrant was thus negated by the loss of accuracy.
2   Donnelly (1964).
3   Baiada *et al* (1995).
4   Turner, AJ (2002).
5   Clarke *et al* (1989), 1–10.
6   Bird (1768) and Le Monnier (1774).
7   Holl (2004).

8   America (where Edward Nairne had captured much of the market to supply the new colleges), Scandinavia (where Bird and Graham had already supplied instruments) and Iberia (which relied on Magellan) made little call on his services.

9   Matulaityte (2004), 51–3. The makers of these telescopes, and thus the dimension to which they were made, is not known.

10 Anon (1999).

11 Sužiedėlis. *Encyclopedia Lituanica*, extracted in the Biographical Archives for the Baltic States, mf 275, frames 207–9.

12 Odlanicka-Poczobutt (1979).

13 VUL, MS F 16–13 f. 5. I am much obliged to Romualdas Sviedrys and staff in the Manuscripts Room of Vilnius University Library for their help in deciphering the documents relating to these dealings.

14 In 1754, while he was in London, Stanislas Augustus was invited to the Royal Society's Dining Club. From 1764, when he became king and was elected FRS, he was customarily toasted at the Club's dinners.

15 VUL, MS F17-9, f. 16.

16 Romualdas Sviedrys, personal communication, 2005.

17 Bernoulli (1779), 4.30–35. 'Mais on sait en Angleterre & en France que M. l'a[bbé] Poczobout, en faisant un grand voyage, il y a une 10e d'années, pour faire fabriquer sous ses yeux d'excellens instrumens par les meilleures artistes, n'a rien negligé pour acquérir une connoissance parfaite de leur usage; son élève et aujourd'hui son digne collegue, M. l'abbé Strzlecki, a pu aisement faire des progrès sous un tel maitre ... lui vient de faire un long séjour en France et particulierement en Angleterre, & qu'il a apporté de ces pays, vers la fin de l'année passée, une nouvelle provision de conoissances & de bons instrumens.'

18 For Poczobutt see the Biographical Archives for the Baltic States, mf 275, frames 207–9. Paris: Institut des sciences, Archives de l'Académie des sciences, Dossier Poczobut. Carl Sommervogel, *Bibliothèque des écrivains de la Compagnie de Jesus*, (1890–1909). vol. 6, cols 9–12; vol. 9 col. 776.

19 Buczek (1980), *Polski Słownik Biograficzny* vol. XXV/4, 638–640, art. 'Perthees'.

20 Sužiedėlis, *Encyclopedia Lituanica,* art. 'Jan Sniadecki', Chamcówna (1968).

21 KCL, MS 1493, 'Catalogus Instrumentorum Mathematicorum Serenissimi Stanislaus Augustus, Regis Poloniarum, 1782', section 'Quadrantes Astronomicae'. The Paris inch, one-twelfth of the old 'Pied du Roi' is equivalent to 2.7066 cm, slightly larger than the English inch, 2.5399 cm.

22 SML, Court MSS, folder T.

23 VUL, MS F 17-3 f. 3. 'Jeurat a l'honneur d'envoyer à Monsieur l'abbé Strecki [*sic*], une lettre pour M. Maskelyne, ministre protestante qui lui fera voir avec plaisir le superbe observatoire de Greenwich, et aussi une lettre pour M. Magellan, prêtre catholique lequel procura vraisemblement à Monsieur Strecki, l'entrée de la Société Royale et aussi l'entrée de la Société des Arts.

Jeurat a donc l'honneur de souhaiter un bon voyage et un bon retour à Monsieur l'abbé Strecki, qu'il desir de voir à Paris lorsqu'il sera de retour de Londres.

À M. Maskelyne il faudra lui parler en latin. À M. Magellan, en latin, ou en françois, ou en anglois, ou en espagnole parcequ'il entend toutes ces langues.'

24 VUL, MS F 17–3 ff. 20–21.

25 VUL, MS F 17–4 f. 17.

26 Buczek (1982), 98–110.

27 The first Duke, whose decisive victory at Blenheim in 1703 was rewarded by a peerage and a gift of land, built the grand house properly known as Blenheim Palace, the only palace not in royal possession.

28 [Pigott] (1792). 1. Preface.

29 'D-ke of M-B-R-GH' [Pigott] (1792) vol. 3, 163–9.

30 *Complete English Peerage,* 1932 edition, vol. 8, 500 n.(e), citing *Complete Peerage,* 1775 edition.

31 Farington (1978–98), 4.1432.

32 Wallis (2000).

33 Lalande (1803), 705–6.

34 Le Monnier (1774), plate 10.

35 CUL, Fitzwilliam Museum Library, Perceval Coll. H-174. Blagden to Banks, Paris 10 July 1783.

36 RAS, MS Radcliffe A2.10.

37 RAS, MS Radcliffe A2.42.

38 RAS, MS Radcliffe A2.44.

39 RAS, MS Radcliffe A2.50.

40 RAS, MS Radcliffe A2.85. Ramsden was summoned to Hounslow by General Roy (see Chapter 8).

41 RAS, MS Radcliffe A2.87.

42 RAS, MS Radcliffe A2.89.

43 RAS, MS Radcliffe A2.123.

44 Reid (1990), 3, citing Private Coll. 'The Professor' is Thomas Hornsby; 'the great equatorial' had been commissioned by Greenwich Observatory, and the 'mensuration' refers to instruments being made for Roy.

45 Mandrino *et al* (1994), 123 and 190–1.

46 Cassini (1810), 25.

47 Piazzi (1788). 'Le mural de mylord Marlborough à Blenheim, qui a six pieds, est, pour ainsi dire, un autre instrument que vous avait admiré comme moi; il tient à un assemblage de quatre colonnes qui tournent sur deux pivots de façon que l'on peut mettre l'instrument au nord et au midi en une minute. Cet instrument est aussi beau qu'il est parfait; mais personne n'étoit plus digne que mylord Marlborough d'en être le possesseur; les astronomes de profession n'ont pas plus de zèle, d'assiduité ni d'exactitude. … C'est pour ce bel instrument que M. Ramsden a imaginé un moyen pour rectifier l'arc de quatre-vingt-dix degrés sur lequel un habile astronome avoit élevé quelques difficultés; mais avec un fil horizontal et un fil à-plomb formant un espece de croix qui ne touche point au quart de cercle, il lui montra qu'il n'y avoit pas un seul seconde d'erreur sur 90 degrés, et que la différence venoit d'un mural de Bird où l'arc de 90 degrés contient plusieurs secondes de trop, et qui n'avoit jamais été verifié par une méthode aussi exacte que celle de M. Ramsden.'

48 RAS, MS Radcliffe A2.136.

49 RAS, MS Radcliffe A2.138.

50 RAS, MS Radcliffe A2.139.

51 BL, MS Add. 72847, ff. 12–28.

52 Vince (1790), 119.

53 I am obliged to Anthony Turner and Ileana Chinnici for a sight of a copy of this document.

54 MHS, MS Radcliffe 39.

55 RAS, MS Radcliffe A2.139.

56   MHS. R.T. Gunther noted that together with a transit of 5½ feet by Dollond, these instruments were sold by Squibb & Son in 1820.

57   Ferrighi (2000), 159–71.

58   Pigatto (2000b), 5–100.

59   Lorenzoni (1922), 67–9.

60   Lorenzoni (1922), 73. Bozzolato *et al* (1986), 77–114, 146–8.

61   Lorenzoni (1922), 81.

62   CUL, RGO, MB III, 4.

63   Lorenzoni (1922),84–7.

64   Santini (1819, revised ed. 1830. 1. 39–74.

65   Mangani (2000); Pedley (2000).

66   Pedley (2000), 58.

67   Padova, Biblioteca del Seminario Vescovile, cod. 798.c.76. 'Je fis donc faire à Londres un excellent quart de cercle de deux pieds de rayon garni d'un micromètre.'

68   Anon *Catalogo generale descrittivo degli strumenti geodetici e topografici dell'Istituto Geografico Militare al 22 ottobre 1922*. (Florence) 33–5, 77–8.

69   RGO 4/307 letter 3.

70   RGO 4/307 letter 4.

71   RS, Council Minutes, 1769–82 (vol. 6, copy).

72   RS, Council Minutes, 16 January 1783.

73   RS, Council Minutes, 1769–82 (vol. 6, copy).

74   Ximenes successfully proclaimed his support for Newton's ideas without upsetting his Jesuit superiors. In 1773, when the Order was suppressed, he convinced the authorities who were taking over Jesuit properties that he had the right to continue occupying the observatory and to keep the instruments.

75   Triarco (1999), 'The observatory and instruments,' Chapter 3, pp. 96–164; abbreviated version in Triarco (2000).

76   Triarco (1999), 162. This may have been an electrical machine for electrical experiments were on the syllabus.

77   Knoefel (1984).

78   Knoefel (1982), 399–422; ASF, MS Misc. Finanze 438; Miniati (1984), 209–20.

79   Bickerton and Sigrist (2000), 204–10. '[Ramsden] voudrait determiner les astronomes à employer des *Cercles* au lieu de *Quarts de Cercles*, soit pour les muraux, soit pour tout instrument employé à des observations delicates, et ou un Cercle peut être admis. Ainsi pour exemple, au lieu d'un Quart de Cercle de 4 pieds de rayon, il voudrait que vous prissiez un Cercle de 3 pieds de rayon; et voici ses motifs.

1. La monture d'un Cercle (*the framing*) étant tout composée de parties ou concentriques ou se dirigeant du centre à la circonference, promet un conservation beaucoup plus grande de la figure de l'instrument (par toutes les causes qui agissent sur le metal) que ne la promet *the framing* d'un quart de cercle, quelque bien qu'il soit calculé pour cet instrument. 2. Un Cercle de trois pieds de rayon portera une lunette de 6 pieds, puisqu'elle diviendra un diametre de cercle se mouvant sur le centre. 3. La division d'un Cercle est sure. Six fois le rayon donne six points incomparablement plus sures, que les meilleures Règles divisées en parties égales ne peuvent donner l'angle de 90°, et ces six points donnés, tous les autres se trouvent par continuelle bisection. 4. La ligne de foi est infaillible dans un pareil instrument, dont la lunette portera un Nonius a chacune de ses extremités. 5. Toutes les variations que pourrait subir l'instrument, toutes les erreurs mème de division, seroient devoilées par ce double Nonius, et en lisant toujours les observations aux deux côtes, et étudiant les causes des differences, on se mettra surement à l'abri de toute erreur.'

80 RGO 14.3, p. 19.
81 'L'altro lavoro è un cercio murale astronomico di dodici piedi di diametro che il famoso inglese Ransden [sic] non aveva ardito di fare nè per l'Amiraglia inglese, nè per il Museo di S.A.R. benchè egli no prendesse l'impegno fino da otto anni passati un tale strumento e per l'enorme grandezza, e per la costruzione totta nuova, e per la sua grande utilità, e per la perfezione che domanda, sarà unico in Europa, e spero di condurlo a fine in due anni, o poco più ....'
82 RGO, MS 35/AC I, 64, 66–7.
83 Knoefel (1982), 417–18.
84 Thrower (1982), 1. 268–9.
85 Masefield (1906).
86 Sharp (1970); Golder (1922).
87 Woolf (1959).
88 Betts (1993).
89 Turner, A.J. (1998).
90 RS, MS MM/7/21.
91 RS, MS MM/7/22.
92 RS, MS MM/7/23.
93 RS, MS MM/7/25.
94 Magellan (1775), 120, n.E.
95 NLW, MS Banks to Lloyd, October 1775.
96 RS, MS/90. Phipps's log.
97 RGO, MS 14/5. After Phipps's voyage, his astronomer Israel Lyons returned all the surviving instruments (some had been broken or lost at sea), at which time the Admiralty paid the instrument makers' invoices for instruments for this voyage, including that from Ramsden for £139.1.0.
98 Bugge (1997), 185–9.
99 Dunmore (1994), xci, cvi–cx.
100 Gaziello (1984), 202.
101 Dunmore (1994–95), 196.

Chapter 6

Observatories and Expeditions after 1786

That wretched Ramsden is really making me wait for the circle which he promised me so many years ago …

Ernst, Duke of Saxe Gotha to Count von Brühl,
16 March 1794, SRO GD157 3379/93

There are few lively persons with such an open countenance as Ramsden, but one can also say that not withstanding this same countenance he has no scruples about failing to meet his promises.

Johan Lexell, May 1781, p. 201 in N.I. Nevskaya,
'Correspondence between astronomers', 177–216.
USSR *Academy of Sciences: Scientific Relations with Great Britain* (Moscow: Nauka, 1977)

Dunsink Observatory

At his death in 1774 Dr Francis Andrews, the late provost of Trinity College Dublin, bequeathed £3,000 to build, equip and staff an observatory for the College. In 1783 Henry Ussher (1741–1790) was elected first Andrews Professor and he decided the observatory should stand on the hill of Dunsine (now Dunsink), five miles north-west of Dublin and on high limestone ground with good all-round visibility.[1]

Ussher was sent to England, probably in 1784, 'to order from Mr Ramsden the best instruments, without limitation of price'.[2] His wish-list included a transit with a 4-foot axis, bearing a 6-foot telescope of 4¼-inch aperture, capable of magnifying up to 600 times; an equatorial instrument with circles of 5 feet diameter; and an achromatic telescope, mounted on a polar axis and carried by an heliostatic movement, for occasional observations. Ussher had intended to acquire a mural quadrant, but Ramsden persuaded him that observations made with a circle would be more accurate, without increasing the cost. According to Sir Robert Ball, a later occupant of the Andrews professorship,

… the enthusiasm of the astronomer and the instrument-maker was conveyed to the Board, who agreed to the purchase of a tremendous circle, ten feet in diameter, on a vertical axis, for measuring meridian altitudes.[3]

The transit arrived promptly. Ussher made some observations in 1785, and was very satisfied with the results. He wrote about certain improvements which Ramsden had made, particularly in the method of illuminating the wires.

> The candle or lamp is here made to describe a circle equal and parallel to that described by the reflecting surface placed before the object glass, the centres of both lying in the same right line parallel to the horizon. This movement is effected by a simple apparatus consisting of a strong ring of wood, which surrounds the axis of the instrument at a small distance, and is made fast to the pillar; this ring supports and confines another, which can revolve and carries an arm bearing, at one end the lamp, and at the other, a counterpoise. However the arm revolves, the lamp hangs downwards, and its smoke and tremulous hot air escape at the back, away from the instrument.

In Ussher's transit, the entire light of the object glass was preserved by a simple contrivance: the pivot of the axis which rested upon the plate regulating the motion in azimuth was perforated with a small hole bearing a convex lens; the plate and pillar were also perforated in the direction of the axis. Inserted into the perforation in the pillar, 3 inches in diameter, was a tube carrying another large convex lens, with the lantern attached, its candle-flame kept opposite the axis of the tube by means of a spring socket.

> The candle light is thus focussed beyond the small lens in the pivot, and diverging within the conical axis, meets a diagonal silvered brass plate with a central hole, though which the light from the object glass passes undiminished. This light is tempered according to the brightness of the observed star by interposing a green glass tinted gradually; this glass is set in a brass frame which slides in a dovetail between the azimuth plate and the pillar.[4]

Ussher comments that he had to prevail on Ramsden to adopt this method of illuminating the instrument; Ramsden supposed that the heat of the candle might affect the axis, but after the two men spent hours in experimenting, this was seen to be not so, and experience proved Ussher right.

Ramsden's 'Ghost'

Ussher's new transit also hosted the first appearance of what became known as 'Ramsden's ghost'.

> I now come to the principal improvement in this instrument, an invention that does high honour to Mr Ramsden, and is a most valuable acquisition to all astronomical instruments where plumb lines are introduced. Although the spirit level is the most convenient method of levelling, it is sluggish in winter and prone to false reading if the tube and support are at all imperfect or the glass is unevenly heated. The plumb line is also imperfect as usually attached, with supports at each end and the line passing over two fine points, one at each end of the tube; the tube then being reversed, seeing if the line again bisects the two points.

Ussher made some modifications but was still dissatisfied, when Ramsden came up with an elegant solution, which Ussher recognized as potentially valuable to all astronomical instruments levelled with plumb lines.

> In one side of the telescope tube, 12 inches from each end, is a small hole in which is inserted a small semi-pellucid slip of ivory with a black dot in the centre; in the opposite side of the tube is a convex lens, in which is focussed an image of each dot. The pillars support a brass frame carrying a plumb line and two microscopes, focussed to the images of the dots, the plumb line being made to swing through these images. The adjustment is thus totally independent of the instrument, and, hanging in the images, cannot be deflected by corpuscular attraction. No additional weight is added to the axis.[5]

Among the instruments which Ussher ordered from Ramsden was an equatorial telescope driven by clockwork. Perhaps owing to Ramsden's alleged feud with Ussher it was never made.[6] In the 1860s Ramsden's transit was replaced by a transit circle made by Pistor & Martin of Berlin.[7]

Interruptions to the Circle's Construction

Meanwhile the non-appearance of the great circle became an increasing cause for concern. Thomas Romney Robinson (1792–1882), astronomer at Armagh, suggested that Ramsden, having quarrelled with Ussher, resolved that the latter should never have the circle. Robinson gives no reason for such a falling-out; Ussher died in 1790; his successor John Brinkley (1763–1835), was authorized to go to London and commission instruments from Ramsden, and it was Brinkley who oversaw the actual construction of the observatory buildings.

With Brinkley's arrival, Ramsden set to work to complete the great circle, but found, to his dismay, that the sulphurous air of London had begun to rot the thin tips of its eighteen radial arms.[8] Whatever the cause of delay, the observatory's visitors arrived in London in 1792 unhappy with the lack of progress, but were permitted to borrow the Greenwich 40-inch quadrant, 'to be returned in three years, or when the instrument now in hand by Mr Ramsden is delivered to them'.

In 1793 Trinity College settled its debt to Ramsden's agents for instruments bought over the preceding years, including the transit, air pumps and other instruments. The overall total of £330, 17s included £55, 3s for packing under Ramsden's supervision, insurance, and payment for the chairmen transporting the cases from Piccadilly to the London docks. This sum did not include shipping from London to Dublin, nor customs dues payable at Dublin.[9] Ramsden gave his assurance that the circle would be completed within the year. But alas for such promises; another seven years passed, with the circle's intended place still unoccupied. Ramsden was by this time in poor health, and the College Board considerately directed that 'enquiries should be made'. With no progress the following year the Board threatened Ramsden with a lawsuit; but the threat was never made good, for Ramsden's health worsened and in November 1800 he died.

The Board's concerns overlapped with the situation in which Ramsden found himself with the part-built circle now decaying. Perhaps he had been driven to buy English-made brass rather than the fine brass manufactured near Aachen and favoured by horologists and instrument makers. In London it was known either as 'Dutch brass', having been shipped from the Low Countries, or as 'Hamburg brass', having arrived from that port. As originally constructed, the circle was ten feet in diameter. Ramsden removed the rim and about six inches from each of the arms, back to sound metal. But as he was doubtful about the stability of the brass, he let the structure lie several years longer, and found his concerns justified. A further six inches was removed from each arm, reducing the diameter to eight feet, and Ramsden awaited the result, ignoring Brinkley's pleas. Shortly before his own death, he was satisfied that no further change was likely and he completed the structure. But in another version of the story, Robinson wrote:

> The materials were actually formed and put roughly together to suit such dimension; but the structure was considered too bulky and heavy, and was therefore reduced to nine feet; in which state, we understand, it was actually divided; but on second consideration, it was reduced a second time, to eight feet, in which state it was not completely finished when Ramsden died.[10]

Robinson could not explain why this destruction was confined to the ends of the arms.

> To judge from the analogy of the Palermo Circle, the diameter of these arms at the outer extremity was very small; and if they were of cast brass, the molecular condition of the metal there, in consequence of the more rapid cooling, may have been different from that of the more massive portions.[11]

Ramsden's death worried the College authorities, who had made interim payments during these 15 years. Maskelyne reassured them that Ramsden had left property and that they were in no danger of losing both their money and the instrument. Berge took over the business, but the great circle was only one of several items which he was left to complete. After four years he promised the instrument in the following August but it did not come. Two years later (1806) Brinkley complained that he could get no answer from Berge. In 1807 word came that that it would arrive in a month. It did not; but in 1808, about 23 years after it had been ordered, the circle, now considerably reduced in diameter, was erected at Dunsink.[12]

Taylor provides a description and the only known illustration of the circle as installed, though his text does not exactly correspond to the engraving (see Figure 6.1) It was suspended between solid pillars and supported in a frame which turned on a vertical axis in the form of a double cone, eight feet high overall. The weight of the circle on its axis was completely relieved by a very ingenious application of a lever assisted by friction wheels. Taylor asserts that 'this great circle and its frame could turn on their axes by the slightest touch of the fingers'.[13] The circle was graduated into intervals of 5 minutes. Three micrometer microscopes, one at the base of the circle, the other pair at the right and left of the horizontal, allowed the observer to read to parts of seconds. With such a large circumference,

Figure 6.1   Dunsink Observatory, engraving of Ramsden's astronomical circle in use.

the temperature could vary around the rim. To counteract any error introduced by this variance, the suspension point of the 10 foot plumbline, and its reference point below, were both formed of bimetallic brass-and-steel bars.

Sir Robert Ball tells us that once the circle was set up, Brinkley, stimulated by having command of so perfect an instrument, attempted the very highest class of astronomical research, resolving to measure with his own eye and with his own hand the constants of aberration and of nutation. He also strove to solve that great problem of the universe, the discovery of the distance of a fixed star.[14]

Dunsink Observatory closed in 1937. The instruments were neglected until 1947 when Professor Hermann Brück (1905–2000) was appointed as director. He modernized the observatory and removed the obsolete Ramsden instrument, retaining the circle as a decorative feature in the Observatory Library (see Figure 6.2). Following a disastrous fire in 1977, the circle was put in an outside storehouse to await restoration. In November 1981 thieves broke in and stole it for the scrap metal value of its brass. Frantic attempts to recover it were made by the police and through unofficial enquiries but to no avail.[15]

Mannheim, Cartography and Astronomy

Prince Carl Theodor (1742–1799), Elector Palatine, and from 1777 also Elector of Bavaria, patron of the sciences (he was elected FRS in 1784), established an observatory in 1764 on the roof of his summer castle at Schwetzingen, outside Mannheim, a city located on the eastern bank of the Rhine.[16] It was directed by

Figure 6.2   Dunsink Observatory, remaining portions of Ramsden's circle and telescope as preserved in the 1950s. Photograph: H.A. Brück.

the Moravian-born Jesuit Christian Mayer (1719–1783), educated at German and Italian universities, who had been a lecturer and curator of the Prince's cabinet of experimental physics at Heidelberg. Mayer and two of his fellow Jesuits had visited Paris in 1757, primarily to study that city's provision of potable water in order to provide a similar network at Mannheim. At this time Paris had a dozen or so observatories, associated with various colleges. Mayer met Lalande, Nicolas-Louis de La Caille (1713–1762), Pierre Bouguer (1698–1758), and Cassini de Thury, and developed an enthusiasm for astronomy and geodesy. He was able to purchase a clock by André Lepaute, and from Canivet an astronomical quadrant of about 2 feet radius and a copy of the standard measure of length known as the 'Toise de Peru'.

Mapping the Palatinate

The quadrant and clock served for observations of the Transit of Venus in 1761, the toise for the cartographical survey of the Palatinate which commenced in 1762.[17] A baseline for the survey was laid along the straight highway between Heidelberg and Schwetzingen, and Mayer's triangulation ultimately extended north to Worms, south to Basel, and across the Rhine to join the French network. The first Ramsden instrument appears to have been a cartographical protractor of 12 inches diameter, acquired in 1770, followed in 1776 by a Ramsden mountain barometer, costing 5 louis d'or (see Figures 6.3a and 6.3b) and a 2-foot refractor costing 12 louis d'or. Ramsden also supplied a 2-foot spirit level, with brass stand and mahogany box, and a theodolite, 9 inches in diameter, with two 7-inch achromatic telescopes and a spirit level, with a tripod and mahogany box. In 1796 the spirit level and theodolite were surrendered to an Austrian officer and not returned.[18] The first provisional 'Basis Novae Charta Palatinae' appeared in 1773; others followed until the turmoils of the Napoleonic war brought the existence of the Palatinate to an end.[19] Bernoulli was an eager visitor to Schwetzingen, lured by its reputation for fine instruments, which he described in a letter of 8 December 1768, but criticising the observatory on the castle roof as 'unstable'.[20]

Mannheim Observatory and its instruments

In 1772 building commenced on a new observatory at the end of the castle gardens, on the western margin of Mannheim and adjacent to the large Jesuit church, but otherwise with clear views (see Figure 6.4). With walls 7½ feet thick, its stability was beyond question. Completed in 1775, the 100-foot tower provided accommodation on the first floor for the astronomer, above which was the 'great instrument room', housing the telescopes, clocks, and an 8-foot mural quadrant by Bird, delivered in 1775. The library and any visiting astronomers were accommodated on the third floor. There were another two instrument rooms on the fifth floor, where in 1778 Sisson's 12-foot zenith sector was set up.

Mayer learned that Jeremiah Sisson's price for a 5-foot equatorial sector was 250 guineas, for an 8-foot mural quadrant, like that at Greenwich, 400 guineas (plus 25 guineas for Dollond's lenses and a further 2½ guineas for four thermometers) and for a transit, 118 guineas. The Prince made funds available for all three instruments, but Sisson had other orders in hand and undertook to construct only the 6-foot transit. Early in 1782, after ordering the transit, Mayer heard from his ex-Jesuit colleagues at Padua about the admirable qualities of their new Ramsden mural quadrant, and in June he made contact with Ramsden.[21]

According to Sisson's contract, countersigned by Maskelyne and the Oxford astronomer Thomas Hornsby and received by Mayer on 31 March 1783, the transit was to be delivered complete for 145½ guineas (it is not clear whether the

Figure 6.3a Mountain barometer signed *Ramsden*. Mannheim, Landesmuseum
für Technik und Arbeit. Photograph: Klaus Luginsland,
Landesmuseum für Technik und Arbeit.

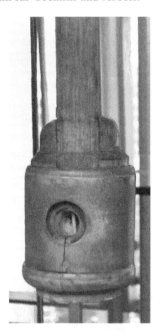

Figure 6.3b Ramsden's mountain barometer, detail of the cistern.

Figure 6.4 Mannheim Observatory today

extra cost was for the lenses, or for the delivery). Mayer died two weeks later. By this time, however, Sisson himself was in no shape to deliver. At his death the following year, only the transit frame had been built. It is unclear how and when the contract was passed over to Ramsden, who now promised this transit in the remarkably short time of four months for the same price.[22]

When its delivery was due Mayer's successor Karl (Charles) Josef König (Koenig) (1751–1809) wrote to Ramsden on 17 April 1784 and again the following week, eliciting from Ramsden a surprised response that König had not received his letter of February stating that work had begun on the transit which was expected to be ready within two or three weeks; he had hoped to have finished it sooner but the weather had been extremely cold and it had been impossible to progress the work. On receipt of this apparently good news, König sent Maskelyne a draft for 100 guineas drawn on his London bankers, writing (in Latin) that he hoped that Maskelyne would examine the finished transit prior to its despatch. The same day he notified Ramsden that the money had been sent. Neither man responded. On 19 June König wrote again to Maskelyne, and on 10 July again to Ramsden, again without effect, although he knew that his draft had been paid in as his Mannheim banker had debited him the equivalent of 100 guineas.[23] Probably he also wrote to Hornsby as an undated draft letter from Hornsby seems to respond to a similar enquiry:

> Your very obliging letter arrived here when I was absent. Very soon after my return I applied both to Dr Maskelyne and Mr Ramsden and have the pleasure to inform you that the draught [sic] for the money is very safe in the hands of the former, and that there is a very good prospect of the transit instrument being finished and sent to you in a very short time. Mr Ramsden indeed assures me that it will be finished and sent away before the end of this month but I know him too well to think that it will be sent away before the end of the present year. Your complaints of that excellent artist are but too well grounded but if you receive your instrument in a few months you may congratulate yourself on your good luck …

Hornsby mentioned the Greenwich equatorial and the Duke of Marlborough's mural quadrant which were under construction and assured his correspondent:

> When these two instruments are out of the way I will lose no opportunity of requesting him to expedite your instrument; and hope in a few weeks to have the pleasure of informing you, unless a letter from Dr Maskelyne should render mine unnecessary, that the instrument is absolutely completed and forwarded …[24]

At a loss as to how to proceed, König obtained Magellan's address from his friend Jöns Mathias Ljungberg at Kiel, and on 21 August 1784 wrote begging Magellan to contact Maskelyne or Ramsden, and so enable him to receive the transit, or recover his money.[25]

The transit was finished in 1785. Transport from London to Rotterdam, then along the Rhine, cost 105 guilders.[26] Though Sisson was known for his precision, J.L. Klüber (a later Mannheim astronomer) thought that Ramsden's transit surpassed it in quality. The 6-foot telescope tube had a triple objective of 3 inches aperture. Three eyepieces gave 90, 130 and 200 magnification. The double-cone axis was 2 feet 9 inches long, the bubble-level 2 feet 7 inches. The vernier allowed the semicircle to be read to single minutes of arc. Six years were to elapse between the start of its construction and the day when this admirable instrument was finally pointed to the heavens.

In May 1786 the astronomer Barnaba Oriani, on his way from Brera to London, spent three days with König, whom he dismissed in his diary as pleasant enough, but 'not a great astronomer'.[27] Apart from brief visits to the Prince's castle in Mannheim, and to Schwetzingen, most of the time was spent in the observatory, and Oriani's diary provides a useful snapshot of the situation. He found the observatory an elegant building but unsuited for observations, and with Bird's quadrant installed in too narrow a space. Sisson's 12-foot zenith sector impressed him less than the two portable astronomical quadrants, by Canivet and Sisson. He also saw a Ramsden theodolite which had cost 20 louis. He noted a small Hadley sextant by Dollond, two Dollond achromatic telescopes of 12 feet and 10 feet, the latter with a heliometer – probably all three from Schwetzingen – and Ramsden's 'really beautiful' transit telescope. He saw a Ramsden 2-foot achromatic telescope which had its eyepiece dented from striking the wall.[28] Since crossing into German territory Oriani was keeping details of his expenses in 'soldi tornesi', and his 'prices of instruments in London'[29] were probably those paid by the Prince. The Ramsden passage instrument was priced at 3,000 tornesi, the Ramsden theodolite at 600 tornesi.[30]

The following morning Oriani returned to the observatory where he observed sunspots with Dollond's 12-foot telescope. In the afternoon they called on Johann David Beyser, the elderly court clockmaker and mechanic. König and Oriani spent the morning of 29 May exploring the vast palace gardens at Schwetzingen; in the afternoon they were back at the observatory where Oriani took a closer look at the transit, sketching its suspension and remarking on the illumination of the micrometer and the way in which the entire apparatus could be raised, to relieve the weight on the pivots.[31]

König subsequently left Mannheim for Munich; the Prince's next astronomer was Johann Nepomuk Fischer (1749–1805), another ex-Jesuit and former professor of mathematics at Ingolstadt. Fischer wanted to install the transit at ground level in a new observatory, but early in 1788 he too departed, to be succeeded by the Lazarist priest Peter Ungeschick (1760–1790), professor of mathematics from Heidelberg.[32] Appointed astronomer at Mannheim, Ungeschick started work on an annex for the transit, alongside the entrance of the observatory. He journeyed to Paris, meeting Lalande, then to London, where he met Maskelyne, Herschel and the Cambridge astronomer Professor Anthony Shepherd, and called on Ramsden. He set out on his return journey in 1790 but died during a stop-over at his home town of Hesperingen in Luxemburg. Ramsden's transit was eventually installed in 1792 by Roger Barry (1752–1813), another Lazarist, and former pupil of Lalande.[33] Barry enjoyed only a short period at his observations for in that year the Napoleonic troops invaded Germany and by 1794 the land west of the Rhine was under French control. Lalande wrote of bombs and bullets flying through the air, the observatory being struck nine times.[34] Under gunfire, Barry packed his instruments into boxes, before the French crossed the river in September to capture Mannheim, followed in November by Austrian troops. Lalande, on his way to Gotha in 1798, was saddened to see the observatory he had visited in 1791 now deserted. It seems that some instruments were sent to Munich, others, according to Lalande, were stored in the cellars of the observatory.[35] The Elector died in 1799, leaving money to build a meridian mark but due to the war it was not erected until 1810, by which time the observatory was once more functioning, under Johann Ludwig Klüber (1762–1837).

Gotha, Seeberg Observatory

In the late eighteenth century Duke Ernst II von Sachsen-Gotha-Altenburg (1745–1804) occupied Friedenstein Castle, near the town of Gotha, in Thüringen. On taking up his inheritance in 1772, Duke Ernst, who had studied in Göttingen, was already familiar with other parts of Germany, Italy, the Netherlands, France and England. Meetings with astronomers stimulated the Duke's interest in the subject and he constructed a small observatory on the castle roof and acquired a good range of instruments.[36]

Duke Ernst's astronomical ambitions were fostered by his friendship with Count Hans Moritz von Brühl, the Saxon minister plenipotentiary in London and himself a passionate amateur astronomer (see Plate 4). Brühl moved in the circle populated by Joseph Banks and other members of the Royal Society, and employed as tutor to his son the young Hungarian Franz Xaver von Zach (1754–1832) (see Figure 6.5a). It was Brühl's suggestion that Duke Ernst should employ Zach as his astronomer, but as we shall see, Brühl himself continued to act on Duke Ernst's behalf in London, watching over the construction of his instruments.[37]

In 1786 Duke Ernst and Zach travelled together to Gotha where the Duke intended to rebuild his rooftop observatory but he was persuaded by Zach to construct a splendid new observatory outside the town, modelled on the Oxford Observatory, known to both men. Meanwhile, Zach continued to observe from the castle and the Duke returned to London, staying there from July until September. Sir William Herschel gave him a 7-foot reflecting telescope, Brühl handed over two of his own clocks and a 2-foot transit instrument by Ramsden.

After another tour of observatories in Germany and France, Duke Ernst and Zach settled down to plan their new observatory, to which Zach was now formally appointed as astronomer. The site was on the summit plateau of Seeberg, a hill several miles outside the smoky skies of Gotha and giving clear horizons (see Figure 6.5b). The design provided two wings for Zach's own accommodation and a workshop, also for stables, coach-house and servants' quarters. In the centre of the observatory was a small turret with a moveable dome, intended to house an 8-foot astronomical circle. One side room was to house an 8-foot transit and its attached pendulum clock, another was for two mural quadrants, mounted on north and south walls. These instrument rooms were unheated, but a stove

Figure 6.5a Franz Xavier von Zach

Figure 6.5b Seeberg Observatory

was installed in the fourth room which Zach intended as his study-workroom and where the observers could thaw out in cold weather. The observatory was ready by 1789. Outside the main building a circular shed with moveable roof was constructed for a large equatorial sector.[38]

Delays and Frustrations

The suite of instruments for which Zach had planned came in from a variety of sources over the years. Several were ordered from Ramsden, but due to his failure to deliver the larger items, after his death Zach had to turn to Stancliffe, Cary and Troughton. The sad tale of the circle that never arrived emerges from the correspondence between Duke Ernst and Brühl. (Although both were German speakers, etiquette required that they corresponded in French.) Duke Ernst wrote to Brühl:

> You were kind enough to mention several instruments which I should seek to acquire, particularly a transit and a mural. You proposed an 8-foot whole circle, which Ramsden has already commenced and which would cost 400 guineas. I wish that on top of all your other favours, you would contract with this artist for the instrument in question, for if I was forestalled by another enthusiast, I would have to wait for the work to be finished and even longer, since he has a reputation for promising much and being slow to deliver.[39]

By 1 September the transit had been delivered to the Duke, who at this time was still in London and could let Brühl know that he was paying for goods received.

> I have at last received the transit from Ramsden and hope this morning to have the Hadley sextant. I shall pay Ramsden for the achromatic telescope and the sextant, with other small items, so that £50 remains in the hope that this encourages him to complete the large apparatus he claims to have begun.[40]

Duke Ernst's hopes that Ramsden would hand over the Hadley sextant before his departure from London, or that it would be dispatched to reach him in Gotha before he and Zach left for Provence, were not fulfilled. In the first of several letters expressing growing frustration at Ramsden's broken promises – and in terms reminiscent of the Duke of Marlborough – he poured out his sentiments to Brühl:

> It is all too true that Ramsden failed to keep his word, up to the last minute of my stay in London. I called on him at eight o'clock at night. He showed me the piece almost ready, almost entirely assembled, claiming that in two hours all would be done. He came to my house between eleven o'clock and midnight, making poor excuses for failing to keep his word and putting the matter off until ten in the morning. I admit that for an instant I was tempted to believe him and to defer my departure. But reason and experience triumphed and I left without my Hadley – not without sorrow but duped by the false hope that he would send it at least within the week and I had taken steps to ask Messrs Rougemont & Co to send it via Brussels and to advise me as soon as it

reached them, and to spare no expense to get it to me before I left for Provence. But alas! My hopes are yet in vain and no doubt I shall leave next Saturday without having it with me. I entreat you Sire, if it is within your power, to protect me from that sublime lying artist, and make him keep his word ...[41]

The Hadley was awaiting the Duke when he reached Frankfurt, in November, on his way to Lyon. But the laborious business of extracting instruments from Ramsden's workshop began again in 1787, this time over the large transit, as the Duke wrote to Brühl:

> I am much impressed, my dear Count, by your good news regarding Ramsden's transit – I very much hope to receive it shortly, and that the circle is also in hand. For apart from these instruments already ordered, my head is full of grand ideas for other items from this artist. My heart is set more than ever on M. Bergeret's mural quadrant; having been its legitimate owner for twelve hours, I cannot allow it out of my mind, and in one or other manner, I must have one by Ramsden, and even were I to own one by Bird, I would never surrender the pride of owning one by Ramsden. And sadly there is but one Ramsden in this world, because if he was not there, and if there were another craftsman able to make one as perfect as that Arch-liar is capable of, I assure you my dear Count, I should long ago have resorted to that man.[42]

The Duke here refers to John Bird's last mural quadrant, made in 1772–74 for the amateur astronomer Pierre-Jacques Onésime Bergeret de Grandcourt (1715–1785) who, reluctant to install it at his private observatory until he could recruit an observer, lent it in 1778 to Lepaute d'Agelet. Bergeret and d'Agelet both died in 1785, at which time it was bought by Lalande. In his next letter the Duke expands on this mural quadrant. Meanwhile, Brühl has obtained for him a most excellent octant by Stancliffe.

> I was delighted to learn from Zach that Ramsden has the transit well in hand and has progressed so far that it might be ready around next spring. I venture to convince myself that you, my dear friend, may be willing to encourage him and spur him on to keep to his word. I have the greatest interest in it, more than ever, since I have a renewed hope of obtaining the Bird mural, to which Lalande gave me access last year, the same which belonged to Bergeret.[43]

The years passed; Duke Ernst suffered the humiliation of a nobleman with a superb but almost empty observatory, plus having to pay a court astronomer to kick his heels while awaiting new apparatus worthy of his situation. Not that Zach was totally unemployed: he produced a stream of useful telescopic observations but the impotence felt by both men when confronted by Ramsden's tergiversations, plus the fact that Ramsden had pocketed substantial deposits, was exacerbated by their sense of distance. Like most of Ramsden's foreign clients, they were unaware of the other major orders passing through his workshop.

By May 1789 the Duke's blood pressure was rising fast. He poured out his woes to Brühl in a lengthy and petulant diatribe, made almost tearful by his belief that Piazzi had succeeded in getting his circle for Palermo manufactured in six months, while he, Duke Ernst, was still waiting for the apparatus ordered several years previously. He was sure that Ramsden had shown Brühl various pieces of brass, which looked like parts of a transit but which were in fact never destined for Gotha. He was probably correct, this being a well-known Ramsden ploy to calm angry customers who called in to thump on his shop counter.[44]

Eventually Brühl politely let Duke Ernst know that he was tired of these continual railings against Ramsden, a man who he much admired and with whom he maintained friendly relations. The Duke's letter of 14 October 1789 was as apologetic as could be expected from one of his superior rank. Its antique formality is untranslatable into modern English, and I simply express what I take to be his sentiments.

> I am very well aware, Sire, that Mr Ramsden's bad faith has led to misunderstandings between us, and has frozen on both sides our warm relationship. For my part I much regret it, my dear Count, but while hoping to correct any offence, I beg you not to lose sight of this Devil of a Ramsden, who has deceived me over so many years. I have been assured that the objective of my 8-foot telescope is in your hands: I trust that you will not forsake our long and valued friendship, and that you will keep pressing that famous artist to complete his contracts with me, so that one day this precious instrument will be mine.[45]

But in March 1790 the sun shone at Gotha, for news arrived that the transit was finished and ready for packing.

> Your friendly letter of the 9th of this month, which I had the pleasure of receiving on Wednesday past, 17th March, allows me to express the most lively and sincere thanks for all the care and trouble that this fine instrument has caused, and which you have kindly told me is entirely finished and ready for packing and embarkation for Hamburg. Please present my compliments to the artist and give him some inkling of the wrongs he has done me. However great these may be, he could wipe them out if he would agree to make me an instrument for measuring declination for he must be aware that the right ascension of a star is in general a secondary matter and that the principal is declination. … I would definitely prefer a mural to the circle which he promised me long ago. I would even promise in advance that if he followed my wish for a mural, the circle would not be forgotten; it is his pet design and I would agree to that.[46]

When the transit reached Seeberg, the Duke joyfully unpacked his new treasure. It would be set on wooden pillars before the stonework was put in place for its final positioning. Meanwhile … there were just a few small matters to be sorted out, for Ramsden had not provided instructions concerning the guard for the plumblines, and it would be helpful to have some spare silver wire for the micrometer, in case of breakage.

> Zach hastened yesterday to let you know of the happy arrival of this superb instrument which is a masterpiece of its art. But he can hardly have conveyed to you the delight which overwhelmed me as I unpacked it. We have just set it on two wooden pillars which serve as a model for its arrangement and direction on the final supporting pillars. … Zach will have told you, my dear Count, how much trouble Ramsden has caused us by failing to give instructions about the fitting of the wire guard. I requested him to ask the artist to make this matter clear. Today he thinks that he has guessed how

it should be and I believe that his ingenious arrangement will be entirely satisfactory. However, to be more certain, it is desirable that the artist should be asked about the plumb-wire. It may well be that both arrangements yield the same result and that we can choose between them. I beg you to ask the artist for a small portion of silver wire for the micrometer. It is so easily broken and as wire of that quality is unobtainable here, we shall be in want, as the artist did not include any spare in the package. I don't know how Maskelyne manages, but his observations continually record that the wire has broken, that an insect has broken it, etcetera, and it is better not to find ourselves in difficulty. ... To you I owe the happiness of possessing one of the most beautiful and most perfect instruments to be found in the most celebrated observatories in Europe. If it were possible to persuade this man unique in his craft to fulfill his other agreements with me, I could boast of seeing my observatory become the equal of any other in existence.[47]

But inevitably this was not to be. In December 1790 Duke Ernst again put pen to paper to bemoan the non-appearance of his circle.

I truly envy, my worthy friend, the circle which you acquired recently and which, although far smaller than that which I await from the heartless Ramsden, so well supports your declination measurements. I congratulate you most sincerely Sire, since you know full well the value of your possession, and that I sense in my own impatience how hard it is to be 200 leagues from a man who fails to keep his word for his engagements, to provide novelties for the first-comer, who importunes him by dint of perseverance; I am fearful that the establishment which I have begun will never be finished. ...

    PS. Please, from time to time, think of Ramsden's circle and remind him with a word from you, that he gave me his word to finish it. – Oh if only that artist unique in his craft could be made to hear the voice of his own honour![48]

Worse was to come: Duke Ernst learnt of 'a full circle ordered by the Duke of Richmond'. He took this to be an astronomical circle, though it was probably the geodetic theodolite purchased by the Duke of Richmond for the Ordnance Survey:

What you tell me concerning Ramsden on the circle ordered by the Duke of Richmond certainly allows me to hope that the cruel but matchless artist may consider finishing it for me. I base this entirely on your valuable friendship for me since it is clear that without your good offices I shall never obtain from him an instrument to measure declinations, without which all our other observations will be defective. To send someone to England solely to supervise and encourage Ramsden to keep his word, so often broken, would doubtless be effective in obtaining work from this artist – but it would be far too costly for me without a certainty of success.[49]

Brühl demonstrated his friendship for Duke Ernst by lending him Cary's circle.[50] By 23 March it was on its way to Seeberg.[51] In January 1793 the Duke learnt more about the Duke of Richmond's instrument, and recounted to Brühl an amusing episode when Ramsden had sent Zach a letter intended for Hornsby. This was happily followed by a letter from Ramsden to Zach concerning Duke Ernst's circle.

What you tell me, my dear Count, about the unspeakable Ramsden, caused me both pain and pleasure. Competing with the Duke of Richmond terrified me; on the other hand his letter to Zach reassured me. On the subject of this last there was a most amusing mistake, which he corrected soon enough. Zach received a letter from Ramsden which was probably addressed to Dr Hornsby and which had items which he had doubtless ordered. A few days later he received another in which the artist informed him of progress on my great circle, which seemed to promise some goodwill on his part.

Whether this perceived goodwill encouraged Duke Ernst to dispatch money to London for his circle, or whether the transfer had been made some years earlier, on 1 May 1793 Brühl sent him Ramsden's receipt for £200, which further boosted his hopes.[52] But by March 1794 he was back in the pit of despair.[53] In 1800 Zach put a pessimistic note into his journal:

We have no hope of receiving the beautiful transit instrument from Ramsden, which for thirteen years has been wanting for our national observatory and on which we have paid one thousand small thalers.[54]

Later that year he reported 'We hear that that Ramsden lies without hope on his death bed'.[55] Ramsden died on 5 November 1800. Zach later noted, 'The famous Ramsden used to say in joke that he would assess the skill of a practising astronomer by how he wielded his knife and fork at the table'.[56] Being undated, it is unclear if this is a post-mortem comment.

The Duke turned elsewhere and ordered a 3½-foot circle from Edward Troughton. Zach explained to Maskelyne:

Seeberg Observatory 4 May 1802
I was only so lucky to get from late Mr Ramsden an 8 feet Transit Instrument, Aperture 4 inches. This instrument, of a most perfect kind, can certainly rival with any other existant, but Mr Ramsden had not finished my 8 feet whole circle, for which he has got an advance of £200, which money is very likely lost now. I entertain no hopes at all, to get this instrument done, by his successor Mr Berge, so I commanded a whole circle to Mr Troughton, of which I hope to come in possession the present year.[57]

Seeberg Observatory languished after the early death of Duke Ernst in 1804. Zach was not favoured by the new Duke August and he departed from Germany in 1806. Ramsden's transits and his refractor are in the Deutsches Museum.[58]

Palermo Observatory

The Bourbon Kingdom of Naples covered the southern portion of Italy, from Gaeta on its northern border with the Papal States, and including Sicily. There were no observatories attached to the venerable universities of Naples, Salerno and Catania, but some private observatories existed elsewhere in the kingdom. With the expulsion of the Jesuits from Spain and Sicily in 1767, it fell to the Bourbon government to rebuild the teaching system, and to create, in 1779, a

Royal Academy – from 1805 the University – of Palermo. The situation was made more favourable by the appointment of two enlightened vice-regents: Domenico Caracciolo (1715–1789) and Francesco D'Aquino, prince of Caramanico (1738–1795), the latter having served as Neapolitan ambassador in London in 1784–86, and by the presence of the mathematician Giuseppe Piazzi (1746–1826).[59] Piazzi had taught mathematics in Rome, Malta, Ravenna and Cremona before being called to the Royal Academy at Palermo where on 1 January 1787 he was appointed professor of astronomy. This was a new subject for him and he asked to be allowed to spend two years in Paris and London where he might learn to make observations before entering on his professorial duties. His request granted, Piazzi left Palermo in March, heading first to Paris, where much of his time was spent with Lalande, and afterwards to London where he joined the French group involved in the cross-Channel surveying operations described in Chapter 8.[60] Piazzi kept a diary of the years between 1787 and 1790, but it was last seen in 1938 and cannot now be found.

Piazzi arrived in London in late September 1787 and headed for Greenwich with his introduction from Lalande. There his awareness of the limitations of Bird's mural quadrant, and probably Ramsden's persuasion, convinced him that a circular instrument was essential if he was to observe to an accuracy greater than one second of arc.

Interlude – the Specola Caetani

Earlier that same year, Ramsden had in fact started to construct an eight-foot diameter circle for the Specola Caetani in Rome. This observatory had been erected in 1775, at the family palazzo in Via delle Botteghe Oscure, by Francesco Caetani, Duke of Sermoneta (1738–1810), and his brother Onorato Caetani (1742–1794) who of the two, probably had the better scientific training.[61] The variety of scientific apparatus which the Duke provided led to his observatory becoming one of Rome's cultural centres. Among its instruments were Dollond achromatic telescopes, and Ramsden gregorian reflecting telescopes.[62] The observatory was sacked by French troops in July 1798 but despite this and other political upheavals in the Papal States the succession of astronomical and meteorological observers continued into the early nineteenth century. It is, however, unclear whether the Duke had actually placed an order. Ramsden had allegedly abandoned work on it, but the 'Caetani circle' maintains a ghostly presence in Ramsden's workshop, and part of its framework may have been incorporated into Piazzi's apparatus while Ramsden was under pressure to deliver surveying apparatus for William Roy, as detailed in Chapter 8.

Piazzi's Astronomical Circle

When Piazzi placed his order, Ramsden had already confronted many of the problems arising from the construction of large instruments. Since 1770 he had employed the hollow-cone construction adopted much earlier by Jonathan Sisson and John Bird for the axes of their transit telescopes. Flat brass plates were joined by the lightweight structure known as 'Jacob's ladder'. The bearings were bellmetal turning in steel. These techniques were incorporated in Ramsden's great theodolites as were the low power microscopes fitted with cross-wires which magnified the graduations nine times, permitting remarkably precise readings. Ramsden claimed originality for this feature, but had probably adopted it from the Duc de Chaulnes. Piazzi's instrument would be fitted with one such microscope to read the horizontal azimuth circle and two more for reading the vertical circle.[63] Ramsden also provided specifications for the domes covering the circle and the transit which was manufactured later. The Sicilian minister Count Lucchesi negotiated with Ramsden for the payment.

While he was in London, Piazzi was a frequent visitor to the Royal Society, often introduced by Ramsden, who had been elected FRS in 1786. The Royal Society's manuscript Journal records his attendance at meetings throughout 1787, then again from November 1788 until June 1789. On 4 December 1787 Maskelyne gave a dinner at Greenwich for the Parisian astronomers Cassini, Méchain and Legendre at which Piazzi and Ramsden were also guests.[64] In his obituary of Piazzi, De Angelis states that Piazzi attended every day at Ramsden's workshop. If that is true, he was privileged, for Ramsden seldom admitted even customers into the works. We are also told that to accelerate progress, Piazzi had the idea of flattering Ramsden by writing a letter to Lalande on 1 September 1788 in which he gave an account of Ramsden's life and work. It was published in the *Journal des savans* in November that year.[65] A translated version of De Angelis's obituary of Piazzi relates that the ploy succeeded, and adds that Ramsden thereafter 'wrought with unwearied zeal', which I find unlikely.[66]

There were always several orders for large apparatus in hand, some of which were perhaps waiting for Ramsden to find a source of brass manufactured to sufficient length, a difficulty which plagued other makers of large apparatus. He was frequently called away from Piccadilly to attend to his instruments at Greenwich or Blenheim or to assist Roy on the survey. Ramsden was also suffering from various ailments, as his friends anxiously noted in 1788, and development work on new apparatus probably ceased during his absences. Matthew Berge would have had his hands full dealing with the production of lesser instruments, and the refurbishment and repair of old items, both of which contributed to the cash flow needed to pay the workmen. Piazzi himself reports that Ramsden twice started work on his new circle, and twice abandoned it, before in January 1788 he started again, and completed it by 16 August 1789. Not discouraged by this situation and enthusing over Ramsden's character, his honesty over prices and his abilities, Piazzi secured his circle within the remarkably short period of 18 months and departed from London in August 1789, the proud proprietor of the first large astronomical circle to have been completed (see Plate 2).

This admirable piece of precision engineering – far too majestic to be spoken of as a mere 'instrument' – comprised an azimuth circle three feet in diameter, with a vertical circle five feet in diameter, bearing the telescope. The entire assemblage stood over nine feet high, and being free-standing, brought an extremely heavy load on the base of the vertical axis. The brass cones of the horizontal axis ran on steel wheels, but the entire weight of the structure rested on the base of the vertical axis. The lower bearing was probably steel in bronze. A curious statement in Piazzi's description relates to the square panels to which the four supporting columns are fastened; He says they are made of 'what the English call prince's metal which is the hardest it is possible to make'.[67] Prince's metal, also known as Prince Rupert's metal or Pinchbeck, is an alloy of 75% copper with 25% zinc. It is rather soft and having a gold hue was used to make cheap jewellery. There may have been some misunderstanding here; it is more likely that the plates were made of the hard bronze consisting of 75% copper and 25% tin.[68]

The circle carried a 5-foot achromatic telescope of 3 inches aperture. Piazzi described it as one of Ramsden's 'good' but not 'excellent' productions. The frame carrying its micrometer wires was not well fitted and Piazzi had difficulty in moving it to position the wires. It was not the only detail which was less than perfect, but Piazzi considered that these defects were due to his urgent demands on Ramsden to finish the instrument.[69] An indication of the problems besetting opticians at this time emerges from a letter written in July 1788 by Magellan, reporting the news that Ramsden had purchased 900 pounds of flint glass without finding a single piece serviceable for the lens of a telescope of 30 inches focal length.[70]

Finished at last, the various parts of the apparatus were crated for dispatch by sea. While awaiting a vessel, the crates were stored at Lucchesi's house. At the end of September they were loaded into a ship bound for Naples, arriving towards the end of 1789.[71] The excise office is said to have claimed duty on the export of the circle, on the grounds that it was an English invention but Ramsden maintained that any novelty the instrument possessed was due to Piazzi, and the claim was abandoned. The statutes in operation at this time prohibited the export of machinery connected with the textile and metal-working trades but in 1786 the Custom House issued a list defining prohibited articles which included brass metal and perhaps the excise officers inspecting the crates containing the various parts of Piazzi's circle were reacting to this regulation. Piazzi meanwhile returned overland via Paris where in cheerful spirits he showed drawings of the circle to members of the Académie des sciences, and exulted over its accuracy before hurrying on to Milan. Oriani, writing to Brühl in late November, reported that Piazzi was so eager to reach Palermo ahead of the circle that his stopover had been brief.[72] Piazzi had begun by informing Oriani that no observatory had a circle to match his new acquisition, then, after looking closely at Oriani's instruments, admitted that he had wasted his time going elsewhere since Brera possessed the best possible apparatus. Piazzi was also carrying Ramsden's models of the revolving roof for the circle and the shutter for the transit. He discovered that Oriani's shutter and his four revolving roofs were in fact less likely to go wrong. He assured Oriani that the circle had errors of less than a half-second, but, says Oriani, he had so little experience that it was hard to believe him. After a couple of years of observing, Piazzi's claims would be more modest.

Piazzi brought to Palermo a young craftsman from Hanover named Drechsler, who had been employed by Edward Troughton and was probably related to Georg Drechsler who spent five years with Ramsden before moving on to Hamburg. In addition to a salary and accommodation the younger man was employed as mechanic to the University of Palermo and was paid for any apparatus he made.

Tempted by the greater rewards of commerce, he stayed in Palermo and died a rich man.[73]

The Circle in Use

By registering altitude and azimuth simultaneously, Piazzi believed that the circle was ideal for single observations of a comet or suchlike temporary phenomena, but the Astronomer Royal, John Pond, contested this claim in his article 'Circle', published in a part of the *Edinburgh Encyclopedia* in 1813. In a blistering denunciation of the Palermo circle Pond declared that:

> It owes its celebrity certainly more to the industry and sagacity of the observer into whose hands it has fortunately fallen, than to the soundness of the principles on which it is constructed; for though it contains numberless examples of beautiful invention and contrivance, yet upon the whole it is an instrument that no judicious artist would ever wish to make again: ... the plumb-line by which it is adjusted, referring to two fixed points in the limb of the circle, cannot be examined at the moment of observation: From this cause, and from the effect of expansion, to which, from its faulty construction, it is much exposed, a single observation cannot be relied on, nearer than to five seconds, though a mean of several will probably be much within that limit.[74]

So much for Pond's opinion. We should not, however, forget its predecessor, the as-yet unfinished Dunsink circle, on which Ramsden had already spent considerable time and thought.

When he was composing his account of the observatory and its instruments, Piazzi wrote to Ramsden asking him to clarify the method of adjusting the micrometer wires of the vertical circle to align them with the plumb-line. Ramsden had explained this in London, but when the instrument was set up, Piazzi realized that he had not entirely understood what was required. There were other concerns, reminiscent of those elsewhere: the wires in the telescope focus having broken when the instrument was being mounted, Piazzi had been obliged to substitute thicker ones; he asked Ramsden to send some fine wire. He was also unclear as to how the prismatic ocular inverted, and sought instruction on how to remove the dust which had penetrated between the two objectives of the telescope, and how to clean and care for the instrument itself. Piazzi closed with a hesitant request for an ocular with a filar mocrometer to set on the circle.[75] A draft of Ramsden's letter deals with these points. Apologizing for having forgotten to describe in detail the adjustment of the microscope wires, a lengthy explanation followed, together with some bluestone powder for cleaning and a drawing to explain the working of the prismatic ocular. Ramsden declined to send an ocular with a wire micrometer as it would need matching to the eye end of the telescope. In 1792 he sent Piazzi some fine wire, which, on examination, turned out to be coarser than wire available to Piazzi locally.[76]

The Ramsden circle achieved fame on 1 January 1801 when Piazzi turned the telescope of his circle to the heavens and discovered a star-like body, the first of the minor planets orbiting between Mars and Jupiter to be identified, now named

Ceres. Competently restored in all its impressive glory, Ramsden's circle is now installed in its original setting in Palermo Observatory.

In June 1801, perhaps before Ramsden's reply arrived, Piazzi wrote to Oriani:

> The circle, either from my experience or the nature of the apparatus itself, lacks the precision which I had praised it for having. The wires in the telescope already having broken, I was obliged to substitute thicker ones. Also I have had to clean the limb three times, damp, dust and flies having rendered it nearly invisible.[77]

In return, Oriani wrote that the immobility of the telescope wires (presumably the subject of an earlier complaint) was an unpardonable lack on such a great craftsman as Ramsden; he also advised Piazzi on cleaning and lightly oiling the graduated scale, to preserve it from corrosion.[78]

The passage of time mellowed Piazzi's memories of any problems he might have had with Ramsden. In July 1813 the author Thomas Smart Hughes (1786–1847), in Palermo during his tour of Sicily, called on Piazzi and was shown the circle – 'probably the finest in the world', reports Hughes, with which Piazzi had published 'more numerous and accurate observations than all the astronomers of Europe together in the same time'. Piazzi informed Hughes that he had spent two years in England with Ramsden, 'during which time he [Piazzi] hardly stirred out of the house'.[79]

Meteorological observations started in May 1791, soon after the building works had finished. Ramsden supplied various thermometers, a De Luc type whalebone hygrometer, and a fine precision barometer, with a cistern float (see Chapter 7), of which only the last item survives.[80] At the same time, he supplied a transit with a 5-foot telescope, a dynameter, a 13-inch radius Hadley sextant and a 35-inch achromatic telescope.[81] After Ramsden's death the observatory acquired other instruments from the Piccadilly workshop, for which see Chapter 12.

Sir George Shuckburgh's Observatory

The largest equatorial ever made was the great observatory instrument commissioned by George Augustus William Shuckburgh FRS (1751–1804), sixth Baronet, (later Shuckburgh-Evelyn,) for his observatory at Shuckburgh Hall, Warwickshire. A man of considerable experience with instruments, Shuckburgh appears to have shared in its design. Intending to use his equatorial for positional astronomy, he wished it to combine the properties of a transit with those of a meridian quadrant. The equatorial and a clock by John Arnold & Son were the sole occupants of his observatory. Ordered in 1781, it was delivered ten years later, and probably included parts originally intended for Greenwich Observatory, Shuckburgh being the person who had capped the Greenwich price by offering 500 guineas, as mentioned in Chapter 5. Shuckburgh described it to the Royal Society on 21 March 1793; his description was published, with illustrations, in *Philosophical Transactions* in 1794 and issued as a separate monograph.

The circles, four feet in diameter, had been divided to 10 seconds of arc, reading by micrometer to one second. Shuckburgh praised the 'unexampled diligence and care, with which Mr Matthew Berge … has executed them, I feel bound to bear this testimony to his merit'. The telescope was 5 feet 5 inches focal length, 4.2 inches aperture. Its objective was an achromatic doublet, magnifying up to 400 times. The dimensions and weight of the instrument required bearings of steel running in bellmetal, this latter being used for the frame joining the upper ends of the tubes and pivoted to the massive cast iron structure rising some ten feet above the floor, which supported the upper end of the polar axis; as installed at Shuckburgh; this axis was further strengthened by two iron braces at right angles, bolted through the walls of the telescope chamber. The telescope was held in a yoke on the axis – a form which became known as the English mounting.[82] A balustrade surrounding the base prevented anyone moving around the chamber at night from accidentally knocking against the lower circle.

Shuckburgh had a plate inscribed in Latin with Ramsden's name and fame and the fact that the equatorial had taken ten years to manufacture, ending in 1791, which was fastened on the conical end of the axis. He provides the text:

> Hocce Panorganon Uranometricum à Jesse Ramsden, Londinense Optico celeberrimo et omnibus id genus arteficum longe anteponendo, excogitatum, decem post annos nunc tandem absolutum, GEORGIUS SHUCKBURGH Baronettus, in testimonium amoris sui erga res astronomicas, et ad easdem promovendas, fieri curavit, anno 1791. (This measurer of the heavens, conceived and now after ten long years completed by Jesse Ramsden of London, the celebrated optician, who was by far the greatest practitioner of his art, was commissioned in 1791 by Sir George Shuckburgh as a token of his love for astronomy and for the encouragement of that science.)[83]

The swinging level was unground, having been selected from many yards of glass tube in order to get one with a proper flexure. A change of level of 1 second of arc moved the bubble nearly one-sixth of an inch, a greater distance than that in the level on the transit instrument at Greenwich Observatory. However, the bubble was about 7 inches long, so it moved very slowly and its behaviour varied with temperature. The divisions were so accurate that any slight imperfections were probably due as much to inequality of the points, play in the micrometer screw, or imperfect reading off. The telescope had a divided eyepiece micrometer by Ramsden and a micrometer by Dollond.

Unusually, Shuckburgh devotes a good portion of his description to the construction of his dome, designed by George Pope, another of Ramsden's foremen. A truncated cone 11 feet in diameter, with folding shutters, it was formed of iron rings carrying iron ribs and covered with transverse layers of thin deal planks, then copper sheets, then white lead paint. This edifice was in place by 1789. Pope also checked the instrument before its delivery along the Thames then by canal barge to Lower Shuckburgh Wharf, from where it was brought up to the Hall by waggon. Shuckburgh recorded in two logbooks the meticulous examinations and adjustments needed before he could start observing.[84]

In November 1798, on the occasion of the award of the Royal Society's Copley Medal to Shuckburgh, Joseph Banks said of Shuckburgh's description of his equatorial:

> The whole tenor of the paper, however, resolves itself into an eulogium on the deep scientific knowledge and admirable mechanical skill of that unrivalled artist Mr Jesse Ramsden, to whom I am ever happy to offer my small tribute of applause.[85]

This colossus was mounted in the observatory at Shuckburgh Hall from 1791 to 1810. We may hope that its impressive size, if somewhat bizarre appearance, compensated Shuckburgh for the long delay in its arrival, and for its less than excellent performance. For when Shuckburgh's heir, Charles Cecil Cope Jenkinson, Lord Aylesford (1784–1851), offered it to Oxford's Radcliffe Observatory, it was declined.[86] Presented to Greenwich Observatory in 1811 and embellished with another plaque, also in Latin, commemorating Aylesford's gift, it was mounted by Berge in the North Dome in 1816. Pearson said of it in 1829:

> The framework was too slender for the length of the polar axis, and it remains at Greenwich, a proof of its former proprietor's munificence rather than of its maker's success in the strength and stability of its essential parts.[87]

By 1838 Berge's divisions were so worn that the declination circle was redivided by William Simms (1793–1860)[88] and installed elsewhere at Greenwich; Troughton & Simms redivided the hour circle in 1860. In 1929 the Shuckburgh equatorial was transferred to the Science Museum.[89]

Brera observatory

Milan, under Austrian governance, enjoyed an intensive period of intellectual life during the second half of the eighteenth century. Jesuits occupied the ancient palace of Brera in Milan until suppressed in 1773. The sighting from Brera of a comet in 1760 brought pressure for the foundation of a proper observatory, in response to which Federico Pallavicini, the rector, invited from Marseille the renowned astronomer Joseph-Louis de Lagrange (1736–1795). One of Lagrange's first tasks was to establish the exact position of the observatory; in 1763 he also set in train a continuing series of meteorological observations. In 1764 an event occurred which brought the observatory to the front rank; the Jesuit Ruggiero Giuseppe [Roger Joseph] Boscovich, summoned the previous year to be professor of mathematics at the university of Pavia, decided to transfer to the Jesuit College in Milan. Boscovich, born in the Venetian port of Ragusa (now Dubrovnik), was one of the most widely-travelled astronomers of his day. In 1760 he was in England, where he visited observatories at Cambridge and Greenwich and the private observatory of George Parker, second Earl of Macclesfield (1697–1764) at Shirburn Castle, Oxfordshire, and was elected FRS. His arrival in Milan in 1763 persuaded Pallavicini to recreate the observatory; by 1765 the necessary works had been undertaken and it was functioning as one of 30 observatories set up as part of the Jesuit teaching programme.[90]

The observatory was on two floors, the lower having a strong north–south wall in the long hall, capable of bearing mural quadrants in the room above. In 1768 a quadrant of 6 French feet radius by Canivet was erected in the upper octagonal room, but within a few years it became apparent that there were serious errors in the graduation of its arc, and the astronomers began asking for a larger mural quadrant, of English manufacture. A clock by Lepaute of Paris, with compensated pendulum, stood in the upper octagonal room and was later joined by another by the pre-eminent London maker John Arnold (1736–1799), along with many small portable instruments including two reflecting telescopes by James Short and two achromatic telescopes by Dollond for use on field surveys and geodetic work. On the octagon roof, facing north-east and north-west, stood two small turrets with rotating conical roofs with apertures. In Boscovich's time one turret held a large moveable sextant and a 4-foot transit, both by Canivet. In another turret stood an equatorial sector made by Jeremiah Sisson in 1774.

Suppression of the Jesuit order in Lombardy sent Boscovich to voluntary exile in Paris, but the Austrian government continued to finance the observatory, which now had an international reputation, and Barnaba Oriani (1752–1832) was put in charge. Between 1785 and 1786 circumstances favoured the astronomers: the grant for instruments was increased and the governor of Milan hoped that their observations would lead to the much-desired accurate map of Lombardy. The astronomers proposed that Oriani should visit the most highly regarded European instrument makers, and commission a mural quadrant which would serve to determine the coordinates for the map, besides, naturally, leading to productive astronomical research.

Ordering the mural quadrant

Their argument convinced the governor; in April 1786 he authorized Oriani's travels and the acquisition of the quadrant. Losing no time, Oriani left Milan on 12 May 1786. Travelling via Switzerland, Germany and the Low Countries, he arrived in London on 14 July and found lodgings in Westminster. On Monday 17 July the Austrian consul Antonio Songa took him to call on Ramsden; again on 20 July he visited Ramsden's workshop and saw a mounted circle, 5 feet in diameter. After another visit on 29 July, on 3 August, he negotiated with Ramsden for a quadrant of 8 feet radius, to cost £440, with £200 (500 secchini in Lombard coinage) to be paid in advance. It was to match that of the Duke of Marlborough, the only difference being that whereas the Duke's quadrant could turn on its supporting axis, the Brera quadrant, for want of space, would have to be attached to the wall. The contract was signed by Songa and Ramsden on 9 August. On 8 September, shortly before his departure, Oriani settled up his debts for various purchases and services, among which was £36, 15s, to Songa for Ramsden.[91]

Writing to Oriani on 25 June 1787, Zach, at this time still employed in London, informed him that 'Ramsden has made a start'[92] on the quadrant and if he could meet his promises it would be finished before the end of the year. This was not to be. In his letter of 27 February 1789 Zach told Oriani 'I have continually urged Ramsden to get on with your mural'.[93] Shortly before its completion, as Songa wrote to Oriani on 10 September 1790, the journeymen did 'something wrong' (we are not told exactly what) which Ramsden had to correct.[94]

The quadrant was finished in 1790. Ramsden displayed its qualities to several of his friends but whether by intention or forgetfulness failed to invite Charles Blagden, then Secretary of the Royal Society:

> Ramsden has finished the mural quadrant for Milan and invited Messrs Cavendish, Aubert, Smeaton and many others to a sight of it and a trial of its goodness, but did me not the honour of including me in the number; by the account it appears to be really a most excellent instrument, true much within the second, & with all sorts of proofs or verifications. He told them that any common man in his workshop, with good eyes and hands, could, on the same principles, have divided it to equal perfection.[95]

The cross-braced frame and the arch with its two sets of graduated divisions resembled other quadrants by Bird and Ramsden. Whereas hitherto the best quadrants were provided with two counterweights to keep the telescope rigid whatever its angle, Ramsden had for the first time added a third counterweight, constant in position and force, and fastened to an arm which acted on the axis of rotation. The telescope tube was secured at five points to a rhomboidal frame. The crosswires at the focus – five vertical and one horizontal, were originally of silver wire but were replaced with spiders' thread in 1817. Primary levelling was done with the plumb-line, then adjusted with 'Ramsden's ghost', his recent invention (see above, pp. 114–15).

Ramsden's first invoice, dated 20 November 1790 for £453, 14s, itemizing the quadrant, an apparatus for testing the accuracy of the arch, the packing cases and their portage to the Custom House, (see Figure 6.6) was followed by a second, dated 22 November, for a portable barometer, a 2-foot telescope and other items. But at the end of 1790 the instrument remained in London, awaiting a ship with hatches sufficiently wide to admit the enormous case in which the quadrant itself was packed. Furthermore, all the cases had to be stowed in the hold without anything being loaded on top, and the unused space had to be paid for. Songa, who was supposed to see to the instrument's despatch, was rather dilatory but in December 1790 the quadrant left London and was landed at Genoa in late January. It then made the slow overland journey to Milan, probably arriving in March 1791. By 1 June it was already fastened to the south-facing wall, replacing Canivet's quadrant. In 1792 the astronomers completed their determination of its position.

The quadrant remained in use until about 1840. In 1812, when Luigi De Cesaris was measuring the sun's position, he noticed that its plumb-wire had moved from its proper position, indicating that the whole instrument had shifted. De Cesaris attributed this to a movement of the entire building, although no similar phenomenon had been found in any other observatory.[96]

The purpose of this instrument was to determine the coordinates of the observatory, to which the proposed map of Lombardy could be related. Oriani and his astronomers were concerned with this survey from 1788; concluded

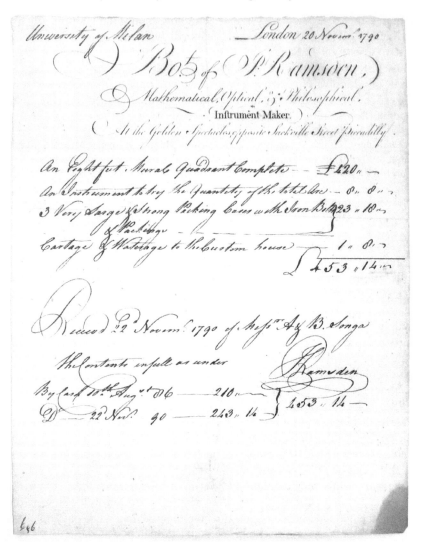

Figure 6.6   Ramsden's invoice to the University of Milan, 1790, for a mural quadrant with its test apparatus, packing and carriage to the Custom House. BOA, MS AAV. 7.

only in 1794, it was the first cartographical operation in Italy with coordinates determined by advanced astronomical methods. Seven sheets were drawn and engraved but the eighth plate was incomplete when Napoleon, having won the battle of nearby Lodi, reached Milan in May 1796 and the seven plates were removed to Vienna for safety. Matters were resumed in 1802 under the Napoleonic Cisalpine Republic.[97]

Ramsden's quadrant is preserved in the museum of Brera Observatory. Many parts are now missing; there remains only the frame and arch, the telescope tube and the vernier.[98]

Instruments for Paris Observatory

The story of the three Ramsden instruments ordered in 1787 is one of French enthusiasm quenched and trust betrayed. While it was played out, a most tumultuous and destructive revolution intervened, politicians came and went, Ramsden died, Cassini IV, the man who had placed the order with such confidence, retired from his post at the Observatoire, and everyone's patience, in Paris and in London, had been tested to the full.

After the instruments installed around the time of its foundation in 1672, the next major items ordered for the Observatoire, all from Claude Langlois who was then official constructor to the Académie royale des sciences, were a 6-foot mural quadrant in 1730, a mobile quadrant of 3½ feet radius in 1738, and another of 3 feet in 1756, which remained in service until 1793.[99]

Paris was well-provided with skilful horologists, opticians and makers of small everyday instruments but, as historian Anthony Turner explains, from the 1730s the fruitful interaction which had formerly existed between these craftsmen and members of the Académie des sciences became increasingly rare and by the mid-century had virtually disappeared. Unlike their leading British counterparts, several of whom were elected to the Royal Society, no French instrument maker was admitted to the Académie, while their increasing isolation from scientific advances impoverished their trade, and discouraged ability, ingenuity and innovation in a manufacturing area of which the Académie had particular need.[100]

The French astronomical instruments business was severely hampered by the long-outdated guild regulations, which restricted craftsmen to the one material deemed appropriate to their guild – be it glass, wood, brass or iron – whereas by this time scientific instruments incorporated several of such materials. Anthony Turner describes how guild officers would enter workshops and destroy items containing 'prohibited' materials.[101] Nor did the French craftsmen have ready access to credit or funds which would allow them to purchase sufficient materials and then pay their workmen for what could be a long period before the instruments were finished and – after further delay – paid for. Consequently they sought full payment before starting work, and avoided introducing any improvements which might have increased the cost. Although it was customary to ask for half the price as down-payment, the London makers were left to carry the remainder of the cost until settlement, and many had bank accounts and access to credit, together with a cash flow generated by small sales, second-hand trading and repairs. When César-François Cassini de Thury (1714–1784), Cassini III, took over the directorship of the Observatoire in 1771 he was painfully aware that French craftsmen were being eclipsed by John Bird, the Sissons and Jesse Ramsden, who had carried the art of optical and mathematical instrument making to far higher levels.[102] 'A Frenchman to the depths of his soul', as Devic tells us,[103] Cassini did not have to look far for confirmation of this sad state of affairs: within Paris the military academy and many of the colleges had acquired instruments by Sisson and Bird.[104]

Cassini III was the astronomer initially involved with Joseph Banks and General Roy on the accurate survey of the longitude distance between the observatories of Greenwich and Paris, as described in Chapter 8. The cross-channel sightings were being planned, when Cassini III, whose health had been declining since 1776, died of smallpox on 4 September 1784. His place was taken by his son, Jean-Dominique Cassini (1748–1845), Cassini IV, (see Figure 6.7) and it was the latter who came across with the mathematician Adrien-Marie Legendre (1752–1833) and Pierre-François-André Méchain (1744–1804), astronomer and geodesist. Cassini's minister, Louis-Auguste Le Tonnelier, Baron de Breteuil (1730–1807), sanctioned his visits to major observatories and calls on Dollond and Ramsden. Cassini III had compiled a long list of instruments needed for the Observatoire, giving priority to a mural quadrant of 6 or 8 feet radius, an equatorial telescope with 5-foot circles, and a 3-foot transit circle. As he had anticipated, only the first three were approved and even these had to be ordered one at a time as funds became available.[105]

Cassini meets Ramsden

So, when Cassini IV reached London with his shopping list, he swallowed his shame at having to buy English-made instruments, and entered into a debate with Ramsden who, he felt certain, would be flattered at receiving his order for a large free-standing transit circle, like that sent to Palermo. Additionally, he had permission to offer to send two men to work under Ramsden, to assist in its production. Cassini had in mind a trained man who would be sent under the pretext of 'dessinateur' or draughtsman, with instructions to look around him and to sketch innovative apparatus and tools, the other on a genuine two or three-year apprenticeship.[106] Both men were given lessons in drawing, geometry, practical astronomy and English, but nothing came of Cassini's proposals.[107] Cassini and the Baron de Breteuil were certainly bent on industrial espionage and its likelihood may have coloured Ramsden's reaction.

Friendly relations were established, but as Cassini reported:

> I must admit that however great an opinion I had formed of Ramsden, after keeping company with him for a few days, and having in the course of two or three conversations, seeing his extensive knowledge and perception, I became discouraged to realise that we would never have such a consummate artist in France.[108]

In fact, Cassini came to wonder if there existed anywhere a man with such a breadth of knowledge, encompassing geometry, astronomy, mechanics, optics and physics. Ramsden answered all his questions, resolved all his difficulties, and always led him to believe that his wishes could be met. Meeting one day a similarly enthusiastic visitor at the shop, Cassini remarked, 'This man is an electrical machine which has only to be touched to emit sparks'.[109] The other

Figure 6.7   Jean-Dominique Cassini. © Observatoire de Paris, Bibliothèque

man (who was it?) declared that he had spent two years in London waiting for his instruments. Ramsden had never ceased to welcome him, had spent entire mornings chatting, showed the visitor his instruments in progress, promised them repeatedly, and never completed them. He assured Cassini that it was not due to indifference or laziness, but the search for perfection; he had seen Ramsden take an instrument which was virtually ready, one of which any other craftsman would be proud, and for one single small defect would break it up and throw the pieces back into the melting-pot.[110]

Thus forewarned, Cassini determined to be resolute, but at their next encounter, Ramsden's affability, his conversation, and his air of good faith, completely overcame Cassini's misgivings. Ramsden declared that the finest transit to leave his workshop should be for Paris; he had one under construction, destined for another buyer; he would adapt a good objective lens and Cassini should have it.[111] It would be in Paris shortly after Cassini's return. Ramsden's manifest sincerity, coupled to French pride, swept away all Cassini's doubts. On another occasion, discussing his plans for the Observatoire, Ramsden proposed sending a new instrument he had in mind, which would allow a single observer to ascertain the difference in right ascension of any two stars, to 3 or 4 seconds of arc. Cassini said that he would make a place for this instrument; they discussed dimensions and even the proposal to send two workmen was apparently accepted without difficulty. Learning that Méchain and Cassini intended to visit Blenheim Palace, Ramsden arranged this with the Duke of Marlborough and himself introduced them into the observatory. They were, of course, immensely impressed with the beautiful rotatable pillar quadrant.

From the Observatoire, Cassini wrote to Ramsden on 6 January 1788, confirming that he had funding to commission two instruments, first and most urgently, the transit which had been promised for the coming August. Secondly, a mural quadrant of 8 feet radius, mounted so as to rotate, like that of the Duke of Marlborough. He proposed that Ramsden should put two instruments on this rotating support, by replacing the counterweight with the entire circle. He asked if Ramsden could deliver these three instruments within about two years from the date of writing, and ended by asking if the French authorities might send two workmen at their expense, to learn a craft from someone who possessed it in such high degree. A copy of this letter in the Observatoire archives bears Cassini's later annotations.[112]

Ramsden's reply, dated 25 January, expressed his honour at being asked to make these instruments, and assured Cassini that he would endeavour to meet the delivery date specified, but this sort of work demanded the greatest care, and nothing would be overlooked on his part. He was however unhappy about the idea of putting a mobile quadrant and a circle on the same framework as the circle stood directly over its vertical axis and needed no counterweight. Cassini made a long annotation on Ramsden's letter and clearly had not understood what was involved. As for the apprentices, Ramsden assured Cassini that his fifty workmen were adequate for his trade; he would have been happy to accept the Frenchmen but feared that his own men would be jealous and join those who had opposed him in past years, disrupting his business. (This was disingenuous; Ramsden had for many years been employing foreign workmen.) Lastly, he could not say what the instruments would cost – his cash flow sufficed to undertake the commissions, and as work progressed he would be able to give a price. Should this prove too high, it would not trouble him to cancel the order.[113] Possibly as a result of this refusal, Cassini then wrote to Magellan asking if he thought that the Troughtons would take a young Frenchman who wished to work in England.[114]

Unlike Ramsden's 'list' given to Felice Fontana in 1778/9, naming the instruments in somewhat erratic French, this letter in Ramsden's own hand is fault-free. It is known that the well-travelled physician Charles Blagden was called on to translate into French Joseph Banks's letters to Cassini III and we may ask whether Ramsden was guided by Blagden, although Louis Dutens tells us, 'At an advanced age [Ramsden] made himself so complete a master of the French language as to read with peculiar pleasure the works of Boileau and Molière'.[115]

Ramsden's reputation was well-known in Paris, from where the Baron de Breteuil wrote in February 1788 to the Marquis de la Luzerne, ambassador in London, with details of the contract and asking him to assign someone to call at Ramsden's shop from time to time, urging him to complete the order and ensuring that he was paid for the transit circle as soon as it was finished.[116] The ambassador's reply must have depressed Cassini, to whom the Baron conveyed the substance: Ramsden had indeed promised to complete the instruments by the due date, but the ambassador does not guarantee that he will keep his word, being well known for never being satisfied with his products and always seeking to perfect them, making it difficult ever to complete them. Worse still, for all concerned, the ambassador continued, Ramsden had thought of an improvement to the rotating mural quadrant, and would be writing to Cassini on the matter.[117] Even two small items – an opera glass and a field telescope – which Cassini had hoped to receive via the ambassador, were not ready for despatch and Ramsden had still to send Cassini the promised model of the support for the quadrant though he continued to confirm the August delivery date.[118]

In February 1790 Brühl wrote to Cassini with news of the progress of his transit. Ramsden had assured Brühl that the two cones of the axis and its support, the counterpoise, the telescope tubes and eyepiece, and the apparatus for lifting the telescope from its supports in order to reverse it, were finished. Brühl had seen some of these pieces for himself and was confident that he was not being misled. George Pope, the workman in charge of such commissions, confirmed the situation and in Brühl's presence urged Ramsden to progress the work. Nevertheless, Brühl added, there clearly remained much more to be done; he himself had been waiting five years for Duke Ernst's promised apparatus.[119] In response, Cassini informed Brühl that he had asked his colleague, the physicist Alexis Marie de Rochon (1741–1817), to take to London a douceur of 3 gold louis for Ramsden's leading workman.[120] We are not told if this gift went into the pocket of Matthew Berge or George Pope.

Time passed; after his visit to London de Rochon wrote to Cassini to say that he had chivvied Ramsden and indeed had hoped before departing to have seen the instrument finished and packed, but from Ramsden he received only fine words and the suggestion that it would be best to delay sending it until Cassini was at Paris to receive it, when there would also be better weather for the Channel crossing.[121]

Political disturbances in Paris began in the winter of 1788–9, escalated throughout that summer, and rumbled on until 1791, when King Louis XVI tried unsuccessfully to flee the country. In 1792 a republic was declared and France became involved in the war with Prussia and Austria. In 1793 the King was executed, the troubles hotted up, and by 1794 what was appropriately known as 'The Terror' held sway till 1799 and the institution of the Napoleonic era. Cassini grudgingly accepted some duties imposed on him at the beginning of the Revolution; he directed a portion of the new administrative maps and was briefly involved with the new decimalized metrology. But, always a monarchist, he opposed reforms to the Observatoire and eventually gave up his duties in September 1793. Shortly afterwards, he retired to the family chateau of Thury.

By this time Ramsden's health was failing. He was busy with other existing orders for astronomical and surveying apparatus and although he had in fact accepted 3,000 livres advance payment in 1789, he might well have hesitated to consign a valuable piece of equipment to a city in the throes of extreme civil unrest. The story of Ramsden's instruments for Paris continued after his death in November 1800, and can be found in Chapter 13.

Voyages and Expeditions

Malaspina's Circumnavigation, 1789–1794

Ramsden instruments were carried on the Spanish circumnavigation of 1789–94 in *Descobierta* and *Atrevida*, commanded by an Italian officer, Alejandro Malaspina (1754–1810). This voyage was more than a scientific expedition, being seen as a journey of imperial inspection, for Malaspina was to report back on all aspects of Spain's extensive overseas empire. On his return, Malaspina began to edit his journal for publication, but unwisely became involved in politics. He fell from grace and suffered a long imprisonment, being eventually released to pass his last eight years in Italy. A printed version of his journal was published in 1990, an English translation in 2001–04.[122]

Magellan acted as agent for the Spanish authorities, procuring instruments from several makers. Ramsden apparatus taken included mathematical instruments, four 60-toise chains, four theodolites and a 3-foot telescope. Two 18-inch radius astronomical quadrants survive, one in the Royal Naval Observatory at San Fernando, Cadiz, the other in the Naval Museum, Madrid.[123]

George Vancouver's Voyage, 1791–1795

Captain George Vancouver, who had sailed on James Cook's second and third voyages, left England in HMS *Discovery* and HMS *Chatham*, rounding south-west Australia and New Zealand, crossing the Pacific, then making his way along the north-west coast of America. Vancouver took 12 sextants, the majority provided by Ramsden, others by Adams, Dollond, Troughton and Gilbert, and he reported that 'they all agreed exceedingly well together'.[124] Ramsden disappointed him in respect of another instrument: delayed by calms and contrary winds at his departure, Vancouver wrote to Philip Stevens, Secretary to the Admiralty, on 12 March 1791, conceding that one good effect of the delay could be 'that of enabling me to receive Ramsden's instrument as I should be exceedingly distressed were I to sail without it'. Inevitably, writing from Falmouth on 20 March, Vancouver admitted, 'I am much concerned to find from Mr Nepean [then under-secretary of state] that I am to be disappointed in the instrument Mr Ramsden was ordered to supply me with'.[125] Vancouver then set sail, hoping that the missing instrument could be embarked on his following storeship, HMS *Daedalus*. This was the 'universal theodolite', an altitude and azimuth instrument ordered by the Board of Longitude for its astronomer William Gooch, costing £126 and delivered in July 1791.[126]

Gooch receipted the list of astronomical and mathematical instruments which the Board had issued to him,[127] Ramsden's sole contribution being the universal theodolite, which was accompanied by its bespoke wood-and-canvas portable observatory.[128] *Daedalus*, with Gooch on board, made her way into the Pacific and put in at Oahu in the Sandwich Islands for provisions and water. The natives of Oahu proved as hostile as they had been when Captain Cook had been killed there a few years previously and on 7 May 1792 Gooch and two other sailors were captured, and killed. *Daedalus* joined Vancouver at Nootka. Gooch's assistant John Crosley formally took responsibility for the instruments in December 1793.[129]

Whatever his feelings had been at the outset of his voyage, Vancouver had mellowed sufficiently by 27 July 1793, when he was running up the coast of what is now British Columbia, north of Vancouver Island.

> We soon reached the east point of the entrance into the north-north-west branch; which, after Mr Ramsden the optician, I called Point Ramsden, lying in lat. 54° 59′, long. 230° 2½′.[130]

Getting an accurate longitude was still difficult; Point Ramsden's correct longitude is 229°54′ E or, as we would now say, 130°06′ W.

The Macartney Embassy, 1792–1793

For almost a century English merchants of the Honourable East India Company (HEIC) had traded with China, but they were strictly regulated and denied access to Chinese officials, being regarded as unwelcome and troublesome barbarians, with little to offer in return for the tea and porcelain they were so eager to purchase. As the range and quality of British manufactures increased, this situation rankled and plans were made to send a diplomatic mission to China. Its leader would present gifts to the Emperor and his court, display a range of fine merchandise representing the arts and sciences, and endeavour to negotiate favourable terms for the Company's merchants and for Britain generally. The sciences were to comprise an assortment of optical, mathematical and philosophical instruments.[131] Unwisely the chosen envoy was Lieutenant-Colonel Charles Allan Cathcart (1759–1788), who at the time of departure was in the terminal stages of consumption. His death at Madeira obliged the embassy to return to England.

A more determined attempt was planned in 1792–93. George Macartney, first Earl Macartney (1737–1806), was empowered as ambassador extraordinary and plenipotentiary to the Quianlong Emperor (r. 1736–1795) at Peking (now Beijing). In the hope of impressing the Chinese, the East India Company purchased a cargo of the best British products. Wedgwood, Boulton and other manufacturers were keen to gain new markets, and there were many bales of various textiles. The scientific and technical apparatus, some of which had been intended for the Cathcart embassy, included a most elaborate and costly orrery of German manufacture, which the London clockmaker Benjamin Vulliamy (1747–1811) had further embellished, chronometers, telescopes from Dollond, numerous instruments from Nairne & Blunt, and what were termed 'measuring instruments' from Ramsden. The embassy sailed in HMS *Lion*, accompanied by the HEIC vessel *Hindostan*. On board the latter was James Dinwiddie (1746–1815), a scientific lecturer with mechanical skills whose duty was to set up the orrery and to demonstrate other apparatus.[132]

The Chinese were known to be interested in time measurement and astronomy – it was supposed that they would be keen to acquire instruments for navigation and cartography. Ramsden was commissioned by the East India Company to provide a steel chain with its accessories, cost £42, 1s; a portable equatorial, cost £110; a zenith sector, cost £189; together with its oilcloth tent, cost £21; and a telescopic level, cost 12 guineas.[133] A second batch of items included two sextants and two 'best' theodolites, which with packing cases and so on cost £82, 11s.[134] Neither batch-list is dated and it is not clear if this lesser order was for the items which had been sent with Cathcart but then returned to the HEIC warehouse.

Inevitably, Ramsden's goods were not ready when the cargo was being loaded. George Pearson (1751–1828), who was on the Royal Society's Council, wrote to Banks on 22 October 1792 describing the outcome: 'Lord Macartney had a pretty violent dispute with [Ramsden] on account of his instruments not being ready when the ship sailed but they were after this quarrel sent by land to Portsmouth.'[135] On arrival in China, Macartney realized that he had no gifts of his own to present and he therefore purchased for about £200 a Herschel telescope belonging to Henry Browne (c. 1754–1830), factor at Canton, and for £773 'the celebrated lens from Parker' (a burning glass) which Captain Mackintosh had brought with him on speculation.[136]

Dinwiddie reported that Macartney divided the presents into three groups, of which only the first two were offered and accepted. A motley assortment, they included electrical and pneumatic machines, pocket barometers, the large burning glass and the planetarium, besides textiles, prints, assorted hardware and stationery, furniture and coaches. The British hope of establishing diplomatic relations between two nations they saw as equals was bound to fail. The Chinese were amused but not impressed by what they saw as inferior products or mere toys, and Ramsden's equatorial and sector were not presented but delivered to Browne at Canton for the Company's account.[137] In this way, Ramsden's zenith sector eventually reached India, where it was employed by William Lambton. (See Chapter 8).

Notes

1   Taylor (1845), 77, 348–57; Wayman (1987).
2   Ussher (1787), 3–22.
3   Ball (1895), 240–41.
4   Ussher (1788).
5   Ussher (1788), 13–26.
6   Pearson (1824–29), 57 and Plate 7.
7   Wayman (1987), 99.
8   Robinson (1871), 445–6.
9   Burnett and Morrison-Low (1989), 29, citing Trinity College Dublin, MUN/ P/4/67/24.
10   Robinson (1828), 21–38.
11   Robinson (1871), 445–6.
12   Ball (1895), 242–3.
13   Taylor W.B.S. (1845), 354–7.
14   Ball (1895), 244.
15   Wayman (1987), 265–6.
16   Budde (1993a).
17   Budde (1993b).
18   Klüber (1811), 28–33.
19   Neumann (1994), 25–40.
20   Bernoulli (1771), 57–61. His comment was perhaps directed at the portable apparatus.
21   Budde (1993b), 19, 21, 61–82.
22   Klüber (1811), 14.
23   BOD, MS Rigaud 38. Volume of letters of men of science. No. 34, König to Magellan, Mannheim, 21 August 1784.
24   RAS, MS Radcliffe A1.44.
25   BOD, MS Rigaud 38. Volume of letters of men of science. No. 34, König to Magellan, Mannheim, 21 August 1784. 'Je suis convaincu Monsieur, que vous plus justement me trouverez digne de votre compassion et pour cela j'espère que vous m'aiderez tant qu'il vous est possible. J'en vous prie Monsieur par les droits de l'humanité et par l'honneur des Anglois savans et celebres, informez vous tant chez Monsieur Maskelyne, que chez Monsieur Ramsden, et poussez la chose jusqu'à recevoir les 100 guinees, ou l'instrument. J'attend avec impatience votre reponse Monsieur, et le dessein, que vous formerez pour parvenir au bout le plus juste. ...'
26   Klüber (1811), 15.
27   Mandrino *et al.* (1994), 72.
28   Mandrino *et al.* (1994), 73.
29   Mandrino *et al.* (1994), 76.
30   The whole Rhineland Palatinate kept accounts in Gulden or Florins, but Oriani was using the old French system, where 25 'Livres Tournois' was equal to 1 Louis d'or.

New coinage was being introduced at the time of his visit. See Kelly (1813), 1. 214, 416–17.

31   Mandrino *et al.* (1994), 75.

32   *ADB* 39 (1895), 303. The Lazarists, a Catholic teaching order founded in the seventeenth century, followed a rule based on that of the Jesuits. The Lazarists did not have an easy time, being suppressed during the French Revolution, re-established by Napoleon, who then suppressed them in 1809, and re-re-established them in 1816.

33   *Das gelehrte Teutschland* 17 (1820), 31–2. Art. 'Barry'.

34   *Connaissance des temps*, year 7, History of Astronomy for 1794, p. 304.

35   Klüber (1811), 17.

36   Wolfschmidt (2004).

37   NAS, MS GD 157 3379/3. Duke Ernst to Brühl, Gotha, 14 February 1786.

38   Vargha (2005), 37–40.

39   NAS, MS GD 157 3379/3, Duke Ernst to Brühl, 14 February 1786.

40   NAS, MS GD 157 3379/9. Duke Ernst to Brühl, 1 September 1786, 'J'ai reçu enfin le Transit de Ramsden et éspère tenir ce matin le sextant de Hadley. Je payerai comptant la lunette achromatique et le sextant ainsi que d'autres bagatelles à Ramsden afin de son accompte de 50£ reste dans son entier et l'engage à achever ses grands ouvrages commences à ce qu'il prétend …'. The transit is now in the Deutsches Museum, Munich, Inv. DM 67743.

41   NAS, MS GD 157 3379/10, Duke Ernst to Brühl, Gotha, 26 September 1786. 'Il n'est que trop vrai sans doute que Ramsden m'a manqué de parole jusqu'au dernier moment de mon séjour à Londres. A 8 heures du soir avant mon départ je fus encore chez lui, il me le fit voir presqu'achevé – presqu'entièrement assemblé prétendant que dans deux heures tout seroit fait; il vient entre 11 heures et minuit chez moi me faisant de mauvaises excuses de son manque de parole me remetter du lendemain à 10 heures du matin. J'avou qu'un moment je fus tenté de les croire sur sa parole et de déferrer mon départ. Mais la raison et l'expérience triomphèrent et je partis sans mon Hadley — non sans de vives douleurs mais dupé de fausses ésperances qu'il me l'enverroit au moins dans la huitième et j'avois pris le parti de faire prier Messrs Rougement & Co de l'envoy par la voye de Bruxelles et de [illeg.] si tôt qu'il leur seroit remis et de n'épargner aucuns frais pour me le faire parvenir avant ma départ pour Provence. Mais helas! tous mes ésperences ont été vaines jusqu'ici et sans doute que je partirai samedi prochain sans l'emporter avec moi.

      J'ose vous conjurer Monsieur, si c'est en votre pouvoir, de m'accorder votre protection auprès de ce sublime menteur artiste, et de le faire [?presse] de tenir parole …'

42   NAS, MS GD 157 3379/15, Duke Ernst to Brühl, 23 July 1787. 'Je suis bien sensible encor mon digne Comte aux bonnes nouvelles que vous voulez bien me donner du Transit de Ramsden – je souhaiterois bien le recevoir bientôt et que le cercle fut aussi en train. Car independant de ces instrumens déjà commandés, je roule encor de grands projets dans la tête pour la construction d'autres objets du même artiste: l'histoire du quart de cercle murale de M. de Bergeret me tient à coeur plus que jamais, en ayant été le legitime possesseur pendant l'espace de douze heures je ne puis consentir à en avoir le démenti, et de manier ou d'autre, il m'en faut un de la façon de Ramsden, et supposé même que j'eusse en celui de Bird je n'aurois jamais renoncé à la vanité d'en posseder un, de sa main. Et malheureusement il n'existe qu'un seul Ramsden au monde, car à son defaut et s'il y avoit un autre artiste capable d'en construire un, au degré de perfection dont cet Archimenteur est capable, il y auroit longtems que je vous aurois conjuré, mon très cher Comte, d'avois recours à cet autre lui même.

43   NAS, MS GD 157 3379/18, Duke Ernst to Brühl, 20 December 1787. 'Je suis bien charmé d'apprendre par de Zach, que l'instrument de passages de Ramsden est en bon train et avance beaucoup si bien qu'il pourra être achevé vers le printems prochain, j'ose me flatter de votre part mon digne ami que vous voudrez bien l'encourager et le talonner afin qu'il tienne parole: j'y ai le plus grand interet et cela d'autant plus que j'ai un nouvel espoir d'obtenir encor le mural de Bird, que De La Lande m'as [co … é] l'année dernière, le même qui a appartenu à Bergeret.'

44   NAS, MS GD 157 3379/20, Duke Ernst to Brühl, 29 May 1789.

45   NAS, MS GD 157 3379/21, Duke Ernst to Brühl, 14 October 1789. 'Je suis infiniment sensible, Monsieur, que la mauvaise foy du Sr Ramsden, aye occasionné jusqu'ici quelques mésentendus entre nous, et ait refroidi de part et d'autre les sentimens de confiance que nous nous avions voues mutuellement. J'en suis pour ma part très fachè mon digne Comte, que d'être à même, de pouvoir vous donner les preuves les moins equivoques de la sincerité de mes sentimens distingués envers vous même. Agrées en attendant les assurances les plus vraies et les plus sincères et continues je vous en conjure à ne point perdre de vue ce Diable de Ramsden, qui me leurre depuis tant d'années. L'on m'assura que l'objectif de la lunette meridienne de 8 pieds de foyer qu'il doit me faire est entre vos mains: j'ose me flatter de votre ancienne et toujours très précieuse amitié, que vous voudrez bien, ne vous point désais, de ce bijou sur lequel repose seules les experiences, qui luisent encor en moi, de posseder encor un jour ce précieux instrument, et de presser souvent ce fameux artiste d'accomplir ses engagemens envers moi.'

46   NAS, MS GD 157 3379/22, Duke Ernst to Brühl, 20 March 1790. 'Votre aimable lettre du 9 de ce mois que j'ai eüe le plaisir Monsieur de recevoir mercredi passé, 17 mars, me pérmettre de la plus vive et de la plus sincère reconnoissance pour tous les soins, toutes les peines, les desagréments que vous a causé le bel instrument que vous avez l'amitié de m'annoncer comme étant entièrement achevé, et prêt d'être emballé et embarqué pour Hambourg. Je vous prie d'en faire mes complimens à l'artiste, et lui faire sentir un peu ses torts vers moi. Quelques grands qu'ils soient, il pourra les réparer s'il en a envie par me faisant un quart de cercle mural ou quelqu'autre instrument propre à des observations en déclinaisons, car il doit sentir que l'ascension droite des astres en général n'est qu'un ouvrage secondaire et la déclinaison le principal. … Je préferois sans doute un mural au cercle qu'il m'a promis anciennement; je lui promets même d'avance que s'il veut entrer dans mes vues pour un mural le cercle entier ne lui échapera pas, c'est sa marotte et je m'y conformerai.'

47   NAS, MS GD 157 3379/23, Duke Ernst to Brühl, 29 May 1790. 'M. de Zach s'est empressé dès hier de vous annoncer mon très cher Comte l'heureuse arrivé de ce superbe instrument qui est un chef d'oeuvre d'art: mais il ne peut vous avoir décrit que très foiblement le plaisir et la satisfaction dont j'ai été rempli en le deballant. Nous sommes après de la placer sur deux piliers en bois qui serviront de modèle et de direction pour les piliers effectifs qui le porterent pour sa destination réelle. … M. de Zach vous aura instruit mon digne ami de l'embaras que nous a causé Ramsden en ne nous donnant pas de direction pour l'emploi du garde fils qui a accompagné cet instrument, il vous a prié à ma requisition de demander des éclaircissemens à l'artiste à ce sujet. Il croit aujourd'hui l'avoir deviné et il me paroit que son explication très ingenieuse remplit parfaitement son objet. Néanmoins pour plus de sûreté et de certitude il seroit à souhaiter que l'auteur nous communiquat ses instructions relativement au fil à plomb. Car il se pourrait fort bien que deux methodes différentes produisent le même resultat. Et dans l'execution nous aurions le choix de la méthode. J'ose vous prier mon digne ami de vouloir nous procurer par l'artiste une petite provision du

fil d'argent qui sert au micromètre filaire; il est si aisé qu'un fil cassat par accident quoique jusqu'ici il ne nous en soit pas arrivé de pareil, que nous serions au deparvu si'il avoit lieu en effet: l'artiste n'en ayant pas ajouté en reserve à son envoy. J'ignore comment M. Maskelyne s'y prend, mais ses observations disant à tout moment qu'un fil s'est rompou, qu'un insecte l'a cassé, etcetera: et il vaudrait mieux de ne nous point trouver dans l'embaras. …

[À vous] je dou enfin le bonheur d'être en possession d'un des plus beaux et plus parfaits instrumens d'astronomie qui existe dans les plus célèbres observatoires de l'Europe. S'il pouvait être persuadé cet homme unique dans son éspece à remplir ses autres engagemens envers moi, je pourrai me glorifier de voir aller mon observatoire de pair avec tous ce qui existent.'

48   NAS, MS GD 157 3379/24, Duke Ernst to Brühl, 28 December 1790. 'Je vous envie bien vivement mon digne ami le cercle entier que vous possedez depuis peu, qui, quoique infiniment plus petit que celui que j'attends de la part du cruel Ramsden seconde si bien vos observations en déclinaison. Je vous en félicite bien sincèrement Monsieur parce que vous sentez tout le prix de ce que vous possedez, et que je sens par ma propre impatience combien il est dure être à 200 lieues d'un homme qui manque de parole à ses engagemens, pour en remplir de plus nouveaux envers le premier venu, qui l'importune par ses assiduités: à ce comte je crins bien de n'achever jamais l'établissement que j'ai commencé. …

[PS] Voudriez vous bien, vous souvenir quelques fois du cercle de Ramsden, et lui faire rappeler par quelqu'un des votres, la parole qu'il m'a donné de l'achever — Oh si cet artiste unique pour son art pouvait être rendu sensible à la voix de son honneur!'

49   NAS, MS GD 157, 3379/25, Duke Ernst to Brühl, 15 March 1791.'Ce que vous dites, Monsieur, l'égard de Ramsden sur le cercle entier commandé par le duc de Richmond, me fait concevoir sans doute l'éspoir, que ce cruel mais inimitable artiste songeoit peutêtre à le finir enfin pour moi. Je le fonde en entier sur votre précieuse amitié pour moi puis qu'il m'est demontré que sans vos bons offices, jamais je n'aurai de lui quelque instrument pour observer en déclinaison, sans lequel toutes nos observations quelconques resteront toujours defectueuses. Envoyer en Angleterre quelqu'un expressément pour controler et animer Ramsden à faire honneur à sa parole donné tant de fois en vain, seroit sans doute un moyen bien d'efficace pour obtenir quelque chose de cet artists — mais ce moyen seroit beaucoup trop dispendieux pour moi sans me procurer la certitude de la réusite.'

50   Herbst (1991), 337.

51   NAS, MS GD 157, 3379/27. Duke Ernst to Brühl, 23 March 1791. NAS, MS GD 157, 3379/29, Duke Ernst to Brühl, 3 January 1793.

52   NAS, MS GD 157 3379/30, Duke Ernst to Brühl, 1 May 1793.

53   NAS, MS GD 157 3379/33, Duke Ernst to Brühl, 16 March 1794.

54   'Wir haben gar keine Hoffnung, das schöne Passagen Instrument von Ramsden zu erhalten, das wir for drey zehn Jahren für unsere National Sternwarte befellt und worauf wir tausend kleine Thaler vorausbezahlte haben.' Anon, *Monatliche Correspondenz* (1800), 68.

55   'Nach unsweres Nachtrichten liegt Ramsden ohne Hoffnung auf dem Todtenbette.' Anon, *Monatliche Correspondenz* (1800), 373.

56   'Der berühmte Ramsden pflegt in Scherz zu sagen, das er die Geschichtlickheit eines practischen Astronomen schon daraus beurtheilen wolle, wie ihn bey Tische Messer und Gabel führen sieht.' Anon, *Monatliche Correspondenz* (1800), 539.

57   RGO, MS 4/119x, Zach to Maskelyne, 4 May 1802.

58   Wolfschmidt (2004), 90–91.

59   Foderà Serio and Chinnici (1997), 9–10; Anon, 'Memoire of … Piazzi' (1827).

60   Débarbat (2002), 20–22.

61   Fiorani (1969), 88.

62   Calisi (2000), 442–3; [Anon] 'Astronomia' (1778–9), No. 1, pp. 1–3; No. 2, pp. 9–12; No. 3, pp. 17–21.

63   Piazzi (1792), 20 and 24–5.

64   CUL, MS RGO 35/140, p. 35.

65   Piazzi (1788).

66   Anon, 'Memoire of … Piazzi' (1827). The original, by De Angelis, is in *Bulletin des sciences mathématiques, astronomiques, phisiques et chimiques*, (1826), 339–44.

67   Piazzi (1792), 16.

68   I am obliged to Michael Wright for this suggestion.

69   Piazzi (1792), 26.

70   Magellan to Dom L. Garrellon, prior of the Monastery of Molesme, France, 22 July 1788. I am obliged to Rod Home for a transcript of this letter from his work in hand: See also, on the scarcity of good optical-quality flint glass, Turner (2000b), 404–8.

71   Chinnici *et al* (2001), 4–5. I have not been able to identify the ship which transported the circle.

72   NAS, MS GD 157 3386/5, Oriani to Brühl, 29 Nov. 1789.

73   Ragona, (1857), cols 262–3.

74   [J. Pond] (1813), *Edinburgh Encyclopedia* (1808–30), art. 'Circle. 6. 484.

75   Mountstuart, Isle of Bute: Bute archives (in course of being catalogued). Letter, in French, Piazzi to Ramsden, undated. 'To set to zero and in the same line the two wires of the micrometers of the vertical circle, firstly, one brings to the centre of intersection the plumb-line wire from top to bottom; at the top by the screw belonging to the wire, and below by the screw on the chassis. Secondly, one checks to see if the upper division matches that below; if not, one looks to see by how much it needs correcting by turns of the micrometer screw. Thirdly, one corrects half of this error by means of the screw, and the other half by the circle. Fourthly, at the top, with the screw on the chassis, the wire is brought back to the division; if some error remains, further adjustment below, and as heretofore.

Such is the method which I understood you to have shown me, and which I have always followed. But when I left London, I imagined that by this method one would render the microscope wires parallel to the plumb-line. But at the moment when I commenced observing, it seemed to me that one only set the wires in the line which was the diameter of the circle, and that the microscope above the plumb-line served only to set the vertical axis vertical without touching the micrometer.

If I am mistaken, as I may well be, I beg you to set me straight, and to give me the correct rectification with its demonstration, as I await this information before publishing the description of the observatory, of which your instrument is the principal element, and which I have documented with the greatest care, illustrating each part separately. You will see there all the observations I have made up to the present time, and various details which should give you pleasure.

The wires set in the focus of the telescope having broken, I have substituted others which have a diameter of [illeg] ″ [seconds of arc], which prevents me from taking precise measurements of stars which exceed the fourth magnitude. I therefore take the liberty of asking you to include some of your wires in the reply which I hope you will send me.

The prismatic ocular inverts differently from the other oculars. If you are willing to take the trouble of making me a drawing showing its effect, I would be most grateful.

Tell me what I should do to clean and care for the divisions on the circle, and the entire instrument, without spoiling it, for the method I employ at present does not seem to be the best one.

The telescope of the passage instrument lacks the clarity of that of the circle. I fear that dust has penetrated between the two objectives, How should I deal with this?

If I could have an ocular with a filar mocrometer to set on the circle, this would seem to be a great improvement but I hesitate to ask you for one.'

76   Piazzi (1792), 58; Piazzi to Oriani, 7 January 1792; Cacciatore *et al.* (1874), 15–16. Piazzi learnt only in 1795 that Troughton was successfully using spider's web in his micrometers.
77   Cacciatore et al (1874) Letter, Piazzi to Oriani, 23 June 1791.
78   Cacciatore et al (1874), 10–12, Letter, Oriani to Piazzi, 21 July 1791.
79   Hughes (1820), 1. 122.
80   Chinnici *et al.* (2000), 16–17, Fig. 1.
81   Piazzi (1792), 49–55.
82   The terms 'English mounting' and 'German mounting' were coined by Sir George Airy in 1844.
83   Howse (1975), 87.
84   CUL, MSS Add 4529 and 4530.
85   RS, MS Copy Journal, 374.
86   MHS, MS Radcliffe 53.
87   Pearson (1829) 2. 518.
88   YRO, Vickers Archives, William Simms Dividing Notebook.
89   Science Museum Inventory 1929-979.
90   Whyte (1961).
91   Mandrino *et al.* (1994).
92   BOA, MS CS 1787. 'Ramsden a fait quelques preparatifs.'
93   BOA, MS, CS 1789. 'Je n'oublierai non plus de reccomander à Ramsden de s'occuper de votre grand mural.'
94   BOA, MS, CS 1791.
95   BL, MS Add. 33272 f.90[r]. Blagden to Banks.
96   Miotto *et al.* (2000) 43–53; Balboni *et al.* (2000), 94–5.
97   Tucci (2000), 49–50.
98   Milan, Museo Nazionale della Scienza e della Tecnica, Inv. 1025 (Brera being the outstation).
99   Wolf (1905).
100 Turner, A.J. (1998), 83–7; Devic (1851), 100–3; Daumas, M. (1973), 303–5.
101 Turner, A.J. (1989), 1–13.
102 Cassini (1810), 4.
103 Devic (1851), 81. 'Français jusqu'au fond des entrailles.'
104 Bigourdin (1887), 502–3.
105 PAO, MS D 5–37, pp. 2, 7, 15.
106 PAO, MS D 5–37, p. 35.
107 Devic (1851), 110–16. Turner, A.J. (1989), 10. One of these men was the optician Haupois.
108 Cassini (1810), 23.
109 Cassini (1810), 24.

110 Cassini (1810), 24.

111 Giorgia Serio suggests that this was the 8-foot circle begun for the Specola Caetani – seemingly a useful item for Ramsden to display to several of his customers. Serio and Chinnici (1997), 12–13, 58.

112 PAO, MS D 5–37; Wolf (1905), 289–90.

113 Wolf (1905), 290–91.

114 BOD, Rigaud 30, No. 18. Cassini to Magellan, 13 July 1788.

115 Aikin (1813), 450–57.

116 Wolf (1905), 294.

117 Wolf (1905), 294.

118 Wolf (1905), 294–5; Devic (1851), 114, n.1. Devic asserts that Cassini IV put too much faith in Ramsden and perhaps other craftsmen, since English patriotism was by nature exclusive.

119 Wolf (1905), 295–6.

120 NAS, MS GD 157 3390/7/2, Cassini to Brühl, 29 April 1790.

121 Wolf (1905), 296–7.

122 David *et al.* (2001–04), 1. xxix–lxxix.

123 Martinez-Ballesteros (?1999), 94–100, 302–4.

124 Vancouver (1984), 313.

125 Vancouver (1984), 201–2.

126 RGO, MS 35/ Ac II. Maskelyne's loan to Ramsden for making this instrument was repaid on 7 September.

127 RGO, MS 14/6, 64, 175.

128 RGO, MS 14/6, 91.

129 RGO, MS 14/9, 70–72.

130 Vancouver (1984), 993.

131 Pritchard (2000a); London BL OIOC, MSS China, Cathcart Embassy.

132 Proudfoot (1868).

133 Cranmer-Byng and Levere (1981), 522.

134 OIOC, MS G/12/20 ff. 596–664.

135 BL, MS Add. 33797, f. 181.

136 Pritchard (2000b), 393–4.

137 Cranmer-Byng (2000).

Chapter 7

A Miscellany of Instruments

I've spoke a thousand times to Ramsden about his experiments on St Paul's, and could never get any positive answer, but promises of looking and searching his memorandums.

Magellan to Pigott, 12 October 1773. RAS Pigott letters 29

Among the various talents which Ramsden possesses, that of rendering a plain matter intricate, and so confounding those who are less acquainted with it than himself, is one upon which he particularly piques himself.

Blagden to Banks, Naples 29 January 1793. BL MS Add 33372, f.113r.

During his working life, Ramsden invented or modified a variety of instruments, not all satisfactory or made to the standard of production which might be expected. He no doubt retailed other items which he bought in to satisfy his customers, and like most opticians, could supply a range of hand-lenses and spectacles.[1] This chapter looks at the miscellany of signed instruments which were made wholly or partly under his roof, and for which there is some linking documentary evidence.

That enthusiastic amateur collector of telescopes William Kitchiner (1778–1827), summing up Ramsden's approach to the design and construction of the instruments which had claimed his attention over the years, echoed the praise lavished on him by the astronomer Giuseppe Piazzi, declaring:

The highest praise is due to the merit of the late Mr Jesse Ramsden for his ingenuity, liberality and persevering endeavours to invent and perfect the various instruments used in Astronomy, Philosophy and Mathematics, to produce which, he devoted almost all his time and almost the profits of his very extensive trade; in carrying on which, his anxiety was not (like the razor-maker, who merely made his goods to sell,) to study and contrive how cheap he could make an instrument, and how dear he could sell it; his sole care, was to make it as perfect as possible; he spared neither pains nor expense in forming an instrument, or bringing it to perfection, and his insatiable thirst for perfection, almost invariably, produced success. Without the least ostentation, pride or reserve in his manners, he was polite, easy, familiar to all that had business with him.[2]

Ramsden was primarily a mathematical instrument maker. His instruments were constructed principally of brass, and their purpose was to measure lengths or angles, with the help of telescopes or microscopes. From time to time, and perhaps in response to specific requests, he produced philosophical instruments: those which register aspects of the natural world which are not directly perceptible, namely atmospheric pressure, weight, temperature, magnetism, and electricity.

One such typical request came from Joseph Banks, who as mentioned earlier, intended to ship with Captain James Cook on his second voyage but was summarily rejected when Cook discovered how many servants and how much baggage he intended to bring on board. Between February and June 1772 he spent £66, 12s, 6d on a diversity of instruments from Ramsden (see Figure 7.1) and went instead to Iceland.[3]

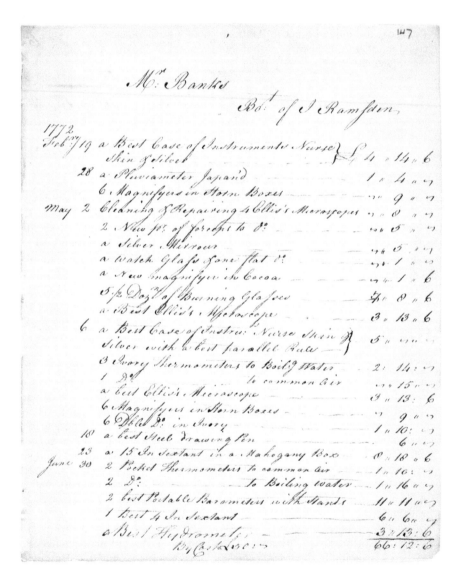

Figure 7.1   Ramsden's invoice to Sir Joseph Banks, 1772, for various instruments and apparatus. State Library NSW, Banks CY3003-172.

The Electrical Machine

Electrical machines were first developed at the end of the seventeenth century by Francis Hauksbee the elder (c.1666–1713), in the course of a long series of experiments on the force generated by rubbing leather pads against glass rods, spheres and cylinders, some of which Hauksbee had evacuated with his air-pump, others being open to the air.[4] The apparatus had at first no practical use beyond demonstration (Hauksbee was operator to the Royal Society), but its effects on various materials, and more dramatically, the sparks and charges which could be generated, made it an object of great interest to natural philosophers across Europe. A battery made up of connected Leyden jars and capable of holding a charge, was devised in 1745, opening up a wider and more impressive range of experiments. The electrical apparatus was modified to augment its generating powers and to enable it to perform a greater variety of tricks, and also to minister to patients suffering mental and physical disabilities, for whom electrical shock was seen as a new medical treatment. Contemporary writers were unclear about the origins of the single-plate machine, where a heavy glass disc replaced the globe or cylinder.

Joseph Priestley in the first edition of his *History and present state of electricity* (1767) wrote 'Mr Ramsden, mathematical instrument maker in the Haymarket, had lately constructed an electrical machine on a very different plan ... where friction is given to a circular plate of glass about nine inches in diameter'.[5] In the second edition (1769) Priestley revised this passage, stating that he had been wrong in crediting Ramsden as the inventor of the plate machine, a mistake 'which he himself led me into', and that the Netherlands physicist Jan Ingenhouz (1730–1799) was the inventor.[6] In the third edition (1775), Priestley credited both men, 'each independent of the other' as co-inventors.[7] Le Roy, writing in 1785, may have drawn on Priestley when he declared that Ramsden had been the first, in 1766, to employ such plates.[8]

In 1769 Charles Blagden, then in medical practice in Gloucester, asked the London clockmaker Alexander Cumming (1731/2–1814) to purchase a machine for him; Cumming wrote on 7 November , 'I have ordered your electrifying machine from Ramsden and I expect it will be ready and forwarded to you in a day or two'.[9] This turned out to somewhat optimistic, for on 23 November Ramsden wrote to Blagden:

> I have been out of town for some time ... I have several electrical machines finished by me at present such as you describe except that there is not any method to insulate the cushion. I believe it has not been done as yet to those made with plate glass – it might be done but it would be a little difficult on account of the number of cushions.
>
> If you will please to favour me with a line, if those made at present will be agreeable, one shall be sent down immediately, or if it is absolutely necessary that the cushions be insulated, I will endeavour to make one in that manner in about a fortnight, but I am afraid it will come a little dearer on that account.[10]

Ramsden supplied a machine with an 8-inch diameter plate to Joseph Banks, who took it on board when he embarked with Captain James Cook in HMS *Endeavour* in 1768. Banks interpolated a description of the machine and its behaviour into his *Endeavour* journal, comparing it to a similar machine supplied by the London instrument maker Francis Watkins to *Endeavour's* astronomer, John Green.[11]

Describing its performance during *Endeavour's* stay at Madeira Banks relates:

> One day we had a visit from the Governor ... and were obliged to stay at home. We however contrived to revenge ourselves on His Excellency, by an electrical machine which we had on board. Upon his expressing a desire to see it, we sent for it ashore and shock'd him full as much as he chose.[12]

Both machines suffered from the high humidity encountered on board in tropical latitudes. It was difficult to turn the glass plate against the damp leather cushions, and both failed to hold a good charge, though Watkins's machine was able to deliver the greater shocks. Banks took his machine out of its box on 19 March 1770, while the ship was off the coast of New Zealand, to discover that one glass jar and the plate were broken. Banks fitted his spare plate, but on trying the machine again four days later, its performance was still poor. Apparently Banks did not inflict this new electrical experience on the native population of Tahiti, as he was to do during his travels in Iceland in 1772.

While he was still in Haymarket – that is, in or before 1773 – Ramsden issued a small pamphlet advertising a single plate apparatus: *Directions for using the new invented electrical machine, as made and sold by J. Ramsden, mathematical, optical and philosophical instrument-maker, near the Little Theatre, in St James's, Haymarket, London.* The claim that it was 'new invented' may have simply meant that the machine, and its ancillary parts, comprised a new assemblage or configuration. Referring to the annexed plate, (see Figure 7.2) the pamphlet gives instruction on the assembly of the various parts, the current being created by turning the glass disc between two cushions. The shock can be precisely regulated to give a greater or lesser force, by the distance of the conductor from the knob, leading to one chain held by the patient, the other chain coming from the glass bottle. 'The most useful part of the apparatus being described', Ramsden mentions other demonstrations which could be performed, such as ringing a set of four bells, attaching a plate on which pieces of leaf gold could be made to rise erect, small figures of people cut from thin paper be made to dance up and down or feathers which could be made to wave as if in the wind, and to repel each other. Other more interesting and dramatic demonstrations could be arranged, such as causing a person standing on the insulated glass stool to be electrified, and if touched, to emit sparks.

The auction sale in 1793, following the death of John Stuart, third Earl of Bute, included as Lot 207 'a portable plate electrical machine and apparatus by Ramsden, in case'. It went to an unnamed buyer for £7, 17s, 6d, considerably less, probably due to its small size, than the prices realized by Bute's other electrical machines.[13]

Many of the so-called 'Ramsden electrical machines' which survive are of the single-plate pattern, but not manufactured by him – sometimes, indeed, not

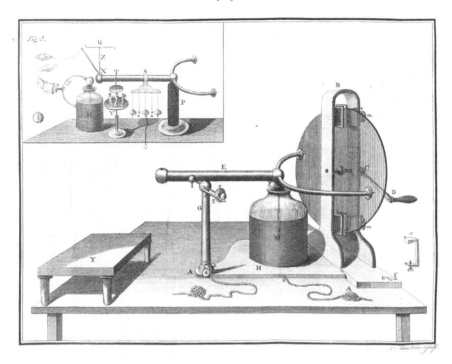

Figure 7.2 **Engraving of an electrical machine, 1770, from** *Directions for using the New Invented Electrical Machine. As made and sold by J. Ramsden ...*

even in England. The machine illustrated in J.-A. Sigaud de la Fond, *Description et usage d'un Cabinet de Physique Expérimentale* 2 vols (1775), vol. 2, plate XX, is very close to that in Ramsden's pamphlet. Gaziello notes that 'the Ramsden plate electric machine was brought to France in 1771 and was rapidly adopted as easier to work and sturdier than globe machines'.[14]

The portable barometer

Domestic portable barometers

Ramsden's undated pamphlet, five pages plus a plate, (see Figure 7.3c) on his portable barometer gives his address as Piccadilly and must have been published immediately after his arrival there but before 1772, when the abbé Jean Rozier published a description in French in his new journal.[15] Ramsden explained that while portable barometers had their uses, hitherto only barometers with an open cistern (see Figure 7.3a), where the mercury level was visible, could be relied on to give a true reading. The new barometer (see Figure 7.3b) was provided with a cistern-gauge, consisting of an ivory peg floating on the mercury and protruding through a small orifice in the top of the cistern. The level of mercury was adjusted

**Figure 7.3a Open-cistern domestic barometer signed *Ramsden*. National
Museum of Science and Industry. Inv. 1893–143.**

**Figure 7.3b Portable barometer signed *J^e Ramsden London*. The scale plate
carries a Fahrenheit thermometer and a temperature correction
table; the ivory float can be seen at the top of the cistern. Palermo
Observatory.**

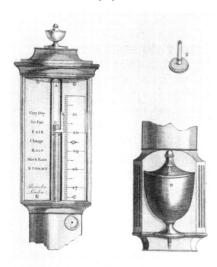

Figure 7.3c Engraving of a portable domestic barometer, ?1772, from
Description of the Portable Barometer as made and sold by
J. Ramsden.

by means of a portable screw, until a graduated ring round the float became visible, when the scale would read true. The height of the column against the scale was read accurately by the index and vernier which encircled the tube (see Figure 7.3c).

Ramsden is generally credited with this double or ring vernier index. The Göttingen professor Georg Lichtenberg (1742–1799), visiting London in 1774–75, noted in his diary that Ramsden's barometer could be read to great precision. The younger George Adams (1750–1810) told Lichtenberg that his father had discovered how to achieve this, but that Ramsden had a simpler method.[16] This would seem to be a reference to the ring index-vernier, which he attributed to Ramsden. He claimed for his father the first application of the floating cistern gauge, adding 'though others since his time assumed the merit to themselves'.[17] This may have been another of those ideas which occurred to several people around the time that the barometer was advancing from being simply a companion piece to the domestic long-case clock, to an accurate instrument consulted by scientifically-minded gentlemen.[18]

The French chemist Antoine Laurent Lavoisier (1743–1794) reported that he had adopted the method first devised by Ramsden of putting a vernier on a toothed rack, to ascertain the true level of the mercury.[19] There are slight differences between Rozier's engraving and that of Ramsden: Rozier shows a thermometer opposite the scale plate, and the method of adjusting the cistern level to the ivory float is less efficient than the version Ramsden describes. The first mention of a Ramsden barometer being transported to distant lands was in 1769, when it is listed among the instruments taken by Maskelyne's assistant astronomer William Bayly to the North Cape of Norway for the Transit of Venus.[20]

Mountain Barometers

The construction of a mountain barometer must allow for the decrease of pressure with height. Even in Alpine regions the column of mercury may fall to around 20 inches. A longer graduated scale is needed, and a larger cistern to hold this extra volume of mercury.

Ramsden was drawn into this topic in response to the enthusiasm of William Roy and his Royal Society friends who were concerned with barometric altimetry. In this respect, they were following in the eminent footsteps of Jean-André de Luc (1727–1817), a citizen of Geneva, whose masterly two-volume *Recherches sur les modifications de l'atmosphère* (1772) had already been widely read and admired by 1773, when he settled in England. De Luc's barometer comprised a siphon tube with the returned portion lengthened sufficiently to allow scales at either end of the column. De Luc had to fit an ivory tap to close the open end when the instrument was being carried but once set up and the tap opened, he could measure the exact length of the column of mercury, and by reference to the thermometer alongside the tube, make the necessary corrections for temperature. It then remained to calculate the fall in the mercury with height – a value then still widely debated but as yet unresolved.[21]

The barometer which Ramsden produced for Roy and his colleagues – Joseph Banks, Sir George Shuckburgh and the army officer William Calderwood (d.1787) – had a cylindrical cistern with a portable screw and a scale extending down to 20 inches or less. As described by Roy, who was using such instruments from 1772, the cistern float was similar to that described by Rozier (see Figure 6.3b). When the barometer was set up and the plug – now in the side of the cistern – was removed, a small ivory float could be seen swimming on the mercury in the cistern. The level of mercury in the cistern was raised or lowered by turning the portable screw until the tip of the float was level with a line scribed within the cistern, whereupon the scale read true. With the help of a vernier, readings could be made to 1/500 of an inch. In early examples, a thermometer was fitted on the case, its bulb reaching into the wooden frame of the cistern. In later examples, the entire thermometer was clear of the cistern. Ramsden also provided a wooden tripod which could be folded round the barometer tube to protect it for transport and he is generally taken to be the inventor of this useful device which was adopted by Magellan and Fortin.[22] Slung in a leather case, Ramsden's barometer must have been considerably heavier than De Luc's outfit.

The wealthy English amateur scientist, Henry Cavendish (1731–1810), writing in about 1777, stated:

> All the portable barometers of Ramsden which I have seen have a vernier division by which we may observe the height to 100ths & 200ths of an inch and have a screw at the bottom by which the quicksilver in the cistern may be adjusted to the proper height[23]

As a meteorological instrument, the tendency of the barometer – that is, the rate at which the mercury is rising or falling – is its most useful forecasting diagnostic.

To measure altitude, however, the observer needs an accurate reading at height, to be compared with a simultaneous reading at the base station. The reading must be corrected – principally for temperature, since the temperature normally drops with height – but also for the average temperature of the atmosphere between the two stations.

Ramsden and others engaged on this work investigated the effects of temperature change on the glass barometer tube and on the contained mercury. Some brief notes in his hand headed 'Mr Ramsden's expansion of the mercury', dated 3 January 1768, describe a test on two open-cistern barometers, first observed in a closet, then alternatively out of doors, then again indoors, showing one-hundredth of an inch change for every 3°F of temperature. Another undated set of observations showed that the expansion was progressive, not arithmetical.[24] The French meteorologist Louis Cotte (1740–1815), describing various mountain barometers, took the illustrated article in *Observations sur la physique* to be 'Ramsden's first'. This barometer has a plug to allow the cistern level to be viewed but its scale is graduated from 26 to 32 (clearly only for domestic use), on which he noted a thermometer with scales for Réaumur and Fahrenheit and a third, whose zero was 55°F, with values above and below showing the 100ths of an inch to be added or subtracted to the height of mercury according to the temperature of the atmosphere. Cotte was more familiar with 'Ramsden's second', or true mountain barometer, having seen examples in Paris, including Shuckburgh's barometer when he brought it for comparison with Cotte's own barometer.[25]

Altimetry in the Field

In June 1772 Joseph Banks bought 'Two best portable barometers with stands … £11 – 11s' from Ramsden. In 1773, when he was needing to make a business visit to Wales, Banks decided to make up a 'philosophical party' to accompany him, and it is likely that barometry was on the agenda, Ramsden, Blagden and De Luc being among those invited. But in the end, Ramsden was kept in London 'by the confusion of his business affairs', and some of the others also dropped out.[26]

Shuckburgh kept two weather journals, recording observations taken with his Ramsden portables. During 1774–76 these were mostly undertaken while he was in France, Switzerland and Italy. In 1777–78 he moved between his London and country houses, and Wales where he climbed on Snowdon. Shuckburgh remarks on a fixed adjusting mark on the ivory in the cistern, noting in his usually precise fashion that the internal diameter of the tube was 0.2 inches; the internal diameter of the cistern nearly 1.5 inches, so one revolution of the portable screw raised the mercury surface 0.09 inches.[27]

In a letter to the mathematician Charles Hutton (1737–1825), written from Lanark on 26 July 1774, Roy wrote: 'Ramsden's barometers do wonderfully well, that is to say, they are uniformly consistent in their own results; though the rule for ascertaining heights by De Luc's method is defective.' Roy spent considerable time and effort on experiments to find the expansion of mercury under different temperatures and pressures but gave less thought to the datum from which his heights were calculated, taking on at least one occasion 'the level of high water

Neap Tide (which I take to be the mean height of the sea) at Glasgow New Bridge ...'.[28]

The Portuguese agent Magellan, took a great interest in barometers for all purposes.[29] In 1772–73, the British astronomer Nathaniel Pigott (1725–1804) and his son Edward (1753–1825) were employed by the Austrian government to measure longitudes of the principal towns in Flanders, in preparation for a cartographic survey of the Austrian Netherlands.[30] Magellan assisted with the supply and repair of their instruments, among which were Ramsden mountain barometers. For Magellan, the difficulties of negotiating with Ramsden were increased by the Pigotts' movements – he was unsure where they were on any particular day, and he had to send letters and instruments to John Needham, director of the Academy of sciences at Brussels, or to one of the British consuls. On 23 March 1773 he wrote to Nathaniel Pigott:

> Mr Ramsden tells me that about ten days ago he did send a barometer to you, but ask[ed] that it was directed for Mr Needham. Therefore please to speak with Mr Needham and see if this already sent was really arrived for him at Oxford, or if he knew anything about it: that I may send one I have already got for him to your friend at Dunkerque, according to your desire: but for fear of my mistake I'll keep it in my hand, till I deliver your answer. For I do not account at all, nor do I rely on what Ramsden says. Please tell me also, where I shall apply for the 5½ guineas I have already paid for this Barometer, and for the expenses of ye box and sending it aboard ye ship &c.[31]

On 30 April Magellan was still wondering if he could trust Ramsden's statements but on 1 June all was forgiven:

> I do believe that Ramsden's assertions on the article of your barometer are true, and at least once he is in the right. It appears to me, as a fact, that he has sent ye instrument tho' in a silly manner, without giving any proper notice by you, or to Mr Irwin at Ostend. I called on him, and shew'd the article of yours relating thereto. But he declares to have lost the instrument and add[s] that in a week or 10 days, you shall hear of it; being at Ostend or Bruges.[32]

Ramsden, like other barometer enthusiasts before him, had climbed to the top of the dome of St Paul's Cathedral at which height the mercury stood half an inch lower than it did beside the tidal Thames at nearby Paul's Wharf. He did provide tables with his barometers, but these did not please Magellan, who wrote to Pigott on 12 October:

> ... I've spoke a thousand times to Ramsden about his experiments on St Paul's, and never could get any positive answer, but promises of looking and searching his memorandums. I have seen of late an engraved plate, where he sets the height for each fall of ye mercury, but this he set aside, saying there were some errors or mistakes which render it useless untill corrected. Now he is very busy with Mr de Luc: and perhaps it is about some schemes of ye kind. This gentleman (Mr de Luc) has met with ye greatest reception in England. Indeed he is very clever in ye barometrical way and must have a very extraordinary patience to go through 2 large vol^s on ye subject. If you have the work you may find there much better tables than those we could get from Ramsden. The last

told me Saturday, that he will make now another kind of barometers, much superior to ye former. But I will wait the sight of them before believing his assertions.[33]

When the antiquary and amateur scientist Sir Henry Englefield (1752–1822) joined the Piggotts in Brussels in 1773, he and Edward Pigott took a Ramsden barometer upstairs to a window close to the great cathedral bell, to see if it registered the pressure wave caused by the effect of sound. With the mercury steady at 29.478 inches, the deep sounding of this 7-ton bell caused the mercury to oscillate between 29.482 and 29.472 inches.[34] In January 1774 Magellan ventured: '… It took into my head to improve Ramsden's barometer, and will have soon some made, according to my ideas by my workman, which [I] hope will answer better.'[35] In his treatise on barometers, *Description et useage de nouveaux baromètres, pour measurer la hauteur des montagnes et la profondeur des mines* (1779) Magellan implies that he has been assisting Ramsden with his 'mountain barometer'.[36]

William Roy relied on Ramsden's mountain barometer to construct his table of heights. In his lengthy account of this exercise, which was read on four evenings at the Royal Society between June and November 1777, he had followed De Luc's three corrections and had laboriously converted his values, made in degrees Fahrenheit and feet, to compare with the European scientists' degrees Réaumur and toises. The toise, (literally 'fathom') was approximately two yards, or just under two metres, but like English measures during this period, it was undergoing small but important revisions.[37] In 1792 Ramsden invoiced Matthew Boulton for 'a best improved mountain barometer and stand £8– 8/–, packing case for do. 2/ 6d'.[38]

Thermometers

It is not known if Ramsden employed a skilled glass-worker, but his instruments served in some of the most demanding experiments – principally those associated with barometry and with distillation. An undated draft from Maskelyne hints at the quality of Ramsden's products:

> A committee of Meteorology meet at the Royal Society house tomorrow at one o'clock to try the boiling point of thermometers made by different artists. We wish to try some by Ramsden & I have applied to him & he has promised to send them. But, as he is a little uncertain, we should be obliged to you to lend one or two that you may have made by Ramsden; if they shall be broke, we shall make them good.[39]

Shuckburgh, in his weather journal of 1774–76, noted that Ramsden, in making his thermometers, had set their boiling points under a pressure of 30 inches of the barometer.[40] Later, writing on the temperature of boiling water, Shuckburgh reported:

> In the beginning of last year [i.e.1778] with the assistance of Mr Ramsden, I procured a most excellent thermometer, in every way adapted for this purpose. It was about 14 inches long, but the interval between freezing and boiling only 8¼ inches* and though every degree was something less than 1/20 of an inch, yet by means of a semi-

transparent piece of ivory, which applied itself close behind the glass tube, sliding up and down in a groove cut in the brass scale for that purpose, carrying a hair-line division, at the extremity of which was a vernier dividing each degree into ten; with, moreover, a lens of an inch focus; this apparatus being made moveable first by the hand and more delicately by means of a micrometer screw; whose head was divided into 25 divisions, each equal to the fortieth of a degree (for so truly cylindrical was the tube, which had been with care expressly selected from a great quantity of glass, that the divisions in the neighbourhood of the freezing point did not differ from those near the boiling point by so much as 1/40th of a degree, and this variation appeared in other parts of the tube strictly uniform, as was found by breaking the column of mercury; by means, I say of this apparatus I was enabled to read off any height of the thermometer to within 1/50th of a degree.

*It may be suggested that a longer tube would yield a larger scale but that size increase brings problems of construction, and the increased size and thickness of bulb lead to distortion and insensivity.[41]

Felice Fontana (1739–1805), purchasing physics instruments, chemicals, and natural history specimens in Paris and London, had seen such thermometers. He wrote ecstatically to the Duc de Chaulnes, 'Ramsden, the famous Ramsden, to whom astronomy and the physical sciences are so beholden' has already adopted the most slender tube which is read against a strip of ivory. But he noted that such thermometers cannot be immersed in acids or certain other substances. Fontana proposed to Ramsden that he should engrave the degrees directly on the tube with a diamond, which he did, but with a very fine file. Ramsden gave Fontana two thermometers divided in this way and while he was in London Fontana used them for his experiments on latent heat, in the presence of the Irish chemist Richard Kirwan (1733–1812) and made similar thermometers when he returned to Florence, claiming to be the first to graduate the tubes with a diamond on his dividing engine. The disadvantage of Ramsden's thermometers was the tendency of the slender mercury column to break in the tube and Fontana advised Ramsden to make the tube large enough for the divisions to be clearly visible, and to expand the bore to prevent this breakage. History does not relate how Ramsden took this young foreigner's advice.[42] The physician John Hunter (1728–1793) carried out a series of trials on the temperature of various animals, using a delicate thermometer made for him by Ramsden.[43] On 4 July 1780 the Board of Visitors toured the Royal Observatory at Greenwich, and ordered 24 thermometers and a barometer from Ramsden.[44]

The Theodolite

Judging from the number of examples in public and private collections, and passing through sale rooms, Ramsden's workshops must have produced many small and medium-size surveying theodolites of various patterns (see Figure 7.4). We seldom find contemporary reports, perhaps because they were entirely satisfactory, as Maskelyne reported in his account of the work done at Schiehallion in 1775.

**Figure 7.4   Theodolite, early pattern, signed *Ramsden*. 8 inches diameter.
National Museums of Scotland. Inv. T.1987.132.**

A theodolite of the best sort was wanting, a necessary instrument for obtaining the
figure and dimensions of the hill. One of Mr Ramsden's construction of 9 inches
diameter was thought fittest for the purpose, on account of the excellence of the plan
on which it was made, and the number of its adjustments, being capable of measuring
angles for the most part to the exactness of a single minute.[45]

The surveys were delayed by bad weather and to speed operations as autumn
approached James Stuart Mackenzie (1719–1800) lent another Ramsden
theodolite of the same pattern.

Theodolites were among the surveying instruments taken to North America for
the frenzy of land measurement associated with land grants. In the US, Benjamin

Henry Latrobe (1764–1820) owned a Ramsden transit and a theodolite, which had cost him 150 guineas,[46] and in Virginia Thomas Jefferson, (mentioned in Chapter 2 as owner of a Ramsden equatorial), employed a small Ramsden theodolite and chain to survey and plan works on his estate at Monticello.[47] This theodolite is at the Monticello Museum.

The Surveying Level

Ramsden's levels also survive in various collections and appear in sale rooms. According to the *Cyclopaedia* Ramsden's was:

> The level which is in the most general use, great numbers having been made by Mr Ramsden, and since his decease, by his numerous pupils. All adjustments to bring its parts into accuracy may be done in the field.[48]

The components of Ramsden's level were stacked with the bubble tube hung beneath the telescope, allowing the bubble to be adjusted independently of the telescope. The compass was inset on the base of the frame carrying the Y-rings for the reversible telescope. A level by Matthew Berge, now in Palermo Observatory, exactly matches one by Ramsden, now at Coimbra Observatory (see Figure 7.5). In Troughton's level which had been designed by Nevil Maskelyne in 1803 the compass was uppermost and the bubble tube was inset over the telescope which was not reversible.[49]

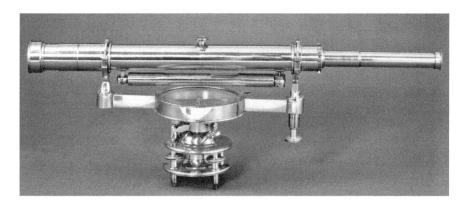

Figure 7.5   Surveyor's level with 16-inch telescope, signed *Berge London late Ramsden*. Palermo Observatory.

Lalande describes Ramsden's arrangement of lenses which served as a 'portable' or 'hand-held' level, to inform the viewer when two objects sighted were at the same horizontal level.[50]

Telescopes

Ramsden's pamphlet on his refracting telescope survives in the French version, dated 1775: *Description d'une lunette achromatique, faite et debité par J. Ramsden, Faiseur d'instruments d'optique, de physique, & de mathematiques, vis-à-vis de Sackville Street, Piccadilly, à Londres.* The telescope was not novel in its design, comprising three objective lenses, the two outer ones being bi-convex, of crown glass, that at the centre bi-concave and of flint glass. Ramsden did however make claims for improved construction by giving the telescope a stand that was lighter and firmer, and easier to use. The pamphlet explains in detail how to set the telescope up for terrestrial and celestial observations, how to clean the lenses, and how to stow the instrument in its case. Berge followed Ramsden's design (see Figure 7.6).

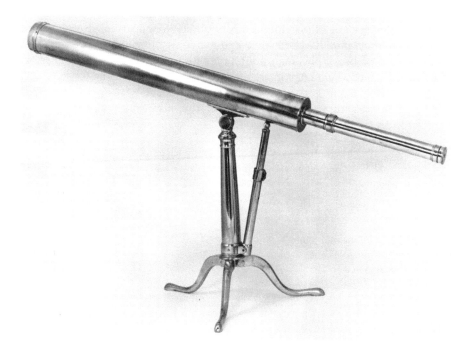

Figure 7.6   Two-foot achromatic telescope signed *Berge London late Ramsden*. Palermo Observatory.

These telescopes were bought by private individuals and by government departments, including the Admiralty and the Board of Ordnance. As with the surveying instruments mentioned above, numerous examples survive in public collections and in private hands, and continue to pass through the sale rooms.

Ramsden's Prices for Telescopes

It was remarked by Giuseppe Piazzi and others that Ramsden would sell instruments at a lower price than his shopmen had been instructed to charge; whatever the truth of this allegation, some prices were noted by direct knowledge or repute. Lalande, in the second edition of his *Astronomie* (1771), converting prices to French 'livres' at 100 livres equal to £3, 17s, 9d, priced 18-inch Hadley octants in wood at 120 livres; 2-feet in brass at 150 livres; and those 'made with care by Ramsden' at 300 livres. Comparing prices of achromatic telescopes in Paris and London in this second edition and the third (1792) edition of his *Astronomie*, Lalande prices achromatic objective lenses of 3 feet at 3 guineas; those of 9 feet, 8 guineas; of 12 feet, 10 guineas; and of 18 feet, 15 guineas, to be found at both Dollonds and Ramsdens. In the third edition he adds that achromatic telescopes of 3½ inches aperture and 3½ feet focal length, which used to cost £24, 5s were very difficult to find at the time of writing. This was probably due to the shortage of optical-quality glass.[52] Prices for the largest observatory instruments which were made only to order must have been time-and-materials estimates, based principally on the weight of brass, plus man-hours; they were usually given in guineas, to round figures and, where known, are mentioned in Chapters 5 and 6.

A document in the Forbes of Callander Muniments gives prices in 1786. (Forbes' concern with wholesale prices arose from his business as a supplier to the East India Company.)

> 8 Aug 1786 Mr Ramsden's prices of refracting telescopes.
> Improved 3ft refracting telescopes sliding tubes
>   wholesale £3, 18s   sell at 6 guineas
>   2ft D°   -- £2, 4s --   4 guineas
>   1 ft D°   -- £1, 3s --   2 guineas
>
>   One foot in one mahogany tube
>   wholesale 16s   sell at 24s
>   Eighteen inch --   20s   -- 31s, 6d
>   Two foot   --   26s   -- 42s
>   Three foot   --   31s, 6d -- 52s, 6d
>
> Improved in one mahogany tube
>   Two foot   38s   3 gns
>   Three foot   52s   4 gns
>
>   One foot telescopes on a stand
>   45s   3½ gns
>
> 30 inch focus with 3 different magnifying powers, on a stand
>   wholesale £7, 17, 6d sell at 10 guineas.[53]

Cassegrainian, newtonian and gregorian reflecting telescopes signed by Ramsden are mentioned, or survive, in old collections; others occasionally pass through auction salerooms. The telescope maker John Watson informed Kitchiner, 'In the year 1780 I made a Newtonian for the late Mr Jesse Ramsden, which was of eight feet focus, and the louge metal 10 inches diameter ... This telescrop was, from its large dimensions, and powerful effect, the finest telescope I ever saw'.[54] Their mirrors were a specialized product.

Ramsden made a unique Pole Star telescope for the amateur astronomer Alexander Aubert. Fixed to the wall of Aubert's observatory north of London, it measured the small apparent movements of the Pole Star, due to the earth's irregular rotation over time, against a spiderline micrometer.[55]

Micrometers for Reflecting and Refracting Telescopes

> Sensible of how much the theory of astronomy is limited by the imperfections of instruments, I always incline to improve rather than to invent, except when repeated examinations convince me that the imperfections arise from defect in principle as well as in the construction.[56]

In this paper, sent to the Royal Society through Joseph Banks, Ramsden dealt with defects in the micrometers applied by Short and Dollond to gregorian reflectors, their cost, and the fluctuations in the image brought about by atmospheric instability. Henry King describes his 'simpler and theoretically more accurate double-image device – a special cassegrainian telescope'.[57] In this arrangement, which had the advantage of needing no additional optical elements, Ramsden duplicated the image by bisecting the small secondary mirror. Turning the micrometer screw gave a sliding action by pushing one half of the mirror in one direction and pulling the other half in the opposite direction. Because of the way the secondary mirror is mounted, the two halves also rotate. The illustration shows that the micrometer could be removed from the telescope and replaced by a standard lens.

Ramsden's preference for the cassegrainian reflector lay in the fact that the spherical aberration of its large mirror is partly corrected by the aberration of the small one, whereas in the gregorian, the small mirror increases the aberration, leaving the observer with a larger residual aberration. Giuseppe Saverio Poli (1746–1825), then a teacher of geography at the Military Academy in Naples, wrote from London to Lalande on 3 November 1779 (the year he was elected Fellow of the Royal Society) with news of this new micrometer which, as he said, overcame the defects of those commonly fitted, whether with moveable parallel wires or split lens types.[58] But his 'catoptric' micrometer, as Ramsden termed it, met with little success, being difficult to construct and easily put out of adjustment.

The second micrometer, for use with refracting telescopes, consisted of two semi-lenses placed in the focal plane of the eyepiece and sliding laterally against each other to duplicate the image produced by the objective. Ramsden noted that any imperfections and optical errors in the glass of a micrometer applied at the object glass were magnified by the whole power of the telescope, but in his design the micrometer was placed in front of the eyepiece. Consequently the image was magnified before it came to the micrometer, and any imperfections in its glass were magnified to a much lesser extent by the remaining glasses.[59]

The eminent amateur astronomer William Pearson (1767–1847) may have confused the unsuccessful catoptric micrometer and the dioptric micrometer; Pearson queried whether any dioptric micrometers were ever made as there were no records of any measurements, and George Dollond was surprised when Pearson told him that the 'new micrometer' which he – George Dollond – had just invented, had been described 40 years previously by Ramsden.[60]

The adoption of the thread spun by the diadema spider in diaphragms fitted in the focus of telescopes and microscopes was a considerable improvement over the fine silver wires previously used. The thread, taken from the radial lines of the web, was thinner than the silver wire and being slightly elastic remained taut while being less likely to snap. Edward Troughton and Ramsden have each been credited with its introduction, but both acknowledged Felice Fontana as the first to employ this natural fibre. Describing eyepieces for compound microscopes, John Quekett (1808–1847) considered 'the cobweb micrometer, the invention of Ramsden', superior to the Huyghenian forms. Ramsden's micrometer had two plano-convex lenses with the field lens reversed. In the focus of the upper lens two fine wires or spider threads were stretched across the field of view, one being moveable by a screw nominally of 100 threads to the inch, the screw head being also divided into 100 parts. The lower border of the field was a comb made of a thin piece of brass with an indented edge, notched by the same screw. Every fifth notch was longer, to facilitate counting, and each notch corresponded to one turn of the milled head, so that the number of turns could be read off in the field of the instrument, and fractions of a turn read on the divided head.[61] If a divided glass scale or fine wires is placed exactly at the focus of the object glass, scale and image are magnified together and every part of the image is in contact with the scale and therefore can be measured up to 1/20,000 of an inch.[62] Such micrometers were an integral part of Ramsden's astronomical instruments and the great theodolites.

Ramsden also made microscopes. They were included in Banks's purchases of 1772 (see Figure 7.1) and a Ramsden microscope is in the Whipple Museum, Cambridge. Jesper Bidstrup, writing to his patron on 31 December 1790, reports having sent back to Copenhagen a compound microscope 'made by Ramsden's method', having received the casting moulds from one of Ramsden's workmen. He explains that craftsmen who make the 'Ramsden model' usually charge 10 guineas, which is the price Ramsden has instructed his shopman to charge. Bidstrup then remarks that if the would-be purchaser speaks directly to Ramsden, the price can be 9, 8 or even 7 guineas. Bidstrup considers 7 guineas ridiculous as that is half the cost of the labour. He adds that Dollond sells smaller versions for 9 guineas, Nairne & Blunt for 7½ guineas, but both these differ from Ramsden's model.[63]

The Dynameter and Other Optical Devices

According to William Pearson, Ramsden was the first to make a dynameter to measure the image precisely and so calculate the magnifying power of a telescope.[64] Hitherto, this was done simply by inserting a graduated slip of mother-of-pearl

Figure 7.7a Dynameter in its case, signed *Berge*. **Private collection.**

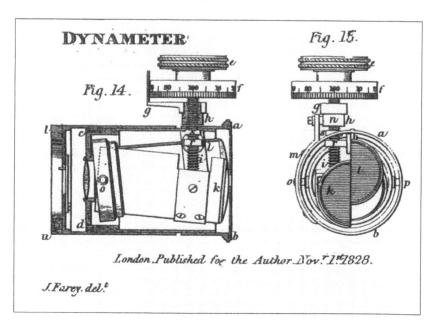

Figure 7.7b Engraving, the dynameter optical system. From Pearson,
       Introduction to Practical Astronomy **(1828), Figures 14 and 15.**

into the common focus of the two lenses, a method which Ramsden found too imprecise. His dynameter worked on the principle of split lenses screwed by micrometer, to gauge the exact size of the image (see Figure 7.7b), an idea which may have been taken from the 'naval coming-up telescope' or rangefinder made by Edward Nairne.[65]

William Kitchiner, that avid collector of telescopes, provided a full description of the dynameter:

> To measure the diameter of the pencil rays with great ease and accuracy Mr Ramsden, in about the year 1775, contrived a clever little instrument, which he called a dynameter, for though when single lenses are used, the power of a glass is readily discovered by dividing the focal length of the object glass by that of the eye-glass, in eye-pieces of the common construction, especially those of a negative focus, it is very difficult to measure in this manner; nor can it be done with any accuracy with those eye-pieces which are made for erect vision with four eye-glasses.
>
> The dynameter is principally composed of a fine plano convex glass; by means of which the image of the pencil of rays is completely separated, and the diameter of it known to the greatest nicety. The wheel or head of the micrometer is divided into a hundred equal parts, and a figure engraven over every fifth division which is cut rather longer than the others: 1, 2, 3, and so on to 20: but adding an 0 to each figure in calculating, it will then read off, 10, 20, 30, and so on to 200. The nonius is divided into 15, 10, towards 0, and 5 on the contrary side.
>
> The revolutions of the micrometer head will bring the edge of the circle round it, and the division on the nonius to coincide at 10: each division, therefore, is equal to the ten thousandth part of an inch.
>
> Applying this little instrument to the eye-glass of a telescope, when adjusted to distinct vision at any distant object, and turning the micrometer head, the emergent pencil will begin to separate: and when the extreme edges are brought into contact, the number of divisions will show the diameter of it in thousandths of an inch; then reduce the diameter of the object glass into thousands and divided that sum by the diameter of the pencil, the quotient will be the real magnifying power. But as it is requisite for the emergent pencil of rays to be in the focus of the divided glass, a thin transparent piece of ivory, precisely one-tenth of an inch in diameter, is set in the sliding cover, to adjust for that distance, which must always be done before it can be used with accuracy.[66]

Thomas Jones may have made most of these dynameters for Ramsden, and after Ramsden's death Berge continued to sell them (see Figure 7.7a). Louis Dutens wrote to Sir Joseph Banks:

> Mr Jones ... will also show you a Dynamometer [*sic*], one of the nicest instruments invented by Mr Ramsden, and in the making of which he was also especially employed.[67]

According to Stuart Talbot's recent account, this was the forerunner of all subsequent telescope dynameters and remains a *tour de force* example of Ramsden's genius. Such a dynameter test confirmed the exact magnification of the telescope objective examined – as precision of the known magnification definitely enhances astronomical observations and calculations.[68] A dynameter, and a refraction machine (see below), both made by Ramsden for the Earl of

Bute, were sold at his death, and bought by Nairne for £4, 2s, 6d.[69] The sale in 1806 of instruments belonging to Alexander Aubert included 'An instrument to determine the magnifying power of telescopes by Dollond', sold to one Greatorex for £2, 16s, and another by Ramsden, sold to Kitchiner for £4.[70] Lalande, writing in 1788, reports that apart from this dynameter, which was available from other instrument makers, Ramsden made another version, also with split lenses, and that it was basically the same as his ocular micrometer. It was however more difficult to make and he had produced only five or six examples.[71]

For those telescope owners unable to afford a dynameter, the *Nautical Almanac* for 1787 offered a cheaper way:

> To find by experiment, the magnifying power of any telescope. Many methods have been contrived to determine experimentally the magnifying power of any telescope. That excellent artist Mr Ramsden showed me, some time ago, a small instrument of his own invention, to measure the diameter of the emergent pencil of rays, at the eye hole, to the utmost degree of precision. By dividing the diameter of the great mirror in a reflecting telescope, or the diameter of the object glass in a refractor, by the diameter of the emergent pencil of rays, determined by that instrument, the magnifying power will then be given. But as that instrument, constructed chiefly upon the principle of Mr Dollond's object glass micrometer, is somewhat* expensive, and therefore may not be found in the hands of everyone who is possessed of a telescope, I shall lay down a plain and easy method ...
>
> *Mr Ramsden informed me that the price of one of his small instruments to determine the power of any telescope would be about 3 gns. Whether any other optician makes these instruments, as invented by Mr Ramsden, I cannot say, never having seen any other than that which Mr Ramsden showed me.[72]

In 'A description of new eye glasses for such telescopes as may be applied to mathematical instruments' *Philosophical Transactions* 73 (1783), 94–99, Ramsden announced his attempt to produce a flat achromatic field for viewing micrometer wires in front of the graduated arcs of theodolites and suchlike instruments, allowing accurate readings to be made with great precision. The eyepiece consisted of two plano-convex lenses of equal power arranged with their plano surface on the outside. I rely once more on Henry King's technical explanation:

> For transverse achromatism the lenses should be separated by a distance equal to half the sum of the focal lengths; but this would mean that the field-lens, and every speck of dust on it, would be in the anterior focal plane of the eye-lens. In practice, therefore, and at the expense of full chromatic correction, the lenses are separated by about two-thirds of this distance. The eyepiece is termed positive because the anterior focal plane of the system is in front of the field-lens; the usable field is about 40° in angular extent.[73]

According to the astronomical writer Francis Wollaston (1731–1815), Ramsden mentioned the idea of reading-off divisions by a microscope 'one evening, at a meeting of our Society in the beginning of 1787'.[74] Applied to Ramsden's pyrometer, constructed to measure with extreme accuracy the expansion and contraction of his surveying chains, and to his great theodolite under construction for William Roy, these microscopes allowed precise readings to be made with confidence (see Chapter 8). Their relation to the pyrometer was also remarked on by Lichtenberg who noted in his travel journal that on 29 September 1790 the Hungarian Miklòs Vay showed him a Ramsden micrometer which could read to 60 divisions in a line (the 12th part of a Paris inch). Ramsden made this micrometer for his pyrometer, and afterwards made them in quantity.[75]

According to King:

> The term 'Ramsden disc' does not appear to have originated with Ramsden, although he was the first to realise the importance of the size and position of this area in telescopic vision.

King explains the optical theory in a more straightforward manner than Ramsden:

> The Ramsden disc, or exit-pupil as it is generally called, is the image of the objective formed by the eyepiece. Its size is therefore obtained by dividing the diameter of the objective by the apparent magnification. Alternatively, knowing the diameter of the objective and exit-pupil, one can, by simple division, obtain the apparent magnification. It had been his understanding of the mathematics involved that had enabled him to design the dynameter mentioned above.[76]

The Long-Beam Precision Balance

Ramsden's large balance, of which a single example is known, was purchased at Banks's expense (see Figure 7.8). Ramsden's invoice, dated 20 May 1788, describes it as 'A large size Balance for weighing Hydrostatically One Hundred Guineas'[77] (see Figure 7.9). It has a 24-inch double cone beam on a four-pillar stand, in a glazed case with numerous drawers. The cabinet stands about 22 inches high by 32 inches wide by 9 inches deep. The balance served for the experiments made around 1790 by Charles Blagden, on behalf of the Royal Society, at Government request, to see if the Excise Office methods of rating spirits were acceptable, and could be imposed on the trade without argument.[78] Ramsden's presence was noted on 29 March 1794, when Excise samples of rum were weighed at various dilutions and temperatures.

> As the specific gravities of liquors was properly adopted for determining their relative strengths, the nicety of the apparatus for weighing was one of the principal desiderata, and fortunately a balance made by Mr Ramsden was obtained which so far surpassed in nicety everything hitherto made of the sort, that the least error in weighing was totally out of the question.[79]

The descriptions – one might say eulogies – of this balance date from 1788, and both come from French sources. In his description in *Observations sur la physique*, Jean Rozier explains that the essential part of a balance was the beam and the manner of its suspension, and that Ramsden, 'whose skill in the invention of

Figure 7.8   Long-beam precision balance, in glazed case 22 inches × 32 inches × 9 inches, signed *Ramsden*. National Museum of Science and Industry. Inv. 1900-166.

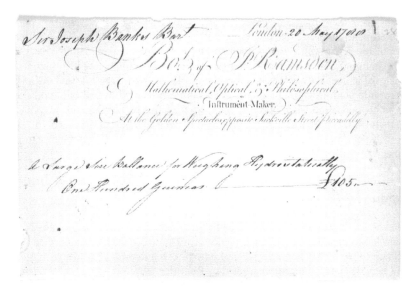

Figure 7.9   Ramsden's invoice to Sir Joseph Banks, 1788, for the precision balance. State Library NSW, Banks 830500.

new instruments equals the great perfection which he gives to everything he does,' had devised a new form.[80] Piazzi, in his encomium of Ramsden's work, wrote more briefly, stressing the importance of the steel knife edges turning on 'oriental crystal', the pan suspensions being exactly in line with the knife edges, and the cones making up the beam. Ramsden also made the thermometer. This may be the instrument described in an undated note among Blagden's papers:

> The thermometer was made by Ramsden. Its ball was 0.22 inch diameter, stem 13 inches in length: below the scale was 3.6 [inches] bare. The scale reached from 15° to 110°: the part made use of, from 30 to 100 was 6.2 [inches]. Scale of ivory, divided to 5th of a degree, of which divisions the quarter could be easily estimated.[81]

The long-beam precision balance had developed in the 1770s.[82] The desideratum was absolute sensitivity extending to very small mass differences, while maintaining the highest resolution of the load capacity to that sensitivity. To achieve this, the balance beam had to be as light as possible without flexing under high loads. The maker had to consider geometry of the points of balance – at the centre of the beam, and at the suspensions of the pans – and the nature of the knife edges on which they turned. The beams were usually frames, but Ramsden adopted the internally-braced cone, his favourite device for combining lightness with strength and rigidity. One argument against this feature was identified by John Pond (1767–1836), Maskelyne's successor as Astronomer Royal, namely: that if the observer's body-heat warmed one or other of the cones and its contained air, the action would be affected. In the article on 'balance' in his *Dictionary of Arts*, the chemist Andrew Ure (1767–1836) wrote:

> A balance made by Ramsden for the Royal Society, is capable of weighing 10lbs, and turns on 1/100th of a grain, which is the 7/1,000,000th part of the weight. In pointing out to me this balance one evening, Dr Wollaston told me it was so delicate, that Mr Pond, the Astronomer Royal [from 1811–1836], when making some observations with it, found its indications affected by his relative position before it, although it was enclosed in a glass case. When he stood opposite the right arm, that end of the beam preponderated, in consequence of it becoming expanded by the radiation of heat from his body; and when he stood opposite the left arm, he made this preponderate in its turn. It is probable that Mr Pond had previously adjusted the centres of gravity and suspension so near to each other as to give the balance its maximum sensibility, consistent with stability.[83]

The later history of Ramsden's balance is related in Charles Weld's *History of the Royal Society* (1848):

> One article however, which belonged to Sir Joseph Banks, the Society possess; and I am led to mention this in consequence of an amusing anecdote connected with it, related to me by Mr Babbage, which I have not seen in print. The article in question is a very delicate balance, constructed by Ramsden.
>   Upon the decease of Sir Joseph Banks, the Secretaries wrote to his widow, appraising her that the balance was lying in the apartments of the Society, and requesting to know her wishes respecting it. "Pay it into Coutts" was her Ladyship's reply.[84]

The inference is that Babbage believed that Lady Banks assumed that this enquiry referred to her husband's bank balance. Another version of events emerges from the Society's Council Minutes, 7 July 1820:

> Ordered that the Secretary do write to the Executors of the late President informing them that the Balance belonging to the late Sir Joseph Banks and now in possession of the Royal Society will be delivered to their order if required, but that it is desirable that the said instrument should remain where it now is, according to the intention of its late owner; the President and Council are willing to purchase the same in order that it may be retained for the purposes of the Society.

The reply from Banks's nephew Sir Edward Knatchbull, one of the Executors, was entered in the Minutes for 16 November 1820:

> It is desirable that this instrument should for the present remain in the custody in which it is now placed, and I have no doubt Lady Banks as soon as the Executorship Accounts are closed will feel happy in the opportunity of showing her high respect of the RS by presenting the balance for its use.

Her donation is confirmed in an inventory titled 'Instruments and apparatus belonging to the Royal Society', dated November 1834: 'Hydrostatic Balance, by RAMSDEN; with weights, by ROBINSON. Presented to the Royal Society by Lady BANKS.'[85] The Council Minutes for 26 June 1824 record: 'Ordered that the Balance Beam belonging to the Royal Society be placed at the disposal of the President [William Hyde Wollaston] for prosecuting his experiments on the corrosion of copper.' The balance, sadly lacking its pans, is now in the Science Museum, London.[86]

Lesser-known Instruments

The Cutting Engine, or Microtome

This small precision gadget was for cutting transverse slices of wood, cork or similar material, for microscopical examination.[87] We learn from John Hill, *The construction of timber ...* (1770) that it had been invented by clockmaker Alexander Cumming who, having perfected the design, passed it to Ramsden for production (see Figure 7.10).

The device, as made by Cumming and described by Hill, consisted of a cylinder of ivory, 3½ inches high by 2 inches diameter; later Ramsden instruments had brass cylinders (see Figure 7.10). The upper plate of the cylinder is of bell-metal, the base plate is brass. A spiral cutter turns to slice the quadrant of wood, which is held against the blade by a screw at the base of the cylinder, turning against an index. As the blade slices, a slender arm holds down the cut slice, to prevent it from immediately rolling up. The sample is held fast in the cylinder by two screws entering from the side of the cylinder. Hill described a more elaborate version, where the sample material was automatically advanced, but in a recent trial with one of these models, it was found, even after the material had been stiffened with

Figure 7.10 Microtome in brass case, signed Ramsden London. Length 6 inches, diameter approx. 2½ inches. Museum of the History of Science, Oxford, Inv. 65389.

wax, that the knife tended to compress rather than slice it. Several examples of these devices, both with ivory and with brass cylinders, are to be found in various museum collections. An example signed by Ramsden, in the collection of the Royal Microscopical Society has a catalogue note that this was the first section cutting engine, incorporating a controllable and measurable fine advance.[88] In the sale of the Earl of Bute's instruments, Lots 99–103 consisted of 1 large and 4 small microtomes, which went to several buyers at prices ranging from £7, 10s to £2, 12s, 6d.[89]

Prisms

Ramsden is known to have sold sets of prisms, but whether these were manufactured in house or simply bought in for retail is not known. One such set, marked 'Ramsden 0-6/0-7/0-5', in a fitted box, is in the Colección San Isidro, at the Museo Nacional de Ciencia y Tecnología, Madrid.[90] A set of prisms by Ramsden, in a wooden frame, was among Joseph Priestley's apparatus lost in 1783 when politically-driven arsonists set fire to his house.[91] Lalande describes a prismatic ocular, but its purpose is not clear.[92]

The Optigraph

This aid to drawing, credited to Ramsden by his former apprentice and employee Thomas Jones (1775–1852), was one of many small portable optical devices, preceding and contemporary with the camera lucida as patented in 1806 by the physiologist and chemist William Hyde Wollaston (1766–1828).[93] Jones' article in *Philosophical Magazine* was probably in response to Wollaston's article in a previous issue of that journal.[94] It was also illustrated in the *Edinburgh Encyclopaedia* and several editions of the *Encyclopedia Britannica*.[95] No example, description or drawing of Ramsden's original simple instrument has been found, and we have to work from Jones's description of his own modified optigraph:

> The late most ingenious Mr Ramsden, so well known for his inventions and improvements in various instruments, considered the present subject an object worthy of his attention and invented the instrument ... described, which is so simple and easy, that a person not possessed of the least knowledge of drawing, may, with less than 3 minutes instruction, be perfectly able to take a perspective view of landscape, building, machinery, or, in fact, any object of any description presented to his eye, with the utmost correctness.[96]

Jones continues:

> Mr Ramsden left this instrument without the means of enabling the operator to enlarge or diminish, an inconvenience which I have obviated, while at the same time I have added some other trifling improvements. This instrument is certainly superior to any hitherto constructed for the purpose; for in this the operator views the object through a telescope, which enables him to delineate minute objects with great exactness and ease, which are often too far from the eye to be delineated correctly.

The improved optigraph consisted of a vertical tube suspended by a universal joint from a fixed mirror at 45°. A small dot on a clear glass part way along the tube was observed by an eyepiece. A wooden handle attached to a pencil at the base of the tube enabled the tube to be moved about the universal joint to carry the dot around the outline of the scene reflected in the stationary mirror. This motion was replicated as the pencil traced over a sheet of paper. The optigraph did not allow the user to view the dot and the point of the pencil at the same time.[97] The National Museums of Scotland has an example of Jones's optigraph.[98]

The Ship's Compass

It is not known what persuaded Ramsden to design a ship's compass with its bowl supported on a spike in the base of its box. On 3 March 1781 he wrote to the Board of Longitude offering to show a new improved steering compass which he believed would be of great use in the navy. The Board sent for Ramsden, and inspected his compass,

> ... which from the lightness of card and the manner of its suspension, appeared to be greatly superior for general use to those hitherto invented and Lord Sandwich informed the Board that upon the same idea the Lords of the Admiralty had directed the Commissioners of the Navy to provide a proper number of them for the use of His Majesty's ships.[99]

Previously to this date, Captain Constantine Phipps had written to Charles Blagden on 22 February 'I left Ramsden's compass in town at my home and beg you will be so kind as to send it directly by the stage to the care of Mr Palmer 126 High Street Portsmouth'.[100]

Admiralty Standing Order of 14 April 1781 instructs: Ramsden's 'New Invented Steering' compasses to be issued in future to all HM ships'.[101] One customer was the Duke of Gordon, for Ramsden's invoice of 21 July 1783 lists 'An improved 12 inch sextant, £10, 10s; A steering compass with a card for a boat and one for a ship, £2, 12s, 6d; and 7 dozen phosphoric matches £1, 8s; total £14, 10s, 6d'.[102]

Ramsden's compass found little favour. It does not appear in histories of the compass, and William Rhind, master of the *Andromeda*, wrote to John Hamilton Moore on 2 September 1788, that he 'preferred Kenneth McCulloch's compass to that of Ramsden, which swings four points where McCulloch's hardly swings half a point each way when the motion of the ship is very quick'.[103] It is a sad fact that landlubbers generally had little idea of the motions of a ship in various conditions, and how to distribute the weight across the card in order to keep it steady in rough weather. There is an example in the Jagiellonian University Museum, Krakow and another, in a glazed box, in the Conservatoire des arts et métiers, Paris.[104]

Apparently Ramsden had spoken about the improvement of other sea-going instruments, but had gone no further. In Aikin's biography we are told:

> Among ideas which he never took to maturity were a very improved log, an instrument for measuring leeway; a most simple and accurate instrument for trimming a ship, or ascertaining her line of floating; and another, equally new and ingenious, for measuring the angle of her inclination either from the pressure of her sails or the action of the waves, or, in other words, her heel and roll.[105]

Nevertheless, Matthew Berge's 1801 catalogue (see Figure 13.2) offers 'Mercurial level for ascertaining the trim of a ship – £3. 5. 0'.

Coin balances

Small folding coin balances were carried by merchants, tradesmen and gentlemen to check the quality and weight of the domestic and foreign coins that passed through their hands.[106] There were four major types: the equal-arm, adapted from the traditional balances into the folding type in the later years of the eighteenth century; the unequal-arm with sliding weight, adapted from the steelyard, which could accommodate a wide range of coins without the inconvenience of loose weights; the single-turn, in which a hinged weight moved from one position to another, nearer the centre of the beam, for weighing the guinea and half-guinea, and in 1798 the double-turn type which could also weigh the seven-shilling piece.

Two coin balances signed by Ramsden are now in the Whipple Museum, Cambridge.[107] The first has a Haymarket label on its box and is dated to c.1770–1773 (see Figure 7.11). It is a folding gold-balance, its beam marked for 10 coins: the half johannis (36 shillings), the moidore (27 shillings), the guinea, and their subdivisions. The second balance, signed 'Ramsden Piccadilly' is a hydrostatic coin-balance, a type which can detect counterfeits where the gold is adulterated with copper, thus weighing the same in air but not in water. The late Michael Crawforth, an authority on coin balances, noted that the few folding gold balances made in France that he had seen were all unequal-arm with sliding weight, similar to the Ramsden type, and that the pattern may have been taken to Paris by Ramsden's former journeyman François-Antoine Jecker.

Figure 7.11 Coin balance, label on box signed *Ramsden Haymarket London*. Whipple Museum, Cambridge, Inv. 630.

Georg Lichtenberg, in London in 1774–75, noted a gold-balance in Ramsden's shop.[108] Thomas Bugge, making his way round the London instrument-makers' shops in 1777, called on a Mr Russel, who showed him a little coin balance by Ramsden. The accompanying rough sketch shows it to be an unequal-arm type.[109] When Gilbert Innes of Stow, in the Scottish borders, who was accustomed to buy telescopes and other instruments from Ramsden, tried to buy a coin balance in April 1780 he was informed by Grant, his agent. 'I was this morning at Ramsdens who has discontinued making the scales you mention above two years being so full of telescope employment, but he has promised to make for me a pair in silver in a week'.[110] It will come as no surprise to see Grant's next letter, dated 9 May: 'I have been coaxing Ramsden about your scales but he fights me at my own weapons.'[111] Nothing more is heard and perhaps Innes found a more obliging supplier.

Atwood's Fall Machine

George Atwood (1745–1807), a mathematician with an interest in mechanics, designed his 'fall machine' as a teaching apparatus, in about 1779. In that year J.H. Magellan wrote to the Italian physicist Alessandro Volta (1745–1827), then teaching at Pavia, inviting him to subscribe to Atwood's forthcoming 4-volume 'Course of Physics'. Magellan had been to see Atwood demonstrating his machine.[112] In later correspondence, we learn that Magellan has ordered an Atwood machine for Volta,[113] and that he has published his letter to Volta, describing and illustrating the machine, which is said to have been made by Adams.[114] Magellan sent an Atwood machine to Portugal in 1780, and another to Martin van Marum in Haarlem in 1790.

Atwood described the machine, which was intended to demonstrate the laws of uniformly accelerated motion due to gravity, in his *Treatise on the rectilinear motion and rotation of bodies* (1784). It enjoyed some years of life, being illustrated in text books in various languages. At an unknown date Ramsden made an Atwood machine for Giuseppe Poli, then teaching at the military academy in Naples.[115] In a letter of 26 July 1791 Blagden wrote to Pictet: 'I have never seen Atwood's machine executed, but I conversed with Ramsden about it, who says he made one with considerable improvements, and that it answered the purpose, but was dear ....'[116]

Notes

1   Volta ((1949–55), 3.146–7. Letter, 31 December 1782, Magellan to Alessandro Volta, 'I found at Ramsden's two sorts of pocket spectacles, silver or silvered, with concave lenses, 20–25 shillings, and with convex lenses, 2½ guineas'.
2   Kitchiner (1818), footnote to pp. 130–33.
3   State Library of New South Wales, Banks 30172/3.
4   Hackmann (1978).
5   Hackmann (1978), 143–4; Priestley (1767–1775), First edition, p. 736.
6   Priestley (1767–75), 2nd ed., p. 300.
7   Priestley (1767–75), 3rd ed., vol. 2, p .111.
8   Le Roy (1785), 53–9.
9   RS, MS CB/1/3/107; Blagden letters in C.72.
10   RS, MS CB/1/6/88.
11   'My Machine': Beaglehole (1962), 2. 276–9.
12   Beaglehole (1962), 1.160.
13   Turner, G.L'E (1967).
14   Gaziello (1984), 163, n. 78.
15   Anon (1772).
16   Gumbert (1977), vol. 1, Tagebuch 1, p. 125.
17   Adams (1790), 8.
18   An accounts book at Mountstuart, Isle of Bute, records an unpriced payment to Ramsden on 18 March 1783 for a portable barometer.
19   Lavoisier (1864–93), 3. 755–6.
20   Bayley (1769), 272.

21 Archinard (1980).
22 Roy (1777), 653–788.
23 Jungnickel and McCormmach (1986), 437. An unsigned mountain barometer by Ramsden is in the Cavendish Collection at Chatsworth, Derbyshire.
24 TNA, MS OS 3/3.
25 Anon (1772), 509–12; Cotte (1788), 1. 509–10.
26 Carter (1988), 121.
27 BL, MS Add 38,481.
28 Close (1969), 7–8.
29 Magellan (1779).
30 McConnell and Brech (1999), 305–18.
31 RAS, MS Pigott 25.
32 RAS, MS Pigott 27.
33 RAS, MS Pigott 29.
34 Englefield (1802), 157–9.
35 RAS, MS Pigott 31.
36 In the collection of the Museu de Física [Physics Museum] at the University of Coimbra in Portugal, there are two barometers of Ramsden's pattern, described in the original catalogue as 'made in London by J. J. de Magalhães' in 1780, which we ought perhaps to read as 'provided by' him as they are signed by W. & S. Jones.
37 Archinard (1980).
38 BPL, MS MB 251.90.
39 RGO, MS 35 No. 40 Draft.
40 BL, MS Add. 38,481, Shuckburgh's 'Weather Journal'.
41 Shuckburgh (1779), 365.
42 [Fontana] (1783), Lettera III, 156–70.
43 Hunter (1778), 7.
44 RS, Council Minutes 1769–1782 (Vol. 6, copy).
45 Maskelyne (1775), 501 and 524.
46 Bedini (2001), 203–4.
47 Bedini (2001), 11, 13, 651–4.
48 Rees (1802–20), Article 'Level', illustration on plate V in 'Surveying'. An example of Ramsden's level, with 50cm telescope, was acquired in 1818 from an unstated source and is now in the Physics Museum at the University of Coimbra, in Portugal.
49 RGO, MS 4/199, 'Account of an improved telescopic level [signed] NM 14 Sept 1803'.
50 Lalande (1790), 34.
52 Lalande (1771), 1. xlix–lii; Lalande (1791), lx–lxiii.
53 NAS, GD 171/4238.
54 Kitchiner (1825), 116–9.
55 Irregularities of the earth's rotation arise from precession, aberration and nutation. Pearson (1829), 558–60 and Plate 12 figs 2, 3.
56 Ramsden (1779), 419.
57 King (1955), 162–4.
58 Poli (1780), 111–18.
59 Considerably more technical detail is given by Randall Brooks in his thesis, Brooks (1989a), 98–9.
60 Pearson (1829), 2.181–4.
61 Quekett (1987), 196.
62 Quekett (1987), 160.

63 CRL, MS Bidstrup Letter 3.
64 Pearson (1829), 47–8.
65 Pearson (1829), 185–6 and Plate 3 figs 7, 8. describes the 'coming-up telescope'.
66 Kitchiner (1815), 87–92.
67 RS, MSS Misc. 8.24. Dutens to Banks, 4 December 1800.
68 Talbot (2003), 8–9.
69 Turner, G L'E. (1967), 213–42.
70 Sotheby's Sale Catalogues, No. 50, 21 July 1806.
71 Lalande (1790), 32.
72 London, Board of Longitude (1786). *Nautical Almanac for 1787*, Appendix, pp. 44–5.
73 King (1956), 164.
74 Wollaston, F. (1793), 133. 'Society' is here probably the Royal Society.
75 Promies (1967–74), 2. 277.
76 King (1956), 164–5.
77 State Library, New South Wales, Banks image 830500.
78 *The Times* reported on 5 May 1791, 3b: 'A series of experiments has lately been made at the rooms of the Royal Society, by desire of Government, to ascertain the best method of proportioning the excise upon spirituous liquors.'
79 RS, MS BLA.6, f. 77; RS MS Journal, 22 April 1790.
90 Anon. 'Description d'une nouvelle balance, construite par M. Ramsden, de la Société Royale de Londres', *Observations sur la physique, sur l'histoire naturelle et les arts.* 2nd series, 33 /2.
81 RS, MS CB4/6/10.
82 Jenemann (1997), 39–42; Stock (1969), 16–18.
83 Ure (1832), 82–3.
84 Weld (1848), 2. 116.
85 RS, Printed inventory (November 1834). 'Instruments and apparatus belonging to the Royal Society', p. 4, No. 47. I thank Dr Robert Anderson for this reference.
86 Inv. 1900–166.
87 Anon., Article 'Cutting engines'.
88 Turner, G.L'E. (1989), 281. Tested in 1974, the microtome produced sections of about 20 μm in thickness. Bracegirdle (1986), 145–7.
89 Turner, G.L'E. (1967), 213–42.
90 Guijarro (1994), 47 and 55.
91 Timmins (1890), 10.
92 Ramsden (1790), 33.
93 Hammond and Austin (1987), 5–6, 102–3.
94 Wollaston (1807), 343–7.
95 Morrison-Low and Simpson (1995), 26.
96 Jones (1807), 67.
97 Hammond and Austin (1987), 102–3.
98 Inv. T.1994.16.
99 RGO, MS 14/6, 21.
100 RS, MS Blagden, p. 44.
101 TNA, MS ADM 106/2508.
102 NAS, MS GD 44/51/432/5. Ramsden was one of several stockists of these matches, or philosophical tapers. An advertisement in *The Morning Post and Daily Advertiser* of Friday 18 April 1783, announces that they were manufactured in Paris. They were

enclosed in sealed glass tubes and on breaking the tip and drawing them out, they immediately caught fire. A tin box of 3 dozen cost a half guinea.

103 TNA, ADM 106/2508.
104 Inv. 0549. It has a 6-inch card on a flat magnet needle with sliding brass counterweight. The card bears the words 'RAMSDEN OPTICIAN PICCADILLY LONDON', the origin of the card is indicated on the fleur-de-lis band 'Cary sculps'. The south side of the card is marked 'SHIP'. Pencilled under the card is the word 'small'. A steel-tipped spike fitted to the base of the box supports the agate centre bearing of the card; there is no provision for gimbals but one exterior lug on the tin compass bowl fits a curved guide inside the box.
105 Aikin, (1813), 457.
106 Camilleri (2001); Crawforth (1979).
107 Brown (1982).
108 Gumbert (1977), Vol. 1, Tagebuch 1, p.125.
109 Bugge (1997), 279.
110 NAS, MS GD 113/4/156/39.
111 NAS, MS GD 113/4/156/61.
112 Volta (1949–55), 1. 338–40. Magellan to Volta, 9 April 1779.
113 Volta (1949–55), 2. 14–5. Magellan to Volta, 21 November 1780.
114 Magellan (1780). According to Millburn (2000), 258 etc., Adams made several of these machines.
115 Segnini and Caffarelli (1990), 124.
116 Sigrist (2004), 96.

Chapter 8

Surveying and the Great Geodetic Theodolites

> No consideration upon earth would ever make me go through the same or such another operation again, merely from the drudgery of having to do with such a Man!
>
> Roy to Banks, 6 March 1790, NHM, MS DTC 7.74–8; 74

> This masterpiece of Europe's greatest artist answered fully ... indeed more than fully ... the demands made of it.
>
> Cassini, *Exposé des operations faits en France ...* (1791), 58

Cartography in Britain during the eighteenth century was generally done on a regional framework under private initiative, with the simple types of instruments used by land and estate surveyors. Angles were taken with the magnetic compass, or circumferentor, and plane table, or where expense allowed, with the more costly theodolite. Distances were measured with the chain or perambulator, or where obstacles such as rivers or marsh intervened, simply by the surveyor's naked-eye judgement. When triangulating over areas such as the larger counties, small defects in base-line measurement introduced by worn chains or rods and the effects of humidity or temperature on their presumed lengths, could accumulate to a considerable inaccuracy. Nor was it thought necessary to take into account the curvature of the earth, where non-Euclidean geometry comes into play and the angles within a triangle do not add up to 180 degrees.[1]

Introducing William Roy

The origins of what grew into the Ordnance Survey are entwined with the career of William Roy (1726–1790), a Scotsman whose familiarity with maps may have been instilled by his father, factor to the lairds of Milton, in Lanarkshire.[2] After 1738 Roy moved to Edinburgh and was probably employed as a civilian draughtsman at the office of the Board of Ordnance in Edinburgh Castle, making surveys and plans. The Board of Ordnance had been created in 1518 to control the work of the ancient body of King's Engineers, and the Artillery, consisting of sappers and artificers; in 1716 the Engineers separated from the Artillery and were established as the Corps of Engineers, being granted in 1787 the title 'Royal Engineers'.[3]

The Jacobite uprising, culminating in the battle of Culloden in 1746, brought home to King George II, his commanders and his government, the urgent need for better maps covering regions of political instability and possible conflict – in this case the Scottish Highlands. The only maps available were piecemeal compass-

and-plane table sketches and a scheme was agreed to survey this region. Roy's talents had been noticed by Lieutenant-Colonel of the Engineers David Watson (1713–1761), the Deputy Quartermaster General of the forces in northern Britain and who had been at Culloden. In 1747, with Roy appointed as Watson's Assistant Quartermaster, to avoid the problems which would arise if a young clerk were put into a senior position over the soldiers working to his orders. The survey commenced at Fort Augustus, with Roy in charge of a six-man survey party. By 1750 the intention was to cover the whole of Scotland. Six parties were in the field, provided with 40- or 50-foot chains and circumferentors without telescopic sights.[4] Speed was of the essence: observations went directly into notebooks, and the teams concentrated on recording roads, bridges and topographical features of potential military use. Matters were brought to a halt in 1755 during the build-up to the Seven Years War and never resumed in that form, leaving a fair copy of the survey only for the region north of the Clyde-Forth line.[5] Roy joined the army and was appointed a Practitioner Engineer, the lowest commissioned rank in the Corps of Engineers. He and Watson left Scotland and reconnoitred the south coast of England in preparation for the anticipated French invasion. Roy then went across to France and Germany where his sketch maps, notes and plans produced under active service, gained him rapid promotion.

The French forces did not invade, but in 1763, when times were more tranquil, Roy moved to London, living first at 32 Great Pulteney Street, and from 1779 at 12 Argyll Street, where he had a small rooftop observatory. He put forward a scheme for a national survey which, as he sensibly pointed out, was best done in peacetime. George II had died in 1760 and his grandson George III was now on the throne. The government considered Roy's proposal for a survey of the whole kingdom at public cost, and inevitably rejected it on the unarguable grounds that 'it would be a work of much time and labour, and attended with great expense to the government'.

In 1766 Roy produced a less ambitious plan with detailed estimates of cost. It was not taken up. The idea of spending money during peacetime in order to prepare for some possible future invasion was still unattractive to the government purse-holders. The start of the American War of Independence in 1776 further distracted the politicians and it was only after 1783, when peace was declared, that Roy began a new approach to his subject.

Roy had not been involved in the American war, and spent his time investigating various ways of increasing the accuracy of topographical surveys. One of his projects was to employ barometer readings as a means of measuring the heights of his survey points. He made some observations for a triangulation between Greenwich (south-east of London), Arthur's Seat (on the outskirts of Edinburgh) and Cotton Hill,[6] but he found a number of theoretical and practical obstacles to accurate calculation. In London he had begun a close association with the instrument makers James Short, Jesse Ramsden, John Dollond and Dollond's sons, acquiring the best available chronometers and instruments. Elected FRS in 1767 he was closely involved with Joseph Banks, and other leading men of science in that body. Between 1771 and 1776 Roy carried Ramsden's portable barometers on his travels to Scotland and Wales, so that he could measure the heights of various mountains, although there was no agreement over the amount by which the mercury fell in the barometer tube for each 100 feet of ascent.

Measuring the Degrees of Longitude between Greenwich and Paris

With Britain and France at peace an earlier question surfaced concerning the relative positions of the observatories of Greenwich and Paris, there being a difference of opinion between French and English astronomers amounting to nearly 11 seconds for longitude and 15 seconds for latitude. Earlier astronomical measurements had failed to resolve the matter, and a proposal was now received from France to triangulate the space between the two observatories.[7] The French government had been setting out meridians and triangulating much of the French kingdom, extending this network into Flanders. In October 1783 Cassini de Thury, director of the Paris Observatoire, sent a formal 'memoir' or proposal, conveyed as diplomacy required, through the French ambassador in London, via Charles James Fox, a principal secretary, to King George III. With the King's agreement, 'by His Majesty's command, the memoir was put into the hands of Sir Joseph Banks FRS accompanied with such marks of royal munificence, as speedily obtained all the valuable instruments, and apparatus necessary for carrying the design into immediate execution'.[8] Banks consulted Roy, and the project was set in motion. The 'royal munificence' amounted to £2,000, which was held by Banks and doled out to Roy on receipt of the invoices for each item.

That summer, Roy had diverted himself by setting out a baseline across the fields between the Jews Harp Tavern, north of Marylebone (the site is now within Regents Park), and Black Lane, (now buried under the railway yards near Kings Cross,) intending to triangulate to the prominent hill-top steeples in and around London. Roy had also produced a draft proposal for his commander, Charles Lennox (1735–1806), third Duke of Richmond, which was passed to the King, setting out the case for a detailed topographical survey of the south coast, which he considered vulnerable to invasion. Maps were available for many parts of this region but while they might be geometrically adequate, they failed to depict the topography essential for planning defensive action. After the south coast had been mapped, the survey could move up the east coast and eventually cover the whole kingdom. A great base of six or eight miles, serving as the foundation for the first triangle, should be laid down on a sandy shore or in some level part of a county such as Cambridgeshire or Wiltshire. A meridian was also desirable; not through Greenwich, as this would be terminated on the north Norfolk coast, but further west, running northwards from Dorset. His estimate detailed the cost of the instruments, and the men, horses and their daily allowances, for the undertaking. The instruments needed would cost about £400:

2½ ft radius quadrant and carriage to transport it	£ 120
Small quadrant	25
Transit and Equal Altitude for tracing the meridian	40
Reflecting telescope	21
Refracting telescopes	30

Spirit level and Tryer	12
3 portable barometers	12
Rods and spirit levels	12
Common theodolites, chains and plane tables	180
Drawing instruments	28 [9]

This assemblage was typical of the time, large portable quadrants and deal rods having been employed on the French surveys. For the Greenwich–Paris triangulation, however, Roy produced a similarly modest estimate:

Mathematical instruments of the best kind	£200.0.0
A carriage on springs (secondhand or otherwise) for	
transporting the instruments during the operation	25
Pay of six men for sixty days, one with another at 2/6d each	45
Hire of horses &c, carriages &c, perhaps	80
	350 [10]

From a measured base-line in the south-east of England a series of triangles would extend towards and across the Channel, to connect with the French triangulation. The instruments employed for lesser surveys were inadequate for such an extensive and prestigious project where all measurements had to be of the utmost accuracy, the various instruments capable of the highest precision, and the surveyors scrupulous in their operations. The challenge was considerable, and beneath their sense of friendly competition each national party strove to put on the best show for its sponsors. Possibly Roy believed that suitable instruments were already available; his unrealistically low estimate was in any case to be overtaken by events. For when Roy discussed the matter of instruments with Ramsden, they agreed that 'the angles should be determined by a large circular instrument', Ramsden having convinced Roy of the superiority of a full circle over the old-style large portable quadrant. Its construction was duly ordered in 1784.[11]

The Hounslow Base

On 16 April 1784 Roy, accompanied by Joseph Banks, Henry Cavendish and Charles Blagden walked the five miles along the proposed base on Hounslow Heath, a desolate area then some miles west of London – one of Roy's terminal points is now within Heathrow Airport. Several small dwellings, one stream, and the London to Staines road had to be passed, but otherwise the only obstacles were furze bushes and anthills. These were cleared and levelled by soldiers from the 12th Regiment of Foot, who were encamped nearby and could be called on for labouring tasks. Operations commenced that summer, a first measurement with a 100-foot steel chain made by Ramsden, constructed on the principal of a watch chain, each 1-foot link made up of one long and two short plates, with circular holes at ends of each, by which they were connected with steel pins.[12] The chain weighed 18 lbs and when folded could be carried in a deal box only 14 by 8 by 8 inches. It performed admirably. Its elaborate end-pieces allowed repeated measurements along what was found to be 27,404 feet. This chain is now in the Science Museum, London.

French surveyors generally measured their bases with deal rods and Roy wished to compare this method with the result obtained with the chain. The Admiralty gave him an old Riga red pine mast, which was cut up to produce three rods, each around 20 feet 3 inches long by 2 inches deep and 1¼ inches broad, with bellmetal tips. The extra three inches allowed both butt and overlap measurements. Ramsden brought the rods to the army camp on 15 July. The rods were trussed vertically and laterally to prevent them flexing, but their lengths varied with the humidity of the atmosphere. On one trial, after a week of rain in what was a very wet summer, the rods were found to have gained 1/15 inch, which over the entire base would have amounted to 7½ feet. In another trial, the rods were left out on grass under heavy dew and next day found to have gained nearly ½ inch over the previously measured 300 feet.

Measurement of the base, with its strange equipment manipulated by the officers and men in their colourful uniforms, was seen as a great spectacle; Banks and his friends came to see its commencement and its termination. Banks set up a tent and provided refreshments for his invited visitors. Plans for the King and Queen to visit on 19 July were postponed due to heavy rain but the King arrived on 21 August and watched the operations for two hours. On 31 August the diplomat and archaeologist Sir William Hamilton (1730–1803), the mineralogist and horticulturalist Charles Francis Greville (1749–1809), Welsh landowner John Lloyd (1749–1815), military engineer Captain Charles Bisset (1717–1791) and astronomer Henry Ussher were among several important personages gathered to watch as the last stage of the base was measured.[13]

A fourth rod was kept as a standard, and this was compared to Roy's own 42-inch brass standard, which he had purchased at the sale of James Short's effects. Its pedigree was impeccable, having been divided by John Bird, then owned successively by George Graham and James Short. It was compared with the Royal Society's own standard, on which the Tower yard, the Exchequer yard and the French half-toise were marked. The comparison was made in August after Roy had laid out the two standards, with thermometers, leaving them for two days to accommodate to the ambient temperature before allowing Ramsden to compare them with his beam compass. They were found to match perfectly. For comparisons in the field, Ramsden made a beam compass able to measure 20 feet and trussed like the deal rods. The standard rod was also measured on the long bench in Ramsden's workshop.[14]

Finally, William Calderwood, a captain in the Horse Guards, suggested glass tubes. Calderwood had collaborated with Roy on barometric measurements and must have been aware of John Smeaton's experiments in 1776 on the expansion of glass tubes at different temperatures.[15] It was found that glass tubes varied less with temperature than glass rods of the same diameter, and less than similar tubes made of steel, cast iron or copper.

Roy procured three 20-foot tubes from William Parker's glass manufactory in Fleet Street. While they were in his workshop, Ramsden was visited by the French natural historian Barthélmy Faujas de Saint Fond (1741–1819), whose account of his travels through the British Isles informs us: 'I found the skilful and modest Mr Ramsden occupied in making an instrument simple in appearance, but which demanded much care and many combinations to render it perfect.'

We can almost hear Ramsden explaining to Saint Fond why glass had been chosen:

> It was required to measure on the ground a base of 4286 toises, so as to avoid the defects of the ordinary instruments of measuring; which whether of wood or metal are liable to be expanded by heat and contracted by cold, and to several other inconveniences that do not permit one to depend on their perfect accuracy, whatever precautions may be taken in using them. To effect this purpose, it was proposed to use rods made entirely of glass; and it was in preparing these that Mr Ramsden was then employed. The glass tubes were executed with all possible care, in the glass manufactory of Parker, to the best of my recollection. [He is probably correct, as it seems that he later inspected Parker's premises.] They were all of the same diameter, and straight as the most perfect ruler. They were very long, fixed on proper supports, with a water level to each; they could be elevated or depressed at pleasure.

Clearly impressed by his visit, Saint Fond adds:

> I had much pleasure in conversing with Ramsden. I went to see him several times and I purchased several instruments at his shop. He possesses all the modesty and simplicity of manners of a man of great talent.[16]

Ramsden devised an elaborate arrangement to allow exact congruity of the rod ends. A description was given in Roy's submission to King George III[17] and published with detailed drawings in the *Philosophical Transactions* volume 75 (see Figure 8.1). These rods have unfortunately not survived. In May 1791, when the base-line had been remeasured, the Duke of Richmond was permitted to borrow them.[18] Informed in January 1792 that the Society wanted everything returned, he wrote to Banks:

> The glass rods were unfortunately at my house at the time of the fire, and Mr Ramsden reports that all the thermometers have been broken, as well as two of the rods, and that the cases are split almost to pieces.
>
> I thought it right to acquaint you of this circumstance, but if the Royal Society should have any use for the Glass Rods, I will direct Mr Ramsden to replace those two which have been broken and to repair the other.[19]

By this time the glass rods were not needed, for Ramsden's chains had proved entirely reliable.

The Pyrometer

William Mudge and Isaac Dalby explained the purpose and construction of this apparatus in their *Account of operations, 1787–96* (1799):

> When measuring such a long base-line, there was a chance that errors would be introduced from the expansion of the glass tubes and the iron chains on hot days. The

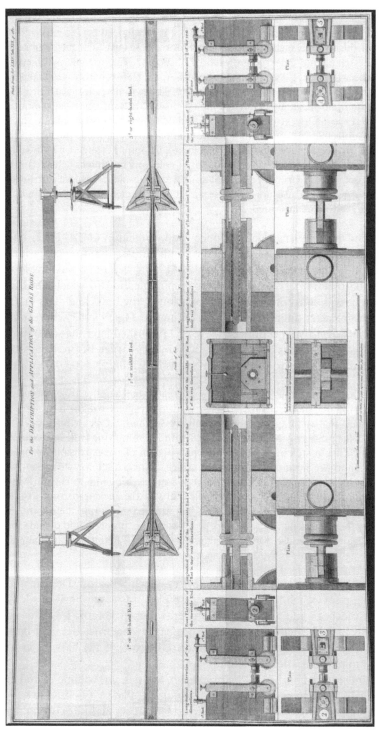

Figure 8.1 Ramsden's apparatus for measuring the Hounslow baseline with glass tubes. Above, the tubes in their cases supported on trestles; below, arrangements to abut the tubes. Tab. XVIII, *Phil. Trans.* 75 (1785).

existing small bench pyrometers, used for testing the expansion of various metals and other substances when heated, were obviously inadequate for this work, and given the uncertain composition of many metals and alloys, it was judged best to experiment with rods of the identical metals and glass used for the base measurements. Ramsden set about designing and constructing what he termed his 'microscopic pyrometer'. Begun in the winter of 1784, it was finished in April 1785, when it was taken to Roy's house in Argyll Street and set up in the yard. The pyrometer, for measuring the dilation of bodies by heat, also exercised the talents of Mr Ramsden. On examining the pyrometer then in use, he observed the radical defect of that instrument, in which the bodies subjected to experiment were not sufficiently separated. But with this microscopic pyrometer he found means to compare the natural state of a body with the same body exposed to any degree of heat or of cold, and by a micrometer adapted to the microscope he measured these variations with an exactness before unknown, and which furnished the measure of a base with a precision ten times greater than in any of those ever before measured.[20]

The instrument maker turned civil engineer John Smeaton (1724–1792) had constructed a small pyrometer for testing the comparative expansion of various metals, glass and other materials as far back as 1754 – before he was involved with Roy on barometry.[21] It was Smeaton's communication to which De Luc referred when he was concerned that the expansion of the brass and glass in his own delicate hygrometer might devalue the readings. De Luc's letter was read at the Royal Society over two meetings in 1778, and published later the same year, but it seems that he began his experiments several years earlier, before Ramsden had constructed his large pyrometer, for he begins by stating:

> I had heard that the ingenious Mr Ramsden sayd, that he had a notion of a pyrometer different from all that had been invented; and knowing his great skill in philosophical and mechanical matters, I applied to him, and pressed him to execute his idea. The multitude of his other engagements prevented with his complying with my request; and he advised me to look no further for the proportions of the expansions of brass and glass than to Mr Smeaton's experiments, which he looked upon, with reason, as the best that had been made. Still, however, upon my desiring him to explain what means he thought of being able to correct the faults of the ancient instruments, he was kind enough to do it, and told me, that he proposed measuring the expansion of bodies by the micrometer of a microscope; by which means he should obviate the greatest difficulties. He added, moreover, that he had made a first trial of his method a long while ago, and was assured of the success. [22]

Ramsden's pyrometer, far larger than the devices which Smeaton, De Luc and others had made for bench testing of small samples, was set up in Ramsden's workshop as the bars being tested were several feet in length. In order to insulate the bar being heated from the reference bar he constructed a twin-bath arrangement. Reading microscopes were fixed to the ends of the bars, so as to read the most minute variations in length.

Detailed drawings, published in *Philosophical Transactions*, volume 75 and in Mudge and Dalby's *Account of operations … volume 1*, Plate V (see Figure 8.2), show a deal frame 5 feet long, 28 inches broad and 42 inches high supporting the

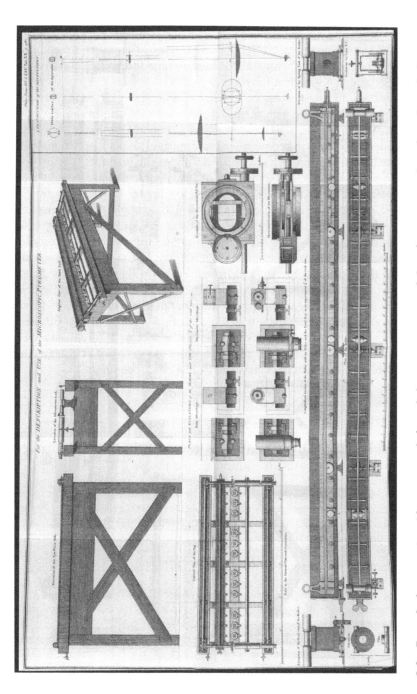

Figure 8.2 Ramsden's pyrometer. Above, general views; below, arrangement of the microscopes and sections through the pyrometer case. Tab. XX, *Phil. Trans.* 75 (1785).

apparatus which consisted of two deal troughs, lined with pitch, each holding a cast iron bar of prismatic section, one end fastened to its trough, the other end free to move. A fixed microscope kept the fixed end of the bar in view, while a moveable microscope sighted the expansion of the free end. Between the two deal troughs was a third trough of copper, heated by 12 lamps and resting on rollers, which could be filled with crushed ice, and raised to boiling point to test the samples through a wide temperature range.

Since the bars under test expanded with heat, the microscopes which measured that expansion had to move with them. To construct such a microscope, where equal parts of an image are to be measured with an equable motion of the object lens, Ramsden could not interpose a second eyeglass (to increase the angle of vision) since that would diminish the size of the image and increase the refraction of the oblique rays of light. His solution was the new system of eyeglasses which he announced to the Royal Society,[23] described in Chapter 7.

Mudge and Dalby give a full explanation of the complicated sequence of operations which enabled the operators to read the expansions to a high degree of precision, adding 'In this manner Mr Ramsden obtains the scale of his pyrometer, in the easiest and most simple way imaginable, without any necessity for knowing the absolute distances of the object lens from the wires of the mark on one hand, and those of the micrometer on the other'. Two observers were needed, one at each microscope. Trials were made with 'Hamburgh plate' brass, the metal of which Ramsden's standard scale was made; with English plate brass; with steel rod from the same source as the chains were made; with cast iron, from which the prismatic bar was made; with glass tube, 'from the same pot of glass as the tubes had been made' and with a glass rod from a clock pendulum.

The First Great Theodolite

With the first base set out, Ramsden's new instrument for taking the angles was eagerly awaited; three years elapsed before it was delivered. A major setback occurred in August 1786, as Banks informed John Lloyd:

> Roy's instrument is not yet done, Ramsden's men suffered a large brass ruler to fall on it when the divisions were nearly all cut – it shook the whole circle so severely that the divisions have been commenced again ab initio so that I have little hopes of much progress this year.[24]

In September 1786 Roy wrote to James Lind (1736–1812):

> It will be yet some days before Ramsden can possibly finish the Division and after that the semicircle for the uppermost telescope is to divide, the levels to adjust, and a number of other small things to do, all of which require time, and it will certainly render it too late in the season to think of taking men into the Field to encamp ... It is hard upon me to have this operation hanging over my head for another year, without any fault of mine; but with such a man as Ramsden there is no help for it.[25]

By early 1787 the instrument seemed to be nearing completion. Banks informed the Foreign Secretary, Francis Osborne, Lord Carmarthen, in February:

> Operations began in 1784, but from that period no further progress has been made in the Business, because the instrument of a particular kind, for taking the angles with great accuracy, was to be contrived, and made by Mr Ramsden, in the execution of which, that very ingenious, but dilatory Artist, has consumed much more time, than was expected. However, the instrument is now in such a state of forwardness, as to leave no room to doubt, that General Roy will be able to resume the operation early in the ensuing summer, or as soon as the weather in this country will permit.[26]

And to Cassini he explained that the instrument was taking longer than expected, but as the divisions and the other parts were done, it should be ready by summer.[27] On 1 June Blagden wrote to Banks, 'Ramsden is hard at work upon General Roy's instrument which it seems may be finished by the end of next week'.[28] This hope was dashed by a sudden deterioration in Ramsden's health. On 28 July Blagden informed Banks:

> Ramsden's illness is a very alarming circumstance. I could not judge from your account what was presently its nature but any sudden discharge of blood, except perhaps by piles, at his period of life, indicates a disposition which most commonly terminates in apoplexy.[29]

The situation seemed to have improved when he reported on 14 August 'I have seen Ramsden, who looks well, but sensibly older'.[30] but six days later the news was again bad: 'Ramsden is said to be very ill.'[31]

Banks's letter to Cassini informed him that the instrument was still with Ramsden, 'who from time to time promises it, but has not yet finished it'.[32] Cassini then wrote to say that after meeting with Blagden in Paris in late July he heard that the instrument was still unfinished,[33] but this letter crossed with the joyous news conveyed by Banks that it was finished and had been delivered, and that Roy was now back in the field.[34] Curiously, Roy and his associates were unwilling to describe Ramsden's creation as a theodolite, preferring to identify it as 'the instrument', a practice followed by the Cambridge astronomer Samuel Vince in a chapter headed 'The new instrument for measuring horizontal angles' in his *Treatise on practical astronomy* (1790), 170–77.

In 1784 Roy had paid Ramsden £340, 10s cash for apparatus delivered; in 1787 he paid Ramsden £444, 11s, 2d for 'his two accounts of the cost of the Instrument and apparatus for the Chain &c'., with a further payment of £37, 9s, 6d for alterations. There were additional payments to carpenters, for alterations to, and subsequently repairs to, the carriage for the instrument, for an oiled silk cover for the instrument, for silver wire, and in 1788, payments to Thomas Milne for a drawing of the instrument – published to accompany the article in *Philosophical Transactions*[35] (see Plate 3 and Figure 8.3).

In the spring of 1789 Roy, in poor health, spent two months in the Welsh mountains, returning, according to Blagden, much improved.[36] He remained, however, sharply critical of Ramsden and his workshop management:

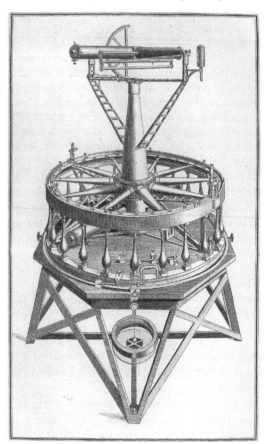

Figure 8.3   Ramsden's three-foot geodetic theodolite, general view, drawn by Thomas Milne. Tab. VII, *Phil. Trans.* 75 (1785).

> For several months of the spring and summer of 1787, Mr Ramsden had been seriously at work in endeavouring to finish the instrument. Not having employed a sufficient number of workmen upon it at the outset, it was now evident, that he had even deceived himself, by leaving too much to be done at the latter end.[37]

He made no concessions for the delay that Ramsden's illness would have on the finer details of the construction of this novel instrument, although by the summer of 1787 its final form was probably decided. It was not, however, the only large instrument of novel design being designed or built in Ramsden's workshop, for between 1784 and 1787 there were, in various stages of construction, the pyrometer for Roy's survey, a transit and a mural circle for Dunsink, a mural quadrant for Blenheim, Shuckburgh's equatorial telescope, two transits and a mural circle for Seeberg, a transit for Mannheim, a mural quadrant for Brera, and a meridian transit for Vilnius. Additionally, from time to time he was summoned to attend the Duke of Marlborough at Blenheim, or was called out by Roy himself, in connection with the Hounslow Heath base.

Roy's annoyance with Ramsden, whom he blamed for the politically embarrassing delay, spilled over into his lengthy account of the triangulation exercise, read at the Royal Society during January and February 1790. Ramsden, who was present for some of the readings, took offence, and sought to have the paper toned down before publication, but Roy had already departed to Lisbon for the sake of his own failing health and did not return until late spring. He died at his home in Argyll Street, Westminster, on 1 July 1790.[38] The dispute is recounted in Chapter 9.

Among the 44 instruments sold after Roy's death were several by Ramsden:[39] a barometer to Lind's design with a thermometer; a 4-inch sextant; two achromatic telescopes, of 1 and 4 feet; two 12-inch brass plotting scales; two portable barometers; a pocket compass, and probably, though not identified by signature, the 'capital large equatorial instrument, which also serves for taking altitudes and azimuths', which sold for £141, 15s. A transit referred to by Ramsden as having been made for Roy had been sold previously to Aubert.[40]

Spanning the English Channel

The first great theodolite, generally known as the Royal Society or 'RS' theodolite to distinguish it from the later instrument, had a 3-foot diameter horizontal circle of brass, divided to 10 minutes of arc and made to read to 1 second by means of two micrometer microscopes. Ten radial conical tubes attached the circle to a large vertical conical axis 24 inches in height. This axis revolved on an inner fixed axis to which the reading microscopes and levelling footscrews were fixed. A flat horizontal bar rigidly mounted across the top of the outer axis supported the telescope axis. The angle to which the telescope was raised was measured against a vertical semicircle of 10½ inches in diameter, divided to 30 minutes of arc and read to 5 seconds by means of a microscope attached to the supporting bar. The telescope axis and supporting bar were adjusted to the horizontal by reference to two spirit levels, 24 and 21 inches in length. There were two achromatic telescopes, each of 36 inches focal length, and 2½ inches aperture, with various eyepieces, one erecting and magnifying by 54. The telescope diaphragm had fixed cross-wires and a horizontal travelling wire controlled by a micrometer screw. For night work the circle could be illuminated through the horizontal axis of rotation. Clamps and tangent screws for fine adjustments were fitted to the horizontal and vertical axes. The probable error of a single observation was about 5 seconds of arc for a distance of 70 miles.

The theodolite came with its ancillary tackle of stand, steps, stools, pullies, ropes, tent and canopy, weighing as much again as the instrument (see Figure 8.4). The whole assemblage was transported in a four-wheeled carriage adapted from a crane-necked phaeton donated by Banks and drawn by two, sometimes four, horses. A housing built on the carriage, covered with painted oil cloth, kept everything dry and secure.[41]

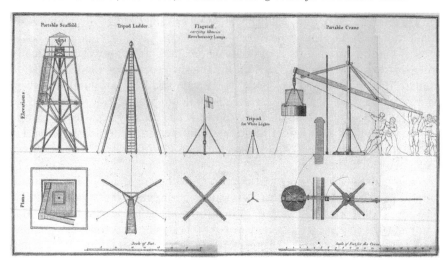

Figure 8.4   Hoisting Ramsden's geodetic theodolite onto its scaffold, and the tripod for the white lights. Tab. XI, *Phil. Trans*. 75 (1785).

Writing in 1856, Sir Henry James, then Superintendent of the Ordnance Survey, remarked:

> Though this instrument has been in almost constant use for the last 67 years, during which time it has been placed on the highest church towers and the loftiest mountains in the kingdom, it is at this day in perfect working order, and probably one of the very best instruments ever made.[42]

It was in use until 1862; for the Principal Triangulation of Great Britain sights were taken over a distance of 100 miles, between Scotland and Ireland.

On 4 September 1784 Cassini de Thury, already in declining health, died. His son, Jean-Dominique Cassini (1748–1845), Cassini IV (see Figure 6.7), who had been closely involved with the French surveys, succeeded as director of the Observatoire and took his father's place on the Anglo-French project.[43] Roy's 'instrument' was set up at the end of the Hounslow base on 31 July 1787 – rather late in the year, but Cassini agreed to be ready at Calais on 20 September.

Greenwich Observatory was one of the first observing stations. Maskelyne jotted down the events taking place during August:

> 1787 August 14 the scaffolding over the transit room for General Roy was begun, & completed Aug 18. Aug 25 General Roy first came here & gave orders about preparing a flagstaff to put over the transit instrument. Aug 27 he came here & put it up and dined here, & went to Norwood to observe it. Aug 28 White lights were fired by the flagstaff.
>
> Aug 29 the new theodolite was brought down by Mr Dalby & Mr Bryce, Aug 30 General Roy dined here. The instrument was set up, the General, Mr Dalby, Dr Blagden & Mr Bryce dined here. The General sty'd and went to town on Saturday & resumed

on Monday morning early & took down the instrument and moved it away at two o'clock to Shooters Hill Tower.[44]

Joseph Banks noted on his copy of Roy's 'Account ...' in *Philosophical Transactions* that the white lights were composed of 28 parts of nitre, 4 parts of sulphur, and 2 parts of trisulphate of arsenic, finely powered, the mixture burnt in copper cups set on tripods.[45]

In August 1987 the Royal Engineers staged a costumed re-enactment of these observations made from the roof of Greenwich Observatory.

The English team intended to set out a second base on Romney Marsh, in Kent, 60 miles east of the first but due to the delay over the theodolite, Roy had to postpone this until after the cross-Channel operation. This base, of 28,535 feet, was measured with a very accurate steel chain which had been compared to the glass tubes, but Roy, his temper perhaps frayed by his declining health, found fault, accusing Ramsden of negligence in its construction. After the reading of his account of the triangulation, Roy wrote to Banks from Lisbon that he had forewarned Ramsden that he would be critical of his actions:

> in regard to the supply of instruments, and particularly the apparatus for the chain; that the blacksmith of Romney Marsh was to be confronted with him, who had found great difficulty in remedying the most glaring defects in the said apparatus, and who would scarcely believe that the famous Ramsden ... could have produced such a blundering piece of business.[46]

At St Ann's Hill, on the way to Wrotham, the box containing the axis level was blown down from the scaffold, shattering the bubble-glass of the level. The quality of the replacement level was also to be the subject of argument between Roy and Ramsden. From Wrotham, Roy's party went directly to the coast. Two principal points, at Dover Castle and Fairlight Down, were chosen to connect with three points on the French coast: Calais, Cap Blanc-Nez and Montlambert. These five points carried four triangles; the French later set out six more triangles, extending north to Dunkerque. The apparatus was taken to Dover to await the arrival of Cassini and his team. They sailed on 18 September, reaching Dover on 23 September for the prearranged rendezvous with the members of the Royal Society. Over two days the programme and timing of reciprocal observations was arranged and the white lights and reverbatory lamps shared out, then the French party with Blagden, who was a fluent French speaker, crossed to France. The operations were straightforward, requiring only clear nights, with good visibility, but bad weather delayed operations until 17 October, when satisfactory observations between Dover and Fairlight Down across to Cap Blanc-nez and Montlambert were achieved.

The French team needed to set up lights on the coast to be sighted from England, and to sight across to Dover Castle; they also wished to improve the accuracy of the old network between the Paris meridian and the coast. Their triangulations were made using a geodetic repeating circle only one foot in diameter constructed by Etienne Lenoir (1744–1832) following a design by Jean-Charles de Borda, a mathematician in the naval service. While admittedly less accurate than Ramsden's great theodolite, it was certainly easier to transport and to raise onto towers or scaffolds.[47] The French also had a good quadrant of 2½-foot radius. The atmosphere was cordial, as Cassini informed his colleagues at Paris:

> When darkness fell we went off to spend the evening with General Roy; we were given tea English style, then a really good supper, which lasted a long time as the Englishmen could not bring themselves to leave the table.[48]

The same conviviality would appear to have reigned on the French side as Blagden reported that they had enjoyed some excellent Burgundy.

The onset of bad weather brought activities to a close. The French minister concerned with the operations instructed Cassini that the French King wanted them to make good use of their time in England by examining instruments by Herschel, Dollond and Ramsden. And so, before returning to Paris, the French party travelled to London and Greenwich, to visit the Radcliffe Observatory in Oxford – as yet incomplete; Blenheim – home to two of Ramsden's masterpieces; and Slough, from where, in 1781, William Herschel had discovered the planet Uranus. At each location, Cassini and his colleagues took note of the ingenious arrangements seen on the instruments: the new means of illuminating the sighting-wires of the telescope, the rotating Blenheim pillar quadrant, which allowed the observer to sight towards the north as easily as towards the south, the simplicity and lightness of Herschel's mounting for his great 40-foot reflecting telescope, with its 4-foot aperture.

Cassini had written his account of the operations by 1789, but out of courtesy withheld publication until the British account should also be ready. Inevitably, this too was delayed – though by Roy rather than by Ramsden. It was published in 1790, not a good year politically, as Cassini ruefully admitted, for advertising his own scientific endeavours in France.[50] The bases having been measured with unprecedented care, and the angular errors virtually nil, the results were deemed to be remarkably accurate. Cassini allowed that Ramsden's theodolite was so perfect that in no triangle did the error exceed 2.8 seconds of arc; the French instrument had a maximum error of 8 seconds, but he could not forego a brief remark on the ease of transporting his own repeating circle, in contrast to the difficulty and hazard of transporting and raising Ramsden's great instrument.

Roy's description of the setting-out of the Hounslow Base, and his account of the triangulation, published in the *Philosophical Transactions* of 1785 and 1790 were both translated into French by the civil engineer Gaspard-Clair-François-Marie, Baron Riche de Prony (1755–1839), who had been in England in 1785 and had watched the survey in progress.[51]

With the cross-Channel sightings completed to general satisfaction, Roy turned back to close the gap in triangulation between the coast and Wrotham Hill. He dealt with five of the remaining seven stations before increasingly stormy weather made observations from steeples or hilltops positively dangerous, bringing operations to a close on 2 November. The theodolite was returned to London where sundry repairs and modifications were carried out: the support for one of the microscopes was improved, another clamp, and an eyepiece and diagonal prism for near-zenith observations were fitted. In January 1788 Roy was elected to the Royal Society of Edinburgh.

The theodolite was sent back in the field on 9 August 1788. The observations being completed, Roy took some additional sightings to other conspicuous points near London, in what was to be his last year in the field. His remaining time was spent on calculations and the writing up of his account of the exercise.

In 1790 King George donated to the Royal Society all those instruments used by Roy for the trigonometrical survey; on 18 November Council decided that the theodolite should go into the library, the three glass rods and chain should be kept in the council room and ante-room. Ramsden was to be asked to estimate for repairs to the pyrometer.[52] (As mentioned above, the rods were later borrowed by the Duke of Richmond and lost to fire.) On 21 February 1793 the Commissioners of the Treasury notified Banks that they wished to examine Roy's accounts, and asked for the vouchers to be produced. These were speedily examined, and returned two days later.[53]

The Survey after Roy's Death

After Roy's death the triangulation so far executed was complete in itself and no move was made to extend it to other parts of England. Eleven months passed, before the Duke of Richmond set it once more in motion. What sparked his initiative seems unclear; although Richmond was a landowner and public administrator with a strong interest in cartography and military engineering projects. Indeed, it seems possible that he was already negotiating with Ramsden for another great theodolite, which might account for the rumour reaching Duke Ernst of Gotha[54] (see Chapter 6), and the report by Giuseppe Piazzi to his patrons in Sicily that Ramsden had such an instrument in hand.[55]

William Mudge and Isaac Dalby thought it probable that the Duke of Richmond seized the chance opportunity of purchasing the instrument, ordered but then rejected by the Honourable East India Company (HEIC) (see below), which, as the property of the Ordnance, could be employed on further surveys, but historian William Andrews suggests that the worsening Anglo-French relations after 1789 stimulated cartographic activity in the southern counties.[56] Whatever fired his decision, the HEIC theodolite became available to Richmond for the price of £373, 14s. To differentiate it from the earlier 'RS' theodolite, it is usually known as the Ordnance Survey, or 'OS', theodolite.

The new chains for the 1791 survey had links of 2½ feet and were strained by 56-pound weights. They were measured with reference to a 21-foot steel bar of trapezoidal section, the base and flanking sides each of 1¼ inches, laid on the 60-foot 'great bench', which ran along the rear of Ramsden's workshop. On this bar, he set off 20 feet from his own scale in six bays of 40 inches, marked by dots on inlaid brass pins.[57] This trapezoidal bar is now held by the National Museum of Science & Industry.

In 1792, with a second-order triangulation under consideration, a smaller theodolite, 18 inches in diameter, was commissioned and delivered in 1795 (see Figure 8.5). It differed from the large theodolite in having footscrews below the horizontal movement which directed the lower telescope. In this way, the whole instrument could be turned without disturbing the levelling. As its telescope did not need to sight towards the zenith, a small index to measure elevation replaced the quadrant. The horizontal circle was divided to 5 minutes of arc and read by three micrometers to 2 seconds, the 8-inch vertical circle was divided to 10 minutes and read by verniers to 10 seconds.[58] With two fitted mahogany cases, and a 100-foot tape measure, it cost £135, 10s.[59] It was delivered for a survey of the coast of the Eastern District, was repaired by Berge in 1801[60], and subsequently employed during the measurement of the Lough Foyle Base in 1826.[61] This theodolite, with one microscope missing but still with its magnificent carrying-cases, is in the Ordnance Survey's Heritage Collection at Southampton.

Figure 8.5   18-inch geodetic theodolite signed *Ramsden*. (One microscope is missing.) Ordnance Survey Historical Collection.

Roy had received the Royal Society's Copley Medal for the Greenwich–Paris survey.[62] Five years after Roy's death, the RS Council awarded the Copley Medal to Ramsden:

for his various inventions and improvements in the construction of the instruments for the trigonometrical measurements carried on by General Roy and Lieutenant-Colonel Williams and his associates and described in their papers printed in the Philosophical Transactions giving an account of those operations.[63]

There is a hint of yet another small geodetic theodolite being in hand during 1798.[64] Pictet, visiting London in September that year, called on Ramsden and reported to De Luc finding him considerably rejuvenated.

He was absolutely in high spirits. He was most friendly with me and told me that over the last six months there had been great advances in the construction of astronomical instruments. He had in hand a theodolite built on the same principles as that of General Roy; it could be operated by a child, having a radius of 7½ inches, and he showed me that it could be read to one second of arc. This value is equal to 1/28000 of an inch. He had abandoned the bubble level and returned to the plumb-bob as more reliable. He fitted a strong microscope and a lateral prism which by bringing the image to right-angles allowed it to be trued up in both directions at once. He seemed to be very much in favour of this method. He was keen to discuss a great many matters with me but sadly I had not the time and to my great regret we had to part company.[65]

On taking command of the Survey in 1799, one of Mudge's first actions was to ask the Royal Society to lend Roy's 3-foot theodolite, this instrument having been laid up in the Society's apartments. His request was granted, and in January 1799 the instrument was returned to Ramsden in order to have new microscopes fitted to it, probably to match the improved arrangement in the OS instrument. Ramsden was also asked to complete the zenith sector which had been ordered by the Duke of Richmond in 1795.[66]

The zenith sector, described in the second volume of the *Account of the Trigonometrical Survey,* served to determine latitudes from the commencement of that survey until 1836. Set vertical by plumb line, its 8 -foot radius carried an arc of about 15½°, divided into spaces of 5 minutes by dots on gold pins, which could be judged by micrometer to about one-tenth of a second. In clear weather, its 8-foot telescope with object glass of 4 inches, was able to sight stars of third magnitude in daylight. Mudge took it across the Channel to Dunkerque in the autumn of 1818 and during the following summer Heinrich Christian Schumacher, astronomer at Altona, observed with it in Denmark.[67] 'But', wrote James:

… not withstanding the general excellence of the instrument, it was not – from the complexity of its construction and the number of its parts – well adapted to constant transport from station to station; the necessary reversal of the instrument on successive nights, and the consequent delay during bad weather, was also a source of considerable inconvenience.[68]

On the night of 30 October 1841 fire broke out in the Tower of London where the Survey had its headquarters. All available fire engines arrived but the 'awful conflagration', more extensive than the fire which had destroyed the Houses of Parliament in 1834, lasted several days. There were some anxious moments while officials tried to locate the holder of the keys to the cabinet where the crown jewels were stored, but eventually he was found, and the regalia rescued. The surveying instruments were then moved to safety, with the exception of Ramsden's zenith sector, which was destroyed. But the Armoury was gutted and the cartographical manuscripts turned to ashes.[69]

The survey's new headquarters was an eighteenth-century barracks in Southampton, which had been vacated the previous year. A century later, the RS theodolite, stored in the basement, was lost during the World War II bombing of that important port, and so cannot now be closely compared with the Ordnance's own instrument, deposited earlier at London's Science Museum, from where it had been prudently evacuated at the outbreak of war. Some major differences in design (see Plate 3) can be discerned from the engraving (see Figure 8.3) in Roy's *Account*.

The Second, or 'OS', Great Theodolite

The improved OS theodolite was described by Henry James in his *Account of the methods and processes adopted for the production of the maps of the Ordnance Survey of the United Kingdom …* (1902), 4–6, with an engraved side view (see Figure 8.6). The instrument stood on three feet, joined by a strong bell-metal plate, 7 inches in diameter and about 8 inches high, co-axial with the instrument. The conical steel axis rose from this cylinder, 2 feet high, with bell-metal base, terminating in a cast steel pivot with tapered cheeks. Originally, the OS probably had two microscopes, but by the time that James described it, it had undergone the same modificiations as the RS instrument, described below. Three arms carrying micrometer microscopes projected from the upper part, with a fourth arm and microscope halfway between two of the others. Each micrometer head was divided into 60 parts, with single seconds marked, and tenths to be estimated.

The horizontal circle was divided every 10 minutes of arc by dots on gold discs, and connected firmly to the lower part of the hollow brass exterior axis by ten hollow cones. Within the base, a collar of cast steel fitted into the bell-metal base of the steel axis. Thus the exterior axis and circle, placed in the interior axis, revolved smoothly and without shake. An arrangement of screws at the top of the axis regulated the pressure, and hence the ease, of revolution. When setting up, a small amount of horizontal movement allowed the instrument to be centred over the station mark. A horizontal bar across the top of the outer axis supported the telescope on Ys sufficiently high to allow the transit axis to pass, and the telescope to sight upwards at an angle. The original semicircle on the telescope axis had also been replaced by 10½-inch circles round the telescope axis by the time that James was writing. The telescope and levels matched those of the RS theodolite.

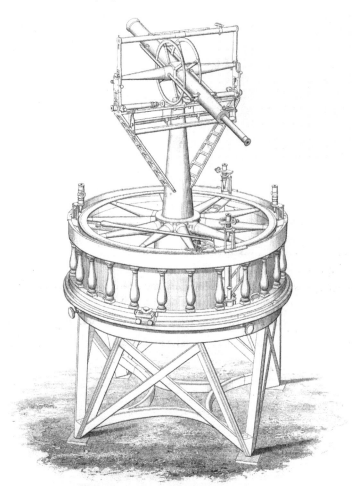

Figure 8.6   Ramsden's geodetic theodolite, as improved by Thomas Colby. H.
**   *James, Account of the Methods and Processes … (1802),***
**   un-numbered plate.**

Like its predecessor, the OS theodolite weighed 200 lbs, and was carried in a sprung four-wheeled carriage. When it was to be moved, the outer axis and circle were let down on the inner axis, this method of locking all motion being one of the most admired features of the theodolite. Indeed, Johnston, by then Director of the Ordnance Survey, believed that no other construction could have withstood for so many years the wear and tear of use, with such a heavy instrument being repeatedly transported in the roughest weather and across the wildest terrain.

As the triangulation proceeded northwards through Britain, Ramsden's steel chain was employed for bases on Salisbury Plain, Wiltshire, in 1794, at Misterton Carr, Yorkshire, in 1804, and at Rhuddlan Marsh, Denbighshire, in 1806. The standard of length for these bases was defined by Ramsden's brass scale at 62°F. Berge compared the chains with Ramsden's trapezoidal bar before and after the measurement of a 5-mile base on Belhelvie Sands, Aberdeen.[70] Ramsden's own brass standard passed to his successors Berge and Worthington but is now lost.

Disputes having arisen again over the exact distance between Greenwich and Paris, it was re-measured in 1821–23, the British team being led by Colby as surveyor, with Henry Kater (1777–1835) as the scientific partner; the French team being led by astronomer Jean-François-Dominique Arago (1786–1853) and his brother-in-law Claude-Louis Mathieu (1783–1875). The RS theodolite was used for the new connection, and to avoid the difficulty and delay caused by changing zero, Kater instructed William Cary to put in four additional micrometers which, with one of the originals, divided the horizontal circle into five parts[71] (see Figure 8.7). This procedure has been attributed to John Pond, then Astronomer Royal, a man with a deep and abiding concern with astronomical circles. It meant that several different parts of the horizontal circle could be used on any one pointing, averaging the results.[72] To some extent this practice did away with the need to change zero and brought it closer in function to the repeating circle, an instrument which had by this time found favour on the English side of the Channel. The OS theodolite was also modified by adding two new micrometers and removing one of the originals. The two great theodolites, and the lesser one, continued in use until 1827, when they were joined by a 2-foot theodolite of somewhat different pattern, from the London workshop of Troughton & Simms. William Simms redivided the 18-inch Ramsden theodolite in 1848, before it was set up on the top of St Paul's Cathedral as the triangulation was carried across London that year.[73]

Honourable East India Company Surveys

In his paper describing the triangulation for the link with France, Roy suggested that the HEIC would be interested in acquiring a 3-foot theodolite for a trigonometrical survey of India, which by determining latitudes and longitudes, would improve navigation to the benefit of its own fleet of ships. Roy also forwarded his proposals to the Court of Directors, who responded by consulting one of the Company's surveyors, James Rennell (1742–1830) and its Hydrographer, Alexander Dalrymple (1737–1808). Their considered reply of 17 June 1787 identified and costed the instruments and the manpower to operate and transport them, and urged that orders should be given promptly, pointing out that dividing, which was done when an instrument was virtually complete, could not be carried out between the end of September and the return of warm weather.[74] This was because in winter, that part of the scale which the workman was dividing and on which his lamp was focussed, inevitably warmed up slightly, and expanded, resulting in irregular divisions. They did, however, add that Ramsden generally rated the prices of his instruments 'considerably lower'[75] than Roy's estimates.[76]

Thus encouraged, the HEIC placed an order with Ramsden for another 3-foot theodolite. It was noted on 19 May 1790:

> As the instruments intended for the observations recommended by Major General Roy, to be made on the Coromandel Coast, are in great forwardness, and as Mr Ramsden

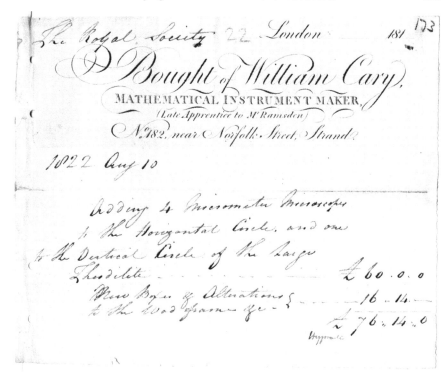

Figure 8.7   William Cary's invoice, 1822, to the Royal Society for improvements to Ramsden's geodetic theodolite. RGO 14/17 f. 173.

gives the most positive assurances that they shall be completed in time to go by some of the ships of this season, we shall postpone till a future occasion saying anything further on this head.[77]

The theodolite was finished in 1791, but Ramsden had incorporated various improvements, based on Roy's experience with the original, and raised his price accordingly. The HEIC refused to pay, and Ramsden was left with this expensive apparatus on his hands and no customer in sight. As recounted above, the situation was salvaged when the Board of Ordnance bought what was thereafter known as the OS theodolite. Meanwhile, the Company placed an order with William Cary for an instrument of the same pattern.[78]

The HEIC also commissioned a zenith sector, the first such instrument that Ramsden had been asked to make, and he asked the Royal Society for a sight of the old Sisson sector in the Society's warehouse. The Society's Secretary George Gilpin (d.1810) explained to Banks on 25 September 1789:

Mr Ramsden being employed in making a zenith sector for the East India Company wishes to have access to the Society's sector which is in their warehouse, and to be allowed to put it up there that he may the better be able to examine it to see what had been done by others. It was made by Sisson, and is the same that Dr Maskelyne had with him to St Helena, and also to Schiehallion – I beg pardon for giving you this trouble, but Mr Ramsden wishing to get on with his instrument is the reason of my not waiting your arrival in town ...[79]

This is probably the sector which had been ordered for the failed Cathcart embassy, then re-embarked with the Marcartney embassy, but not presented. It arrived in India during William Lambton's time with the Great Trigonometrical Survey and served his measurement of an arc on the Coromandel coast, Bengal, in 1802.[80] In 1806 it was sent to John Warren at the HEIC Madras Observatory.[81]

The surveyors delegated for the Indian survey were Michael Topping (1747–1796), who was then at Madras, and Reuben Burrow (1747–1792), mathematical master to the HEIC Corps of Engineers, in Calcutta. Burrow expected the arrival of his instruments in 1789 and wrote several times asking for a zenith sector to be purchased locally, but none was available and it was supposed that the HEIC had one on order from England. In March 1790, on receipt of Lord Cornwallis' order to measure a degree of longitude on the Coromandel coast, near Krishnagar, Burrow started with such instruments as he could buy or borrow from other surveyors. Those by Ramsden were an astronomical quadrant of 1-foot radius, a brass scale, a 50-foot steel chain 'of Ramsden's new construction'. His apparatus included a theodolite, possibly the Ramsden instrument which he had acquired from Captain Garstin, 'an excellent Ramsden telescope' from a Mr E.E. Pote[82] and a sextant, several glass rods, ground to a particular length, long bamboo rods and some 10- and 20-foot rods and stands.[83] Topping, engaged in a survey for the Kistna-Godavari Irrigation Project, reported that 'The levels were taken with an excellent instrument of Mr Ramsden's construction'.[84]

As HEIC surveyors died, or left India, there was enormous demand for their surveying and drawing instruments. Unlike Burrow, who was not a covenanted company servant and could therefore claim for his purchases, the HEIC officers had to provide their own instruments, those by Ramsden, Dollond and Troughton being much in demand. In 1782 the Surveyor General had to beg the directors to send out every two years various instruments, 'and if at all possible to be made by Ramsden', to which, with the usual miserly pruning, they agreed.[85]

While he was in Bengal, Colonel T.D. Pearse acquired a Ramsden barometer which had lost its mercury. He refilled it, but in boiling the mercury, the glass broke, and being unable to procure another tube, he was obliged to return it to Ramsden for repair. His covering letter to Ramsden is critical of several features of the barometer; he asks that the cistern be made more capacious, and that to resist the Indian heat, the boxwood should be baked and treated with oil, or the cistern perhaps made of iron lined with cork, or of glass covered with metal. Pearce was a great admirer of Ramsden: 'I have seen many instruments of Ramsden's in India exquisitely good'. While not appreciating the finer points of barometer construction, Pearse was frustrated by the London craftsmen's ignorance of the high temperatures and extremes of wet and dry weather to which their thermometers and barometers would be subjected.[86]

The 'Berne' Geodetic Theodolite

In 1790 Marc-Auguste Pictet (1752–1825) wrote to Charles Blagden to propose a triangulation within Switzerland, and linked only to the survey of Savoy, thus avoiding political difficulties with France. His letter (of which only a draft survives) must have mentioned the possibility of borrowing Ramsden's great theodolite for on 21 September Blagden's response asserted 'that used by Roy would, I believe, on no account be suffered to go out of the kingdom'.[87]

Pictet, a citizen of Geneva, was a man of many scientific interests, well-travelled, and on good terms with learned men throughout western Europe. Despite Blagden's assertion, Pictet decided to try his luck with the President of the Royal Society and early in 1791 he wrote to Joseph Banks, opening with flattering remarks on the accuracy of the base measurement for the Greenwich–Paris survey. He then proposed that a similar survey be made, commencing from Geneva Observatory, and overseen by commissioners appointed by the Royal Society. The observatory already had a 5-foot transit and a 2½-foot astronomical quadrant, both by Sisson, clocks by Shelton and Lepaute, 10-foot and 3½-foot achromatics by Dollond, a good Ramsden 30-inch achromatic, and sundry other instruments. From England they would need one or two zenith sectors, one or two clocks, and an instrument for measuring terrestrial angles. Pictet then came to the point of the letter:

> I saw in London, about three years ago, that made by Mr Ramsden and employed so successfully in the late measurements from Hounslow Heath to the French coast. It is a matter of doubt to me whether this precious instrument would be allowed to go out of the kingdom; and even should that be permitted, as far as I can judge from the impression it has left on my memory, it would be too heavy and too large of bulk to be conveniently carried to the top of mountains. The instrument ought perhaps to be made of separable parts easily put together and capable of being steadily adjusted; they might be packed separately in a box, not exceeding the common charge of a man. The boxes could be so contrived as to fill a larger one, suspended on springs in a small narrow carriage made on purpose, and in which all the necessary things besides could be collected.
>
> Such a disposition would be very convenient, not only for sending the instruments from England, but also to carry them from one station to another, through cross country roads, with more quickness and less trouble and danger. A convenient tent, to shelter the instruments in the stations, should be added, or might be made at Geneva.
>
> The glass rods for the measurement of the basis [sic] could be drawn of any convenient shape in a glass house not far distant from Geneva. The thermometers, and any additions or repairs the apparatus might require, would be duly performed by Mess. Paul, eminent artists of this town, and we abound with artists of lesser merit.[88]

Pictet was swiftly disabused of his hope that the British government would pay for this survey. Banks's reply of 22 February 1791 explained that Roy's exercise had cost £2,000 and there was no prospect of funding from a government which would not benefit from the operation.[89] Pictet's letter proposing the triangulation was considered worthy of publication in the *Philosophical Transactions* but Council decided to omit the sentences urging the Society to undertake the business, lest it gave the impression that they had more or less agreed.[90] Pictet was elected FRS on the Foreign List in May 1791.

If the appeal for a loan of Ramsden's theodolite had been rebuffed, the plan for a survey remained live. Johann Georg Tralles (1763–1822), a geodesist from Hamburg, was appointed professor of mathematics at the Bernese Academy in 1785. In 1792 he was detailed to undertake a cartographic survey of the canton of Berne. The Economic Society of Berne provided 150 louis d'or to buy instruments, and a great theodolite was ordered from Ramsden in 1792, to be delivered in two years, together with a transit telescope. While waiting for this instrument to arrive, Tralles and one of his best students, the Swiss geodesist Ferdinand Rudolph Hassler (1770–1843) measured a 7¾-mile baseline using a Ramsden steel chain.

The theodolite reached Berne only in 1797, a delay which may not have been wholly due to Ramsden since the first Napoleonic War was hampering British exports to the continent. As delivered, the theodolite cost 6,000 French livres, approximately 1,600 reichthalers or 3,773 Swiss francs, a considerable sum for the authorities.[92] This instrument seems to have matched its two predecessors; Tralles reported that the aperture of the main telescope was 31 Bernese lines (65 mm), magnification × 41. The main circle was divided on brass into 10 minutes of arc, readable with two diametrically fixed drum micrometers, each divided into 20 parts. One turn was equal to 1 minute of arc, allowing the observations to be made to 3 seconds.[93]

Napoleon's military advances and reversals, as he endeavoured to absorb the various Swiss cantons, and to remould their governments as he had done with parts of the Low Countries and Italy, interrupted Tralles's survey with Ramsden's great theodolite. When France's war with Austria ended, Swiss neutrality changed from being an asset to France, to being a danger. The shortest route between Paris and Milan was through the Simplon pass, and to keep in being the Cisalpine Republic which he had created in northern Italy, and which included the former Swiss Valtelline, Napoleon needed to occupy Berne and Valais.

Tralles had funding of 75 louis d'or annually for four years but it seems that little useful work was accomplished, before the French invaded Berne in 1798. Tralles dismantled the theodolite to disguise its value from the French forces, and the French general Schauenberg, glancing into a crate which held a jumble of brass, rejected the parts as worthless.[94]

When Martinus van Marum (1750–1837), founder of Teylers Museum, Haarlem, in the Netherlands, visited Berne, he noted in his diary:

13 July 1802. ... In the afternoon I visited Prof. Trally [*sic*], who teaches here physics and astronomy, and who on his arrival had received for this purpose from the previous Government physical and astronomical instruments to the value of about £900. Of these I saw a 4-ft Dollond telescope with an opening of about 2½ inches, a time-keeper, 2 sextants by Kerry, an astronomical circle of about 10 inches by Kerry, a 3-ft transit instrument by Ramsden, still in the case in which it had arrived. The other instruments were standing all over the floor in the courtyard and were very poorly protected.[95]

For 'Kerry' above, read 'Cary'. We learn from Blagden's brief drafts of his letters to Tralles that in 1791–92 he was overseeing William Cary's construction of these instruments, which apparently made the journey unscathed, and dealing with Tralles's payments to Cary.[96]

After the political situation had stabilized, the French intended in 1802 to establish a map of Switzerland in collaboration with Swiss scientists and to connect with the adjacent surveys of Cassini and Rizzi-Zannoni. The government ordered a new triangulation of the canton of Berne, again under Tralles's leadership, but Tralles declined to cooperate and he resigned his professorship in Berne in 1803, moving to a post at the Berlin Academy. On 8 November 1822, while on a visit to London to buy a geodetic pendulum from Edward Troughton, he fell ill and died. Hassler emigrated to the USA in 1805, to become eventually the first director of the US Coast and Geodetic Survey.

Friedrich Trechsel (1776–1849), Tralles's successor as professor of mathematics at the Bernese Academy and commissioned by the government with the trigonometrical survey of the canton, soon realized that the Ramsden theodolite was not suited to the demanding task of surveying in such a mountainous district. Two members of his team, Maurice Henry (1763–1825), head of the French engineer geographers in Switzerland, and his collaborator François-Joseph Delcros (1777–1865), maintained that the instrument was completely useless, for both astronomical and geodetical observations. They argued that there were too many correction screws, making the whole apparatus unstable, and impossible to adjust with the required accuracy. In 1811 a repeating circle was ordered from Reichenbach in Munich.

After this rejection of an initially prized instrument, it was moved into the observatory and set on a permanent wooden base, to serve as a transit instrument to fix the meridian of Berne. A mire, or signal point, was set on the Gurten, the local mountain; this point, connected with a point beneath the theodolite's axis, defined the meridian line for the triangulation. In August 1812 Maurice Henry wrote from Berne to Tranchot:

> I have just taken the latitude of Bern, and a bearing. I had the benefit of a magnificent Ramsden theodolite, the third of its type, and the only one on our continent, and a smaller but more admirable theodolite by that incomparable artist Reichenbach....[97]

By 1819 Pictet and Trechsel were corresponding on the problems associated with surveying in a mountainous country.[98] Trechsel was trying to find the best theodolite; Ramsden's great instrument was unwieldy but smaller ones were inadequate. In an article in Pictet's *Bibliotheque universelle* he remarks:

> Ramsden's great theodolite, of three feet diameter, which belongs to the Academy, cannot be transported to stations difficult of access; Borda's circle, an admirable instrument for astronomy, is not always so for geodesy, especially in mountainous country: its results are liable to be distorted by refraction and it becomes a *traitor* when the angle is too steep.[99]

By this time the assumption that only large instruments gave measurements to the desired accuracy was past; the unwieldy and costly mural quadrants, theodolites and sectors were replaced by the smaller apparatus being made in German workshops, which delivered equally accurate results at considerably less cost.[100]

The obsolete Ramsden instrument remained in Berne Observatory until about 1835, when it was dismantled. The telescopes were fixed on a simple altazimuth mount and used for student practical exercises. In 1854 a meridian circle from Georg Ertel in Munich was installed at the observatory (now in the Historical Museum in Berne). Parts of the Ramsden circle were then sold as brass scrap. The mechanic Karl Loder bought the screw microscopes and the big circle, with the idea of making a dividing machine, but this proved beyond his capability.[101]

Around 1928 parts of the theodolite were in the Berne factory of Haag-Streit, a precision engineering firm founded in 1858. In 1970 a private collector found in the firm's scrap box an English triplet objective of the late eighteenth century; its diameter and focal length matching those reported by Tralles for the telescope of the Ramsden circle. It seems, therefore, that this object glass, now in his private collection, is the only surviving part of this great instrument.[102]

Notes

1   Seymour (1980); Richeson (1966), 173–7.
2   O'Donoghue (1977); Gardiner (1977), 439–50.
3   Chatham, Royal Engineers Museum, leaflet.
4   The circumferentors and chains were made by Benjamin Cole of London. No observations were made to determine magnetic variation (then 19°W), hence the maps were aligned on magnetic, rather than true north.
5   Whittington and Gibson (1986).
6   I have failed to identify Cotton Hill. It may be a mis-reading from a lost Roy manuscript intended for Calton Hill, Edinburgh, from where Arthur's Seat is easily sighted.
7   Débarbat (1989), 47–76. Mapping in Britain was undertaken by military engineers, whereas in France this task fell to the astronomers of the Paris Observatory.
8   BL, MS Kings 270, 4–5.
9   Fortescue (1927), 1. 328–34.
10   Seymour (1980), 14. Roy's original is in RS, MS DM4.5.
11   Roy (1785), 389.
12   Portlock (1833), i–lxiv.
13   Carter (1988), 205.
14   Roy (1785), 401.
15   TNA, OS 3/3.
16   Saint Fond (1797), 1. 92–4. English translation 1799.
17   BL, MS Kings 270, 4–5.
18   RS, Council Minutes, 26 May 1790.
19   RS, DM 4.36. Duke of Richmond to Banks, 11 January 1792. The circumstances of the fire were described in *The Times* 22 December 1791, 3c. The fire started in the morning of 21 December on an upper floor of Richmond House, Whitehall. Most of the Duke's art treasures and papers were rescued.
20   Piazzi (1788).
21   Smeaton (1754), 598–613.

22 De Luc (1778), 422–3.
23 Ramsden (1783); Mudge and Dalby (1799), 1. 69–88.
24 Cardiff, National Library of Wales, MS 12415.28.
25 Close (1969), 19.
26 RS, DM 4.15 Banks to Lord Carmarthen, 8 February 1787.
27 RS, DM 4.17 Copy of a letter in French to Cassini, 18 May 1787.
28 BL, MS Add. 33272 f.27.
29 BL, MS Add. 33272 f.33.
30 BL, MS Add. 33272 f.35.
31 B,L MS Add. 33272 f.36.
32 RS, DM 4.19 Copy, Banks to Cassini, 29 June 1787.
33 RS, DM 4.20 Cassini to Banks, 27 July 1787.
34 RS, DM 4.21 Banks to Cassini, 6 August 1787.
35 RS, DM 4.23. The artist was the civil engineer, surveyor and illustrator Thomas Milne (1768–1809).
36 RS, MS CB/2/266 Draft, Blagden to Cassini, 8 July 1789.
37 Roy (1790), 112.
38 His obituarist informs us that 'he was transacting business at the War-office till 8 o'clock of the previous evening'. *Gentlemen's Magazine* 60 Pt 2 (1790), 670.
39 Harley and Walters (1977), 13–14.
40 RGO, MS 4/178. In Sotheby's sale of July 1806, after Aubert's death in 1805, it was Lot 335: 'Late Roy's very large and complete equatorial circle by Ramsden with a great variety of apparatus, bought by Bentley for £68.5.0.'
41 Mudge and Dalby (1799), vol. 1, 130.
42 James (1856), 607–8.
43 Provost, *DBF* art. 'Cassini, J.D'.
44 Wiltshire Record Office. MS 1390/2. Maskelyne Memorandum Book III, 33.
45 Seymour (1980), 16 n. 110, citing BL pressmark L.R.292.
46 NHM MS, DTC 7.74–8; 74. Roy to Banks, Lisbon 6 March 1790.
47 Turner, A.J. (1989), 55.
48 PAO, MS D2-41. 'Lorsque la nuit fut venue nous allames passer la soirée chez le général Roy, on nous presenta le thé selon la mode angloise, ensuite un fort bon souper, qui dura fort longtemps car les anglais ne peuvent pas quitter la table.'
50 Cassini (1790), 'Avertissement' [unpaginated prelim.]
51 Walckenaer (1940). The earlier account appeared as *Descriptions des moyens employés pour measurer la Base de Hounslow-Heath dans la province de Middlesex* (1787), the later one in 1795.
52 RS, Council Minutes, 18 November 1790.
53 RS, MS DM 4.32.
54 NAS, MS GD 157, 3379/25, Duke Ernst to Brühl, 15 March 1791. 'Ce que vous dites, Monsieur, l'égard de Ramsden sur le cercle entier commandé par le duc de Richmond, me fait concevoir sans doute l'éspoir, que ce cruel mais inimitable artiste songeoit peutêtre à le finir enfin pour moi.'
55 Chinnici *et al.* (2001), 4, citing Piazzi's letter of 16 August 1789 to the Deputazione dei Regi Studi: '[Ramsden] … having agreed to make another one, worth not less than 600 pounds, for the Duke of Richmond ….'
56 Andrews (1975), 2.
57 Yolland (1847), 73–4.
58 Mudge and Dalby (1799), 409–11; Seymour (1980), 36.
59 TNA, WO 52/88/375, 1795.

60 RGO 14/17 f.107.

61 Seymour, *Ordnance Survey Bulletin* 20, also TNA OS3/59; Yolland (1847), 24–5.

62 The gold Copley Medal was the Royal Society's highest award. Instituted by Sir Godfrey Copley, it was awarded, generally annually, from 1736, to the British or foreign author of some discovery or contribution. Hall (1984), 8.

63 RS Council Minutes, 12 November 1795.

64 Pictet, M.-A. (1797), footnote pp. 286–7. Pictet saw and sketched one of the new theodolites under construction.

65 Bickerton and Sigrist (2000), 474. 'Il était tout a fait *in high spirits*. Il m'a fait beaucoup d'amitiés, m'a dit qu'on avait fait depuis *six mois* des progrès très marqués dans la fabrication des instruments astronomiques. Il a en main un théodolite sur le principe de celui de General Roi; il est [...?] par un enfant, il n'a que 7 pouces ½ de rayon, et il m'a fait voir qu'une seconde pouvait s'y apercevoir. Cette quantité repond a 1/28000 de pouce. Il abandonne l'usage du niveau à boulle d'air et reprend la plomb comme plus sur. Il lui applique un microscope fort, et un prisme triangulaire latéral qui rapportant l'image à angles droits permet de caler a la fois dans deux sens. Il parait grand partisan de cette manière. Il voulait continuer de me developper une foule de choses, mais malheureusement le temps me manquait et à mon grandissime regret il a fallu nous quitter.'

66 Close (1969), 43.

67 Portlock (1833), 41.

68 James (1902), 8.

69 *The Times* 1 November 1841, 5d; 2 November 1841, 5d, 6a; 3 November 1841, e–f; 4 November 1841, 3c, &c.

70 James (1902), 2.

71 RGO 14/17 fols 172, 173.

72 Portlock (1833).

73 *The Times* 17 June 1848, 7e; 29 July 1848, 5e.

74 OIOC, MS E/1/80 Misc, fols 632–5.

75 OIOC, MS E/1/80 f.635.

76 Edney (1997).

77 OIOC, MS E/4/876 Paragraph 69 (p. 410–11).

78 Everest (1839), 20–1.

79 BL, MS Add. 33978 f.262.

80 Lambton (1803), 319–20.

81 Warren (1811), 513.

82 Phillimore (1945), 1. 1 and 204.

83 Phillimore (1945), 1.165.

84 Phillimore (1945), 1.102.

85 Phillimore (1945), 1. 205.

86 Anon (1822) 'Memoir of T. D. Pearse', 154.

87 Bickerton and Sigrist,(2000), 89.

88 Bickerton and Sigrist (2000), 55.

89 Bickerton and Sigrist (2000), 56–7.

90 Pictet (1791).

92 'Auszug aus verschiedenen Briefen des Hrn. Professor Tralles in Bern an den Herausgeber' *Allgemeine Geographische Ephemeriden.* 1/2 (1798), 241–8.

93 R. Willach, personal communication 2005.

94 Graf, (1884) vol. 1 (1884), 526–44; 530–2; *DAB*: 'F.R. Hassler; F. Cajori, 'Swiss geodesy and the U.S. Coast Survey', *Scientific Monthly* 13 (August 1921), 117–29. Zschokke 1882, pp. 13–5. Berthaut, (1902), vol. 1:350–5.

95 Lefebvre and de Bruijn (1969–76), 2. 344–5.

96 RS, MSS CB/2/529, 622, 640, 645, 661.

97 Berthaut (1902), 2. 326–7, citing Archives du service géographique de l'Armée. 'je viens d'observer la latitude de Berne et un azimut. J'ai eu l'avantage unique de pouvoir faire l'usage d'un magnifique théodolite de Ramsden, le troisième de son éspèce et le seul qui existe sur notre continent, et d'un théodolite plus petit mais encore plus admirable de l'incomparable artiste Reichenbach.'

98 Bickerton and Sigrist (2000), 661–2.

99 Trechsel (1819), 78–85. 'Le grand théodolite de Ramsden, de trois pieds de diamètre, qui appartient a l'Académie, ne pouvait être transporté aux stations de difficile accés; le cercle de Borda, instrument admirable pour l'astronomie, ne l'est pas toujours pour la géodesie, surtout dans un pays de montagnes; ses resultats sont susceptibles d'être influencé par la refraction, et il devient *traitre* quand le plan de l'angle est trop incliné.'

100 Zach (1823), 26–8.

101 Wolf (1855), 76–7 n. 28; Flury (1929), 345. I am indebted to Martin Rickenbacher for these references.

102 Rolf Willach, personal communication 2005. I am much obliged to Mr Willach for this information.

Chapter 9

Roy and Ramsden: Dispute Over his Standard of Work

> Where you have written of the inexcusable negligence of Mr Ramsden and [you] might give up the 'inexcusable' on condition of retaining the 'negligence'.
>
> Banks to Roy, 30 January 1790. TNA, WO/30/54 No. 31.

> Nor would it be consistent with common sense, that a Tradesman or Mechanic would suffer his <u>professional character</u> in particular to be publicly traduced in so respectable a place as a meeting of the Royal Society, himself being present, and not to make any reply.
>
> Ramsden to RS Council, 13 May 1790. Royal Society MM 3.30.

Ramsden's delays over the production of the great theodolite for General Roy led to serious recriminations, again aired through the Royal Society. Roy was probably by this time seriously unwell, and resentful of any delay which might prevent the accomplishment of the Anglo–French triangulation, especially where the hold-ups were on the English side. The first part of his paper on the triangulation was read at the Royal Society on 28 January 1790, chosen, as Banks explained in his letter to Roy, 'as the time of year when the Society is usually attended in the most respectable manner'. Banks's letter continued with an account of the audience reaction, and his own sentiments.

> Ramsden, who was present, seemed a little struck by the severity of your language when you described the various difficulties you met with from his advertence and idleness. Aubert, who wishes the world to be ever compounded with water and sugar without any lemon seemed to wish you had forgiven him. I maintained that you had felt infinite more pain and trouble from Ramsden than he could from you and that you was in the right to try to make him sensible that such matters ought not to pass in jest … Instruct me in case anything could be said in the Council re lecture and the words in which you have described Ramsden's conduct. I approve the whole and think he deserves more than he got, but as Humanity is the vice of the age, that is as that virtue ready extended to a vicious extreme. What must I do if I am pressed? A letter from you on the subject would be a helper in my hands of such importance to tell me whether I am to defend the whole with every drop of my blood and not give up till I am beaten in a regular battle that is a bollett or whether may yield to the enemy the outworks; that is, the epithets, as for instance where you have written of the inexcusable negligence of Mr Ramsden and might give up the 'inexcusable' on condition of retaining the 'negligence'.[1]

A Three-Cornered Argument

Following the last of the readings, on 25 February, Roy took his physician's advice and departed to the warmer climes of Lisbon. Consequently, when Ramsden put pen to paper in March, addressing Charles Blagden as Secretary of the Society, the matter was raised in Council: 'A letter from Mr Ramsden to Dr Blagden complaining of some misrepresentations relating to himself in General Roy's paper lately read at the Society, and desiring that he may have leave to peruse it, with a view to prepare a vindication of himself' was read, in Roy's absence. Council resolved: 'That Dr Blagden acquaint Mr Ramsden that as General Roy was out of the kingdom, the Committee of Papers have determined to postpone the printing of the paper till his return and that in the meantime in hopes that on the General's return his difference with Mr Ramsden may be amicably settled, the Council do not think it expedient to comply with [Ramsden's] request at present.'[2]

Roy's letter to Banks, dated Lisbon 6 March 1790, would probably have reached Banks a fortnight or so later:

> I had all along told Ramsden that I would strongly animadvert on the blameable parts of his conduct in regard to the supply of instruments, and particularly the apparatus for the chain; that the Blacksmith of Romney Marsh was to be confronted with him, who had found great difficulty in remedying the most glaring defects in the said apparatus, and who would scarcely believe that the famous Ramsden (whom it seems, right or wrong, everybody but myself is resolved to praise) could have produced such a blundering piece of business. The truth is that, if the French gentlemen had gone down to see our mode of proceeding at the outset, we should have been affronted. No man whatever, even the most zealous of Ramsden's admirers, has praised him more than I have done at all times where praise was due: but in justice to the public in general, and to the Royal Society in particular, whereof he is a member, it appears but right that the reprehensible part of his conduct should not pass without notice. This was and still is my way of thinking; but if you should judge otherwise, I beg you will have the goodness to smooth down Mr Ramsden's back, by removing the whole or any part of the asperity of my mode of expression, in the manner you think best. You have my full leave to do it. No consideration upon earth would ever make me go through the same, or such another operation again, merely from the drudgery of having to do with such a Man![3]

More words must have then passed between Banks and Ramsden, for in May Ramsden felt that he had come to the end of the road in his discussions with Banks, and wrote again to Council:

> Gentlemen
>   In consequence of a misunderstanding between the President and myself, which I am extremely sorry for, but for which I have given no just ground, all future communication between us being at an end I find myself obliged to apply to the Council of the Royal Society, to request that my answer to the accusations brought against my conduct and professional character by Major General Roy may be read at a meeting of the Royal Society as soon as conveniently can be done. I am further led to make this application from being informed by the President, that it rests with you whether my defence shall be read or not.

I am also informed it has been represented to you, that I agreed not to make any reply to those accusations as read at the meeting of the Royal Society. How this mistake can have happened, I cannot imagine, I never gave the least grounds for such representation, nor would it be consistent with common sense, that a Tradesman or Mechanic would suffer his <u>professional character</u> in particular to be publicly traduced in so respectable a place as a meeting of the Royal Society, himself being present, and not to make any reply. It would to all intents be allowing the justice of those accusations but for which in this case there is not the smallest foundation, and I leave it with the Council to judge what must be a person's feelings in such a situation, who has thro life been more attentive to the improvement of his Art, than he has been to the accumulation of wealth.

I do not mean to trouble the Council with the particulars of a private conversation requested of me by the President, because any misconception on either side can never preclude me from being heard in my own defence. But the statement of the result of that conversation in the President's letter of May the 4th conveyed to me a very different idea, from that contained in his letter of May the 6th; in his letter of May the 4th he says "When I said over to you the alterations which General Roy had at my particular request, consented to make in his paper, you declared that you was [*sic*] fully satisfied, and you assured me that you would not offer any reply to those strictures upon your conduct which I had prevailed on him to leave out."

In the letter of May the 6th he says "In the conversation you had with me on the subject of General Roy's paper, you fully gave up your wish that it might be printed in the state it originally stood; and after you had heard my reasons to the contrary, all intention to answering that parts which had been at my request struck out of it, you certainly did persist in your intention of answering such objections to your conduct or your instruments as was to be printed, in which I told you I thought you was [sic] perfectly justified."

I do not mean to insinuate any contradiction in the two letters, but the first I understood as engaging me not to give any answer at all, while the second only pointed out the manner in which that answer is to be given; and although I can never be brought to think, there is any impropriety, in desiring my defence to be read at the same place, and to the accusations as they were read before the Society, yet I have no objection to accept the terms offered in the President's letter of May 6th, provided I am permitted to see the General's paper, which is essentially necessary to enable me to comply with the terms proposed; for my own sake I will take care that my language be <u>decent</u> and my style <u>temperate</u> agreeable to the caution conveyed to me <u>very politely</u> in the President's letter, and tho' I have in those respects been treated with little ceremony.

I have the honor to be with the greatest Respect

Gentlemen

Your most humble servant

Piccadilly   J Ramsden

30 May 1790[4]

Council Minutes for 12 June report:

A paper containing a vindication of Mr Ramsden's respecting some charges brought against him by General Roy in the last account of his trigonometrical measurement was read, and also a letter from Mr Ramsden to the Council desiring that his said vindication may be read as soon as convenient.' Council resolved, 'That the Secretary acquaint Mr Ramsden that the Council do not think the paper in its present form proper to be laid before the Society, but that if he will draw up a plain statement of the facts free from personalities, they see no objection to its being read to the Society.[5]

Ramsden Explains his Design Concept

This 'plain statement of the facts' provides a marvellous account of Ramsden's thinking on this novel instrument, and the technical problems encountered and overcome during its construction:

> When I took in hand this instrument with which General Roy made his survey, my first object was to consider what were the defects, and the improvements to be made in this and other astronomical instruments both as to construction and the application of different materials, a quadrant of three feet radius was the instrument proposed by the General but having for more than 20 years back considered circles as the only form that could be confided in with certainty for all kinds of astronomical observations, I had constructed them with cones, and with plates having pillars between them, which we have been accustomed to call Jacob's Ladder, thereby obtaining strength without additional weight, a most essential part in the construction of the larger sort of instrument, and seeing how much I had improved the method of graduating, a similar improvement became necessary in every part of the instrument.
>
> 1st The center work, which being made hitherto with bell metal, the parts rubbing against each other, could never be brought so close as to be perfectly free from shake, without being subject to gnaw and consequently to spoil itself.
>
> 2ndly I inverted the position of the vertical axis, which preserved the stability of the instrument, and prevented sand or grit from lodging and spoiling the center work.
>
> 3rdly The method of reading off angles by nonius divisions compounded the error of the divisions on the limb with those on the nonius; and by reading off by micrometer screw are lost, in a great measure, the advantage of circles, because, when once the circle is moved by the micrometer to make the coincidence of some divisions on the limb with one on the vernier, we lost the reading off at the opposite vernier, it not being possible to restore with certainty the position of the circle, to that it had at the time of observation, and from thence to read off at the opposite vernier, moreover, the imperfect construction of the micrometer screw, or the vis inertia of the circle moved by it, either of these are sufficient to render such screws inapplicable to their purpose.
>
> 4thly Another very essential object of improvement was that of the micrometer; hitherto they were made to slide in a kind of dovetail, so that from the position, and from the play of the screw, there was no consistency when the screw was turned in opposite directions; it thence became necessary to give the instrument a totally new construction in every part and by the application of a spiral spring they seem now to be sensibly perfect, that is to say, the sensible difference is visible when the micrometer screw is turned in the opposite direction.
>
> These with many other mechanical improvements equally applicable to almost every other astronomical instrument I applied to this instrument among others, illuminating the wires in the focus of the object glass by perforating the axis. This many years before I had invented, and first applied it to a small transit made for General Roy, with which he some time afterwards drew the line of the base on Hounslow Heath.
>
> In General Roy's description it will be seen that to perfect the center work I made the collar of steel, hard as it was possible to make it, and by driving it into a collar of

brass, I could grind it and turn the outside with the same facility as if the whole was made of brass.

The third defect I obviated by the application of microscopes but to do this I was obliged to alter the construction of the eyeglasses, in the present state they were made with two eyeglasses having an image formed between them, by this means pencils towards the extremes of the field were more refracted than those nearer the center; consequently spaces passed over by a wire in the micrometer could not be proportional to those on the object. This rendered the microscopes then made perfectly inapplicable to these purposes. I therefore contrived a new system of eyeglasses, on principles diametrically opposite to those laid down by optical writers who seem not to have considered this part of optics with due attention and which they deemed impossible; yet I have corrected both the aberrations for the different refrangibility of sight and for the spherical figure of the lenses although both the lenses are between the eye and the image by this position. An image becomes similar to the object and a micrometer can be applied with great advantage.

With regard to the glass rods, it was Col Calderwood that suggested the preference of glass to any other material, by reason of its less quantity of expansion, but the mechanics was my own construction.[6]

Countering Roy's Accusations

Ramsden meanwhile sat down to itemize each of Roy's accusations and to counter each with his defence:

Nothing could equal my surprise on hearing the charges brought against me, and misrepresentations contained in a paper presented to this Society by Major General Roy. I was the more affected by it as coming from a Gentleman with whom I considered myself in Friendship, and, who had many obligations to me for my assistance in the business that is the subject of that paper.

Nor is it to this Society alone that I am thus traduced, but also to His Majesty, to whom a copy of the above paper was presented, previous to its being given to this Society. After such public abuse, it becomes me both as a subject and a Member of the Royal Society, to give a true statement of the facts alluded to in the General's paper, which will clearly show the injustice of the aspersions therein contained, and the very different treatment I had a right to expect from its author.

Not having seen the General's paper nor having been present at the reading of the whole of it, and having only my memory to guide me with respect to the parts I did hear, I hope for the indulgence of this Society, if I should not be perfectly correct in stating the General's complaints against me.

The first charge to bring is, that I was the cause of great and unnecessary delay, and instead of assigning the true cause, the terms in which the General expresses himself, imply the delay to have been intentional, and on this head I need not call to the recollection of the Society how liberal he is in his reprehensions of me. The true state of the case is this:

After I had contrived and executed the mechanism of the Glass Rods and of the Chain the General applied to me to construct and make for him the best instrument I could within a limited price, for measuring horizontal angles. In consequence thereof, to the neglect of a more lucrative business and chiefly with a view to the advancement of science and to the honour of this Country, I constructed an instrument on principles very different from what had ever been done before. Among other improvements, that of the center work, a most essential part in these sort of instruments, was new, and more perfect than anything of the kind ever executed. But having made the Instrument ready for fixing on the nonius plate, and for the dividing, it occurred to me that further improvements might be made, and I took an opportunity to mention to the President of this Society and to the General, that the Instrument might be rendered superior in point of accuracy to anything of whatever radius yet made, if the expense necessarily incurred could be allowed. This being agreed to, I contrived microscopes of a new construction, one of which was placed at each end of a diameter of the horizontal circle, and to these I applied micrometers of a new construction also whereby the divisions on this instrument only 18 inches radius, could be read off to 1/2 a second, which is about 1/24,000 part of an inch. Perhaps it may not be improper here to mention the advantages arising from this position of two microscopes: an observer is no longer obliged to rely on the account of his instrument. He is thereby furnished with means to examine the accuracy of it himself and to detect and allow for its errors. Indeed it has been with me an invariable rule, to construct my Instruments in such a manner as to shew their own errors whatever they might be, that an observer may be certain how near his observations can be depended on so far as relates to his instrument. I also improved the mechanism of many other parts of the Instrument besides those already mentioned without which this accuracy in reading off divisions would have been of little use. And I further contrived a new method of dividing equally applicable to circular instruments and to quadrants, whereby a degree of exactness can be obtained beyond what had hitherto been thought possible, and which can be executed by workmen little practised in these operations. But the best judgement of the accuracy of this Instrument may be had, by comparing observations made therewith, with observations made with the astronomical Quadrant at Greenwich. Altho one is of 8 feet and the other only 18 inches radius it may not be improper to mention here as a proof of the durability of the center work that neither the jolting of a carriage on which the Instrument was carried from one station to another, along cross roads, nor its being exposed for months together to all kinds of weather during which the high winds often covered the whole to a considerable thickness with a fine and sharp sand, did it any injury; though it had caused an error of 1/20,000th of an inch, the microscopes must have discovered it.

Those acquainted with the application of new principles in the construction of those more accurate instruments are sensible how much time must necessarily be employed to bring them to the perfection of the one in question. The least accident is often attended with considerable expense and loss of time to rectify it which was unfortunately the case here. After the instrument was graduated an accident happened by one of the workmen that obliged me to take out the whole of the dividing and to begin entirely anew. These circumstances well considered will preclude the imputation of any unnecessary delay. But had an instrument of this sort even been executed before, or could it have been made elsewhere, there might have been some ground for complaint. But surely a mechanic who employs his mind, neglects a constant business, and spends his money for the promotion of science, and after all receives abuse, has much greater reason to be dissatisfied.

The apparatus for the Chain affords the General a second ground of complaint, as if it had been executed in a bungling manner. The truth shortly is the General applied to me to contrive some means to mark the extremities of the Chain with great exactness, when employed in measuring the base on Romney Marsh. Not having the Chain in my possession, nor remembering the width of the handles, the screws in the apparatus for the above purpose happened to be made rather too near for the handles to pass between them, however the General, with the assistance of the Blacksmiths, had the ingenuity to cut off the sides of the handles and thereby correct this error in the manner represented in his drawings.

He further charges me with giving him a bad level, in place of the one that fell from the scaffold on St Ann's Hill. The level I substituted was one I had selected and examined in order to apply it to a large Transit Instrument at that time in hand. I am further confirmed in the goodness of that level by the trials made on it by two of my workmen at the time the instrument passed through London on its way from St Ann's Hill, Norwood, to Greenwich, they both of them found it exceedingly good and very sensible, a property essentially necessary for the Pole Star Observations. But in common with other good levels, care and attention is necessary to adjust the Instrument with it. Unfortunately for me this level was broke before the Instrument returned from Dover which prevents the possibility of any other kind of proof but that of my workmen. On making them a third level, imagining the principal object was accuracy, I lamented that the level was not quite so sensible as the former. However I now find myself mistaken in my opinion of levels for the General's purposes; as despatch was his principle object the instrument could be adjusted with less care, and in much less time, by the later level than by the former.

A 4th complaint is against the microscopes or the fixing of them on the instrument.

When at St Ann's Hill, in consequence of the box containing the apparatus falling from the scaffold, the General mentioned that on a sudden one of the microscopes had become much worse than the other. On examining it I found a brass plate thro which it passed and was fastened, so much bent as to throw the divisions almost out of the field. No doubt the force whether by a stroke or by pressure, must have been very considerable to have bent a plate of brass of that thickness in the manner it was. This fact will be confirmed by the workmen employed to repair this part of the Instrument. Surely an instrument maker cannot be expected to guarantee an instrument of this sort out of his posession against every injury that may happen whether from accident, or from the unskilfulness of persons that may use them improperly.

5thly He complains of delay in making the prism eyetube.

I can with truth say I never actually promised or intended to make him one. They gave me much trouble and were very expensive. The one the General has I had made for a very worthy Friend and it was not till after much persuasion that I let the General have it.

Having thus stated to the best of my recollection the charges brought against me by the General, and shewn, I trust, how little foundation there is for them, I cannot conclude without pointing out to this Society, that the General has done me nearly as great injustice by his silence as by his groundless charges he has brought forward. In no part of his paper does he own or acknowledge that every part of the apparatus is of my invention. A Stranger or a Foreigner pursuing his account would never suspect that I had any other merit than that of finishing with a bad level or making a bungling apparatus for his chain; nay it would seem as if he had some difficulty in instructing me how to make an instrument sufficient for his purpose. Now this is the more extraordinary as through the whole paper there appears the greatest readiness to every even the most minute assistance. Why am I treated in so different a manner I will leave the General to explain. To say nothing of the Instrument itself I might have expected a very different requital, for the assistance I gave him in his description of the mechanism of the several parts thereof, and explaining to him at length the construction of the microscopes and their micrometers. And that in so complete a manner that I may venture to affirm any good workmen may from thence execute a similar instrument.

To the justice of this Society I therefore now appeal, and require that my defence may be as public as the attack made on my character, and this not for myself alone but for the sake of repressing in future personalities of this nature equally prejudicial to the advancement of science and unworthy of scientific men.[7]

If Roy's cutting words had not stirred Ramsden to respond formally and at length to the Royal Society, we might never have learnt so much about his design concept and the way in which it was constructed to become such an admirable working instrument.

Notes

1   TNA, WO/30/54 No. 31. Banks to Roy, 30 January 1790. The paper was published in *Phil. Trans.* 80 (1) (1790); 111–270 and Plates 5–15. For Roy's comments on Ramsden see 111–12; 116–17; 131–2; 155; 263.
2   RS, Council Minutes, 7 and 11 March 1790.
3   London, NHM Archives, MSS DTC 7.74–8; 74–5.
4   RS, MS MM 3.30.Ramsden to RS Council, 13 May 1790.
5   RS, Council Minutes, 10 June 1790.
6   RS, MS DM.4.44.
7   RS, MS MM 3.30, pp. 1–9.

Chapter 10

Final Years

Ramsden has already been ordered to make the trumpet for the Last Day so that it
will be ready in time.

W. Promies, ed. *Georg Christoph Lichtenberg, Schriften und Briefe*.
Vol. 1 (1968), 759.

… yet after my death, which may not be long …

Ramsden to Harrison, 15 March 1800, TNA, CRES 2/730.

Rebuilding 199 Piccadilly

In his final years Ramsden embarked on a major undertaking, to rebuild his
property at 199 Piccadilly, at a cost of £200. A survey dated 18 October 1799
yielded a detailed ground plan and the statement that the house had a basement
kitchen, three stories above, and a garret in the roof. The timber workshop at
the rear had four stories. The negotiations took some time, for the Crown lease
for 60 years and 36 days and a yearly rent of £51, 7s is dated 30 April 1798 and
refers to 199 Piccadilly and workshops to the rear, 'the messuage lately rebuilt'.[1]
Another copy of the lease, still among the Crown Estate documents, is also dated
30 April 1798 and states that the annual rent is to be £55, 11s, with a requirement
to show each year the certificate of £700 fire insurance cover. Ramsden's insurance
policy of 1800 listed the rebuilt house, covered for £700, the brick workshops for
£1,000, with £400 cover for utensils and stock in the house and the same in the
workshop, giving a total cover of £2,500.[2] In his will, written in March 1800,
Ramsden bequeathed these premises to Matthew Berge on condition that he
paid the outstanding bills.

Throughout the 1790s Ramsden's workshop continued to be as busy as ever
and new apprentices were taken on. Among the larger items in hand were zenith
sectors for the Ordnance Survey and for the East India Company, transits for
Seeberg and Paris, and the long-overdue mural circle for Dunsink. It is uncertain
how far the construction of the large instruments had progressed. Ramsden
himself had been ill during 1787 and the dispute with Roy in 1790 must have
exacerbated the problem. He was called to serve on the Second Middlesex Jury at
the Old Bailey session starting 26 May 1790.[3] He was one of the 'artists' elected
to the Society of Civil Engineers when it was reconstituted in 1793, but due to
illness, he seldom attended meetings. Also in 1793 he was one of three eminent
foreigners elected to the Imperial Academy of St Petersburg.[4] Other difficulties
were caused by exceptional weather, with consequent interruptions to outdoor
work, transport, and delivery of goods. In October 1791 a high tide flooded parts

of Westminster, while the series of cold winters continued. In 1788–89 the tidal Thames below London Bridge was frozen from November to January. Jesper Bidstrup's letter of 6 March 1789 to his patrons in Denmark expands on this event: 'We have had an extremely severe winter, the Thames has been completely ice-bound, some thousands of people have taken pleasure in walking on the ice, they have booths and stalls and so big a fire on the ice that they have roasted two entire sheep, the like of which no one can remember happening for quite some time.'[5] Eventually the ice thawed, the loading quays became accessible, so that ships could come up, but the cost of coal and provisions rose considerably and only fell with the advance of spring. In 1796 the temperature over Christmas fell to –6°F, with the Thames above London Bridge frozen.[6]

Political Troubles

News would have reached Ramsden in July 1791 of the Birmingham riots, culminating in the burning down of the house of the Unitarian theologian and scientist Joseph Priestley (1733–1804), and attacks on other members of the Lunar Society. There was equally disturbing news from across the Channel, some of it brought by those seeking refuge in England. The impact of the French Revolution in England was, to quote the historian R.B. Rose, 'immediate, drastic and multifarious' and led to a shifting of other political and religious causes.[7]

   At this time Birmingham was still largely a town of small masters' workshops. The opening of the Birmingham Canal, linked to the Grand Trunk by 1772, was drawing in more heavy industry but Matthew Boulton's Soho works with its 700 employees was still untypical. Dissenters, many of them Unitarians, played their part in county and municipal government and worked alongside Anglicans in the Commercial Committee founded in 1783 to promote Birmingham's trade and manufacture. Religious antagonism was more marked in the lower classes, but the trigger which fired what became known as the Priestley riots, was political, originating in a dinner held on 14 July to celebrate the fall of the Bastille. Priestley himself, sensing trouble, decided not to attend, and the gathering was chaired by the Anglican James Keir, a chemical goods manufacturer and Lunar Society member. A crowd assembled outside the hotel where the dinner was held, voicing hostility although without any defined target. After causing some local damage, the crowd proceeded to Priestley's home at Sparkbrook, there ransacking, burning and destroying an immensely valuable collection of manuscripts and scientific apparatus. Among the losses subsequently itemized was a wooden frame holding several prisms, by Ramsden.[8] In the following hours and days attacks were made on the houses and workshops of other Lunar Society members, both Anglicans and dissenters, causing prudent industrialists such as Boulton, to secure their properties against assault.

   More worrying news came in 1797. News that a small French detachment had landed in Wales triggered alarm throughout the kingdom and a run on the banks, with cash being paid out against notes which were then passed to the Bank of England and immediately cancelled. The Chancellor ordered that no further cash would be issued until the situation was regularised but the Bank of

England's declaration that it was not in financial difficulties inevitably persuaded people that the opposite was in fact true.[9] In the course of a letter to the Duke of Marlborough Ramsden expressed his own concerns:

> Affairs with us common people wear a very unfavourable aspect. Bank notes are at 5/6 percent discount. I don't think [the] company will ever get up again. Everything I have got is there and altho there is no possibility of saving any part of it by being in different annuities secured only on the Bank … I have my health and faculties and am willing and must begin again.[10]

The worst did not happen; there was no major French invasion, the London merchants declared that they would continue to accept notes and Ramsden did not lose his entire savings, as the probate of his will indicates.

Portraits of Ramsden

Alongside these demands on his time Ramsden's interest in and competence with optical devices led to his collaboration in the 1790s with the surgeon Everard Home FRS (1756–1832), later first baronet and first President of the Royal College of Surgeons.[11] It was Everard's brother, the artist Robert Home (1752–1836), who during a brief residence in London, painted the famous portrait of Ramsden, now the property of the Royal Society (see Plate 1). This portrait belonged to Everard, who bequeathed it to his son Sir (James) Everard Home FRS (1798–1853), a captain in the Royal Navy. On 14 June 1850, shortly before he sailed in the emigrant ship HMS *Calliope* for Australia, Home wrote to the Royal Society:

> Sir – I have the honour to offer for the acceptance of the Royal Society, to be suspended in the Meeting-Room, two pictures, one of John Hunter, the other of Jesse Ramsden, optician, both of them painted by the late Robert Home Esq.

At its meeting on 20 June Council resolved that Sir Everard Home's present be accepted, and that the thanks of the Council be conveyed to him.[12] This was a most fortunate occurrence, for after landing at Sydney, Sir Everard went on a voyage into the Pacific, calling at Fiji, Tonga and the New Hebrides. He returned unwell to Sydney on 29 October and died there on 1 November 1853.[13]

A faithful mezzotint engraving of Home's portrait, by John Jones (1745–1797) was published in 1791 (see Figure 10.1). A stipple engraving by Knight, published in 1803, appeared as the frontispiece to the English translation of Piazzi's biography (see Figure 1.1). Another portrait, drawn by H. Estridge was published in May 1801 as a stipple engraving by Charles Turner (1774–1857) who at the time was apprenticed to Jones (see Figure 10.2). Only Ramsden's posture matches; the face is unrecognisable, he wears indoor clothes, and is seated at a table, dividers in hand drawing on a sheaf of papers. Among a collection of portraits displayed at Palermo Observatory is 'Gesse Ramsden nato in Halifax', commissioned, along with those of William Herschel and Lalande, from Giuseppe Velasco (1750–1827) and delivered in the summer of 1791. Beneath Ramsden's head and shoulders, a banner covers a telescope mounted on a circle said to be that of Roy's geodetic

Figure 10.2 Jesse Ramsden, by H. Estridge. Stipple engraving by C. Turner 1801.

Figure 10.1 Jesse Ramsden, by J. Jones after R. Home, 1790. Mezzotint.

theodolite. Understandably neither the great man nor his instrument bear any resemblance to real life, since Velasco could not have seen Home's original nor Jones's engraving of 1791.[14]

Medico-Optical Experiments

Home's eminent elder brother-in-law, the surgeon and anatomist John Hunter (1728–1793), subject of the portrait mentioned above, had announced that he would take the structure of the crystalline humour of the eye as the subject of his Croonian Lectures. Although the several components of the eye in human and certain animal species had been ascertained from dissections, the way in which images were passed through the different substances, apparently without suffering dispersal by refraction, was poorly understood. Hunter's animal dissections had shown that the fibrous nature of the crystalline humour allowed it to adapt the eye to various distances and Hunter believed that the same held good for the human eye. He was delaying his investigations on the human eye until Ramsden could assist him, but on 16 October 1793 he suffered a heart attack while at his clinic at St George's Hospital, Knightsbridge, and died there.[15] Home, who had been trained by Hunter and continued to work closely with him, took over the Croonian Lectureship and announced that he would continue Hunter's enquiries in this field. In his first lecture, delivered in 1794, Home stated:

> In prosecuting this enquiry, I consider myself to have been fortunate in having had the assistance of my friend Mr Ramsden. It was a subject associated with his own pursuits, and one which had always engaged his attention; he was therefore particularly fitted, both by his own ingenuity and knowledge in optics, for such an investigation.[16]

We find in the Society's record of the first of these lectures, given on 10 November 1795, that Dollond (presumably Peter) was present; at the second, on 17 November 1796, we know that Ramsden was present because he is recorded as having introduced a stranger. Home related Ramsden's opinion that the crystalline humour served to correct the aberration arising from the spherical figure of the cornea, producing the same effect that an achromatic object glass delivered less perfectly, by proportioning the radii of curvature of the different lenses. He thought that the lens, with its density decreasing towards the periphery, also corrected spherical errors. Both men agreed that this might be tested on a person with one good eye but lacking the crystalline humour in the other.

The opportunity came in November 1793, when Home was treating a young man who had had a cataract removed. He was tested for his ability to read, with the defective eye, text held at various distances, then assisted with plano- and double-convex glasses, in daylight, and by candlelight. More tests were made on the same subject a year later, this time with the assistance also of the scientifically-minded Sir Henry Englefield. Ramsden then suggested that as the eye changed focus, there would be a visible change in the curve of the cornea. He went to Home's clinic at St George's Hospital to watch Home manipulating a cornea from a recently-deceased body, and said that he would contrive an instrument to examine the cornea on a living subject, while the eye changed focus. Seven months passed; Home was concerned to explain to his Croonian audience that it should not be assumed that Ramsden had lost interest, only that he was fully occupied on other aspects of his business.

In due course, on 31 July 1794 Ramsden, Home and Englefield met in Ramsden's upstairs front room, which gave a prospect down Sackville Street opposite, and where he had set up his test apparatus. This consisted of a board, set up vertically a few feet from the window, in which a square hole had been cut, supporting the patient's face so that his left eye protruded forward of this frame. Outside the board was fixed a short-focus lens or microscope, directed at the corneal apex and free to move forwards, vertically and horizontally. Between the eye and the window Ramsden had fixed a brass plate with a small hole. The subject was asked to focus first on the rim of this hole, then on a chimney located 235 yards up Sackville Street. The experiment was repeated on each of those present and it was always found that the cornea's shape changed with the focus, protruding more for close viewing, flattening for distant viewing. Home concluded that the crystalline lens prevented reflections in the passage of rays through the various substances of the eye, and that it also corrected spherical aberration.

While the crystalline humour displayed a fibrous structure which Hunter and Home had taken as permitting focussing in certain animals, Home now needed to discover how the cornea was able to change shape. In his second Croonian Lecture 'On muscular motion', delivered in 1795, Home explained that Ramsden had again assisted and had made all the optical experiments, which now measured the cornea's convexity. This was achieved by means of an image reflected from the corneal surface and viewed in an achromatic microscope with a divided eyeglass micrometer, the device which he had described to the Royal Society in 1779. This new apparatus was in effect the first ophthalmometer.[17] The first trials were made using convex mirrors as artificial corneas. Two mirrors, one four-tenths, the other five-tenths of an inch focus, had their flat surfaces roughened and blackened, to prevent an image being seen from both surfaces. One mirror was stuck on a piece of wood 12 feet from and directly opposite the window. A board 3 feet long and 6 inches wide was placed perpendicularly against the window sash, and its image reflected from the mirror on the object glass of an achromatic microscope, with a divided eye-glass micrometer. The two images were separated by means of the divided eye-glass until the black line indicating their surfaces of contact was as small as possible. When this effect was produced on the mirror of four-tenths focus, the other mirror took its place and the contact of the two images was then increased, corresponding exactly to the difference between the convexities of the two mirrors. Ramsden then made measurements with Home as the subject.

Ramsden next sought to determine the amount of change which could take place in the cornea before his instrument detected it. He set up two mirrors, one with a focus of four-tenths of an inch, the other four-hundred thousandths of an inch, the difference in size of their images being just visible in the micrometer. But this delicacy of measurement was impossible on the human eye, being spoilt by the subjects' involuntary eye movements and their tiredness brought on by the tests. Further work was undertaken at the clinic by two of Home's students, experimenting on newly-dead eyes, from which Ramsden made his calculations. Home and Ramsden subsequently came to accept that accommodation to changing focus took place simultaneously in the cornea, the axial length within the eye, and in the crystalline lens.[18]

In his next Croonian Lecture, coincidentally delivered on the anniversary of Ramsden's death, Home spoke at some length on Ramsden's virtues, his interest in the behaviour of the human eye (which Ramsden seems always to have considered as equivalent to glass lenses), and his willingness to assist Home with the provision of apparatus. The content of the lecture dealt with the ability of the eye to focus to different distances, even when deprived of the crystalline lens.[19] In the 1790s the physician and physicist Thomas Young (1773–1829) was also investigating the muscular fibres of the crystalline lens. In his Bakerian Lecture of 1800, published in 1801, Young speaks of Hunter's earlier work in this field, 'continued by Home and Ramsden – whose recent loss this society cannot but lament –'[20] Young gave a description of his apparatus for testing ocular movement, for which he is generally credited as the inventor of the ophthalmometer or keratometer.

Ramsden's Dispute with Peter Dollond over the Invention of the Achromatic Lens

Historians past and present have assumed that Ramsden enjoyed a happy relationship with John Dollond senior and, after his death, with his sons Peter and John. This may have been the situation during Ramsden's courtship of Sarah Dollond and their early years together, but the separation extinguished any good feeling on Peter's part. His animosity may have been sustained by Ramsden's growing personal fame, for though the Dollond optical products were widely admired, and indeed Peter's trade card boasted two royal appointments, the firm's output lacked the range and innovation of Ramsden artefacts. Ramsden was elected FRS on 12 January 1786, but Peter Dollond was never proposed for election, and this may have reinforced his hostility.[21]

During the 1780s a feud arose between Ramsden and Peter Dollond over the identity of the inventor of the achromatic lens. Images seen through an achromatic lens were free of the halo of colours which reduced their clarity – though spherical aberration remained to be corrected. This issue of invention had arisen in the late 1750s and entailed a dispute between John Dollond senior and the Swedish professor of mathematics at Uppsala, Samuel Klingenstierna (1698–1765), all of which took place before Ramsden had entered the optical instruments trade. The European mathematicians working on this problem, had difficulty in procuring glasses of the appropriate density for their experiments.

The English patent granted to John Dollond senior on 19 April 1758 was for 'making object-glasses of refracting-telescopes, by compounding mediums of different refractive qualities'. Dollond's letter and example of his achromatic telescope, was produced by James Short, the Scottish maker of reflecting telescopes, for the Royal Society's members to examine and admire. Council awarded Dollond the Copley Medal, and on 28 May 1761 he was elected FRS.

Patents were valid for 14 years, and it was up to the patent-holder to guard his rights during that time, prosecuting those who infringed them. During John Dollond's lifetime, these rights were shared between John Dollond and his partner, Francis Watkins, who had contributed the substantial legal fees needed to secure the patent – about £100 – in exchange for a share of the future profits. Meanwhile, at least two London optical instrument makers had been making compound lenses of crown and flint glass since the early 1750s, and they continued to do so after the patent was granted, without any objection from John Dollond.[22]

Chester Moor Hall – the Other Inventor

According to Ramsden, the earlier inventor of achromatic lenses was Chester Moor [More] Hall (1703–1771), of Moor Hall in Essex, a minor landowner in that county where he sat as a magistrate. Nothing is known of Hall's schooling prior to his legal education in London. He was admitted to Inner Temple in 1723/4, called to the bar in June 1734, and to the bench in 1764. Between 1740 and 1752 he occupied chambers in Tanfield Court, in the Temple.[23] Hall is not known to have belonged to any of the learned or polite societies, nor have any of his letters or papers been traced. He was unmarried and died intestate, his sister Martha being responsible for the administration of his estate; his library and another were sold together, making it impossible to trace his interests through the books he acquired. On his monument in Sutton church in Essex, set up by Martha, Hall was described as 'an able mathematician'.[24]

What roused Hall's interest in the subject of optics is unknown. Brian Gee makes the plausible suggestion that he attended the lectures and demonstrations which were being given in the City of London by various scientific instrument makers and visiting speakers.[25] Presumably Hall did not passively accept Isaac Newton's assertion that the achromatic lens was an impossibility, and from experiments reached the broadly correct conclusion that a compound lens made of two types of glass of differing refractive index would correct this chromatism and throw an image largely free from spurious colours. By 1733 he had approached the London opticians Edward Scarlett (c.1677–1743) and James Mann (c.1685–1756), requesting from one a convex lens of crown glass, and from the other a concave lens of dense lead crystal. It so happened that both opticians contracted the job to the glass-grinder George Bass (c.1692–1768) of Fleet Ditch, who – the story goes – realising that he had been asked to make two lenses which perfectly fitted together, and gave a colourless image, discovered that the buyer was a Mr Hall. A similar story was told in September 1777 to the Danish astronomer Thomas Bugge, who was in London, touring the nearby observatories and shopping at various instrument makers. On 15 September the translation of his diary note reads:

> I went to see the instrument maker A[ddison] Smith, Strand, near Charing Cross … He told me that, a long time ago, Mr Hall, a gentleman and intending purchaser, began to join two pieces of glass in order to avoid miscolouring and that he had the idea from Newton's Optics Book I, VIIth proposition.[26]

In his account given many years later, Jesse Ramsden named the optician James Ayscough (c.1719–1759) as the person who regaled John Dollond with an account of a marvellous telescope in his possession around 1755–56. Watkins, in a letter to Ramsden referring to a case heard at King's Bench under Lord Mansfield in 1763, asserted that Mansfield had established Hall's priority as first inventor:

> It was clearly proved that Mr Hall had several object glasses ground by Mr Bass and others of the trade, many years before the late Mr Dollond began to apply himself to grind any glass. Mr Hall in court produced several object glasses, and the court admitted him the prior inventor ... Mr Bird and myself went to Mr Hall's house near Rochford in Essex. He shewed us a very great variety of glasses, and had some made as long ago as 1732 or thereabouts. [27]

Peter Dollond's Battle with the Opticians

John Dollond died intestate in November 1761. Peter Dollond administered his estate and inherited his father's patent. He fell out with Watkins over the division of the rights, and after litigation at King's Bench, bought Watkins out for £200. He then embarked on a series of actions against those opticians who were making achromatic lenses without permission. The whole tangled history of this matter will be made clear in Brian Gee's forthcoming account of the Watkins family. The London trade, backed by the Spectaclemakers Company, made strenuous objections to Peter Dollond's action, claiming that their members had been making and publicly selling achromatic telescope lenses since Hall's invention became known. In a judgement made in February 1766 Lord Camden found for Dollond, declaring in a phrase which has since passed into patent case law, 'it was not the person who locked up his invention in his scrutoire that ought to profit by such an invention, but he who brought it forth for the benefit of the public.'[28] Peter Dollond's achromatic telescopes, with his correction for spherical aberration, found a ready market within Britain and across Europe.

The Dollond patent expired in 1772. In France and elsewhere in Europe, however, concerns with optical theory and practice continued to generate reams of correspondence and learned papers, most authors acknowledging Euler and Klingenstierna as the founding fathers of the achromatic lens. In 1789 Peter Dollond could bear no more; he wrote a long paper 'On the achromatic telescope' which he sent to Nevil Maskelyne to lay before the Royal Society where it was read on 21 May 1789. Peter Dollond made his position absolutely clear at the outset.

> Seeking to correct inaccuracies or false representations I was led to these reflections by having seen some accounts of that discovery in different publications which were related in a manner that lessened the merit of my late father Mr John Dollond, and gave it to others, who never thought themselves in any manner entitled to claim with him, or even appeared to be inclined to do so ....[29]

Jesse Ramsden, Hall's Champion

Dollond's claim roused Ramsden to a passionate defence of Hall's priority. In a manuscript headed 'Some observations on the invention of achromatic telescopes', he wrote:

> In consequence of a paper respecting the invention of achromatic telescopes read at a late meeting of the Society, (on 21 May 1789) I beg leave to trouble them with the following observations, which I think myself called upon to bring forward, in justice to a person of extraordinary merit who from a love of retirement, and the little thirst he had for public fame, was not so well known to the learned world as he deserved to be. Chester Moor Hall Esq., of the county of Essex is the person referred to, whom I have always been accustomed to consider as the first inventor of achromatic object glass. I knew him personally several years before his death which happened about the year 1767 ... [*recte* 1771] Mr Hall found great difficulty in getting the different parts of his object glass wrought by the opticians of that time and he applied to two different persons to make the concave and convex parts of the object glass probably from some jealousy respecting his invention but they employing the same out of door workman to do their business whose name was Bass it came to be known that the glasses were worked to fit one another and he learnt from his employers that they were intended for the same person, Mr Hall. ... Two or three years previous to the [Dollond, 1758] patent while I worked with Mr Burton I perfectly remember an account being given to him by Mr Ayscough of the wonderful performance of a telescope in his possession ... Bass, who made reading glasses and mirrors for me until the time he died ...

Quoting from Watkins' letter referring to the King's Bench trial of 1763, Ramsden declared:

> It was clearly proved that Mr Hall had several object glasses ground by Mr Bass and others of the trade, many years before the late Mr Dollond began to apply himself to grind any glass. Mr Hall in court produced several object glasses, and the court admitted him the prior inventor ... Mr Bird and myself went to Mr Hall's house near Rochford in Essex. He shewed us a very great variety of glasses, and had some made as long ago as 1732 or thereabouts.

Ramsden also drew on a letter from the optician Addison Smith (d.1795), then retired and living in Charlotte Street, north of Oxford Street.

> 4 June 1789. Dear Sir [to Ramsden], I am honoured with your letter respecting the achromatic object glass, for which the late Mr Dollond obtained the King's patent about the year 1758, and as you desire my sentiments upon this matter, I shall in the most candid and impartial manner lay before you what I know relative to it. In the year 1763 I had the honor to be aquainted with Chester More Hall Esq., a gentleman of great candour and scientific knowledge, who told me that about the year 1729 he had made some prismatical experiments in order to illustrate a new theory in optics, that of correcting the refrangibility of the rays of light in the construction of object glasses for refracting telescopes, and the success attending his experiments had fully convinced him that his theory was well founded, in consequence of which he then applied to Mr Scarlett, optician in Soho (father of the late Mr Scarlett), to grind him one glass of a certain diameter and spherical proportion, and afterwards applied to

Mr Mann, optician in Ludgate Street, for him or some of his workmen to grind him another glass, which was to answer in diameter and spherical figure to that Mr Scarlett had done for him, which compound object glass made of flint and crown glass ground with concave and convex surfaces he gave to me in the year 1763, and told me it was the first that was made and to the best of his recollection was made about the year 1733, which of the two glasses was worked by the two different artists I know not, as I was not curious enough at the time to enquire, he told me his reasons for imploying different persons for grinding his glasses was to prevent his discovery being known; he had many other object glasses ground afterwards upon the same principle, conducted by the same artists, which were of various diameters and different focal distances. Mr Dollond or his friends, yourself or your friends, shall at any time when it is agreeable by calling at my house see the original object glass, which I value as the prime and practical foundation of this useful discovery, and which has never been out of my possession since Mr Hall was pleased to present it to me in the year 1763. Therefore I must beg leave to assure you that I think myself happy in having an opportunity through your means on this occasion to render impartial justice to the memory of Mr Hall and believe me to be with the utmost respect …[30]

The French physicist Alexis Marie de Rochon, visiting London in 1790, called on Ramsden at this time. Like most of his countrymen, Rochon credited Euler with the achromatic lens, but in London he found the matter still in dispute.

Such were the facts which I had gathered in London in 1790: and I made sure to consult scholarly men of great impartiality on this topic. I had often talked about the achromatic lens with the famous Ramsden, yet without entirely trusting in his testimony, taking into account his falling out with his brother in law Dollond; but what ever the case may be I could hardly be persuaded that Euler was not the first one to correct the spherical aberrations, by making up a lens of glass and water ….[31]

The Royal Society declined to publish Peter Dollond's paper and in 1789 he issued it privately as *Some account of the discovery made by the late Mr John Dollond FRS which led to the grand improvement of refracting telescopes … With an attempt to account for the mistake in an experiment by Sir Isaac Newton. By Peter Dollond, member of the American Philosophical Society at Philadelphia.*[32] The manuscript draft of this text, showing how Dollond was searching for the best way to express his feelings and his memory of events, remained with his descendants until the late twentieth century, when it was donated to the Oxford Museum of History of Science.[33] The bulk of the published text seeks to deny claims made by and on behalf of Euler and his assistant Nicolaus Fuss, and by Klingenstierna, and it deals with the comments by Clairaut and Lalande, with a final assertion that these scientists had not, for theoretical or practical reasons, achieved the achromatic lens, this being solely due to John Dollond. Hall's name is nowhere mentioned.

Shortly thereafter, a letter signed 'Veritas' was published in the *Gentleman's Magazine* of 30 September 1790, pp. 890–91:

30 September – The writer of an Introduction to some letters lately published on the Improvement of Ship-building seems to have been misled in saying, "That great discovery in optics, the *achromatic* glasses, was entirely owing to three or four ingenious

men assembling at a public-house in Spitalfields, to amuse themselves in friendly conversation upon mathematical and mechanical subjects."

As the invention has been claimed by Mr *Euler*, Mr *Klingenstierna* and some other foreigners, we ought, for the honour of England, to assert our right, and give the merit of the discovery to whom it is due; and therefore without further preface, I shall inform the author of the above quotation, that the inventor was CHESTER MORE HALL Esq., of More Hall, in Essex, who, about 1729, as appears by his papers, considering the different humours of the eye, imagined they were placed so as to correct the different refrangibility of light. He then conceived, that if he could find substances having such properties as he supposed these humours might possess, he should be enabled to construct an object-glass that would shew objects colourless. After many experiments he had the good fortune to find these properties in two different sorts of glass; and by forming lenses made with such glass, and making them disperse the rays of light in contrary directions, he succeeded. About 1733 he completed several achromatic object glasses (although he did not give them this name), that bore an aperture of more than 2½ inches though the focal length did not exceed 20 inches, one of which is now in the possession of the Rev. Mr Smith, of Charlotte Street, Rathbone Place. This glass has been examined by several gentlemen of eminence and scientific abilities, and found to possess the properties of the present achromatic glasses.

Mr Hall used to employ the working opticians to grind his lenses; at the same time he furnished them with the radii of the surfaces, not only to correct the different refrangibility of rays but also the aberration arising from the spherical figures of lenses. Old Mr Bass, who at that time lived in Bridewell Precinct, was one of these working opticians, from whom Mr Hall's invention seems to have been obtained.

In the trial at Westminster Hall about the patent for making achromatic telescopes, Mr Hall was allowed to be the inventor, but Lord Mansfield observed, that "it was not the person who locked-up his invention in his scrutoire that ought to profit by a patent for such invention, but he who had brought it forth for the benefit of the publick". This, perhaps, might be said with some degree of justice, as Mr Hall was a gentlemen of property, and did not look to any pecuniary advantage from his discovery; and consequently, it is very probable that he might not have an intention to make it generally known at that time.

That Mr Ayscough, optician on Ludgate Hill, was in possession of one of Mr Hall's achromatic telescopes in 1754, is a fact which at this time will not be disputed.
VERITAS

By this time, there were few apart from Francis Watkins and Addison Smith who could have recalled the events during and after the lifetime of the patent. Ramsden's opening remark, to 'some letters lately published on ship-building' is obscure – I was unable to locate any letters in *Gentlemen's Magazine* or any other publication to which this comment might refer. A question arises from the title of 'Reverend' bestowed on Addison Smith, also as to the identity of 'Veritas', but unless the Royal Society had thrown open its private correspondence to non-fellows, which seems unlikely, I am confident that Ramsden wrote this letter.

A Foray into Chemistry

In the early 1790s Ramsden ventured into the domain of chemistry – the measurement of specific gravity, a subject then of interest to HM Excise, the government body responsible for taxing spirits. In the opening remarks of his treatise, published in 1792, *An Account of Experiments to determine the Specific Gravities of Fluids, thereby to obtain the Strength of Spiritous Liquors, together with some Remarks on a Paper entitled The Best Method of proportioning the Excise upon Spiritous Liquors, lately printed in the Philosophical Transactions,* Ramsden explains what has prompted his interest:

> It appears, by a late paper, printed in the *Philosophical Transactions* (vol. LXXX) that government have it in contemplation to adopt new regulations for ascertaining the duty on spirituous liquors; and for this purpose have applied to the President of the Royal Society, to consider on a set of experiments to carry them into execution; and from thence to contrive a practical method, whereby the excise officer may be enabled to ascertain the duties on the different compounds, which may come under his inspection, in the most easy and satisfactory manner.
>
> On perusing the report made by the Secretary of the Royal Society on this subject, the plan pursued, in their experiments, appeared to me not quite so well adapted to the purposes for which they were made, as might have been expected.

Having been concerned in this matter in about 1776, when he was making hydrometers able to measure the proportion of spirit in any compound to one part in a hundred, Ramsden offers a 'suite of very simple experiments' which will deliver more satisfactory results than hitherto. He will describe an instrument for determining the specific gravity of any mixture of spirit and water, and give a practical method of ascertaining the proportion of spirit in what was known as 'proof', namely a mixture which weighed 7lb 12oz per gallon, at a temperature of 55°F. He must have been working on his hydrometer in December 1791 when Blagden wrote apologising that he and Joseph Banks would be unable to call on Ramsden the following day as planned, adding pointedly 'that if Mr Ramsden will let them know when he is really ready, they will appoint the earliest day after Saturday to attend the trial of his hydrometer'.[34]

Charles Blagden, who considered himself something of an expert on this matter, criticised Ramsden's proposed method, which in its verbosity seems anything but simple. In 1793 Ramsden reacted to this criticism by publishing *Animadversions on a Paper entitled Supplementary Report on the Best Method of proportioning the Excise on Spiritous Liquors. by Charles Blagden ... printed in the Philos. Trans. for the year 1792.* It seems that Ramsden was ill-prepared to conduct such experiments on his own, without the guidance of someone familiar with the work. Although he went to the trouble and expense of two publications, neither his proposed instrument nor his 'method' were adopted.

Last Days

By March 1800 Ramsden sensed that the end was nigh, for on 19 March he wrote in his usual neat firm hand to William Harrison at the Land Revenue Office, saying that he had had an offer for the house and workshops at 196, but 'yet after my death which may not be long' he hoped that the Crown would restrict the terms of the lease 'to prevent any unsavoury trade to take over'.[35] Harrison's reply was comforting if non-committal.

At some time that year Ramsden was driven to seek respite in the Sussex seaside town of Brighthelmstone (now Brighton), where he died on 5 November. *The Times*, and the Brighton newspapers ignored his passing, perhaps less newsworthy than the great storm of 9 November – one of the worst in living memory which caused severe damage on the south coast and as far inland as Westminster.[36] His body was brought back to London and on 13 November he was interred in the burying ground adjacent to the chapel of St James Hampstead, which since 1793 had taken the burials from the mother church in Piccadilly where the small graveyard had no more room.[37] The *St James Chronicle* and the *Whitehall Evening Post* noted his death; both mis-identified him as 'Mr John Ramsden, late an eminent optician in Piccadilly'.[38] *Gentlemen's Magazine* printed a long and affectionate notice in its 'Deaths' column:

> 5th [November] At Brighthelmstone, to which place he went for the benefit of his declining health. Although this gentleman's great skill and science in the line of his profession enabled him to improve upon the most eminent of his predecessors, and to take the lead of those of his own day, yet he had qualities independent of his professional merits which rendered him highly acceptable to the most respectable characters. Among those who took the most pleasure in the philosophical turn of his mind, the benevolence of his heart, and the urbanity of his manners, are to be reckoned their Graces of Richmond and Marlborough, whose friendly and tender attentions he had the pleasure of experiencing, at their residences both in town and country; which, no doubt, tended to assuage those infirmities, which intense studies seldom fail to bring upon a life nearly exhausted ...[39]

He was not short of friends, but had no immediate family beyond the Dollonds, with whom he had long ago severed friendly relations. Legally a widower since Sarah's death in August 1796, his only child, John, a captain in the East India Company's marine, was at sea, on his way to China.

Notes

1   Beckenham, Bethlem Royal Hospital Archives.
2   LGL. MS 11936/419, Sun Insurance policy no. 706245 of 1800.
3   www.oldbailey.org. Ref. f17900526–1. 26 May 1790.
4   Modzalevskii (1908), 137. Alexander Aubert and Hans Moritz von Brühl were elected at the same time.
5   CRL, NKS 287 ii. Bidstrup to his patrons, 6 March 1789.
6   Brazell (1968).

7   Maddison R.E.W. and Maddison F.R. (1956–7); Rose (1960), 68.
8   Timmins (1890).
9   *The Times*, 10 April 1797, 3d; T. Fortune, *A concise and authentic history of the Bank of England*, London, 1797.
10   BL, MS Add. 72847 f. 28ᵛ. Ramsden to the Duke of Marlborough, 2 March 1797.
11   Beasley, (2002).
12   RS, *Minutes of Council*, Vol. 2, 1846–58 (1858), 158.
13   Beasley (2002), 124–7.
14   Foderà Serio and Chinnici (1997), 35, 38. Chinnici et al. (2001), Figure 2.
15   Home (1794), 21–7.
16   Home (1795), 2.
17   Mandell (1960), 633–8.
18   Home (1796).
19   Home (1802), 1–4.
20   Young (1801), 23.
21   Sorrensen (1993) suggests that whereas John Dollond and Jesse Ramsden had devoted some of their time and money to the scientific development of their crafts, Peter Dollond was viewed as a tradesman in pursuit of wealth, and thus unworthy of Royal Society fellowship.
22   The complexity of issues and litigation surrounding the Dollond patent will become apparent in Brian Gee's biography (in progress) of Francis Watkins and his optical instrument-making contemporaries, wherein the whole set of issues are reconstructed.
23   Hall's land ownerships and other official local records are in Essex Record Office, Acc. 448 D/Dne. His only known work, a printed table 'to show the daily increase in any sum &c.' was prepared in January 1771, to assist with calculating the rent of chambers in the Inner Temple, listed in the *Calendar of Inner Temple Records* as *Records* vol. 26, No 86, 19 January 1771, 20 ff.
24   McConnell, (2004).
25   A possible candidate is John Thomas Desaguliers (1683–1744).
26   Bugge (1997), 130–31.
27   RS, MS L&P IX.138. 'Some observations on the invention of achromatic telescopes by J Ramsden FRS.'
28   TNA, MS CP 40/3667 f. 626.
29   RS, MS L&P IX.131, 'Peter Dollond's account …'. March–July 1789.
30   RS, MS L&P IX.138, 'Some observations on the invention of achromatic telescopes' by J. Ramsden FRS.
31   Rochon (1801). 'Tels étoient les faits que j'avois recueillis à Londres en 1790; et j'ose assurer que j'ai consulté à ce sujet des savans d'une grande impartialité. Je m'en suis souvent entretenu avec le célèbre Ramsden, sans cependant m'en rapporter entièrement à son témoingage; attendu sa brouillerie avec son beau frère Dollond; quoi qu'il en soit on me persuadera difficilement qu'Euler n'ait pas été le premier qui ait pensé à corriger les aberrations de réfrangibilité, comme Newton avoit imaginé le moyen de corriger les aberrations de sphéricité, en composant un objectif de verre et d'eau. ….'
32   Dollond, P. (1789).
33   [Bennett] 'Peter Dollond answers Jesse Ramsden'. Oxford Museum of the History of Science Newsletter *Sphaera* (1998), 4–5.
34   RS, MS CB/2/712. Blagden to Ramsden, 22 December 1791.
35   TNA, MS CRES 2.730.

36   *The Times*, 11 November 1800, 3.c.
37   WA, MS, St James Piccadilly, PR.
38   *St James Chronicle* 8–11 November 1800, 3c, *Whitehall Evening Post* 8–11 November 1800, 3d.
39   *Gentleman's Magazine* 70 (1800), p. 1116.

Chapter 11

Ramsden's Will and Probate

Ramsden had left a will, naming as his executors his employees Matthew Berge and Edward Pritchard. A romantic story tells how this document was dampened by the tears of those reading it, then charred as it was put to the fire to dry. This nice tale is countered by the two copies and explanation now in the National Archives.[1] The original (copy 'A') had been written by Ramsden. The day after his death Berge produced the will and dictated it to James Allan and Pritchard who, standing together, each wrote copies. Allan's copy (copy 'B') was given to the Dollond brothers. Berge said that sitting alone on 10 November, he made another copy from Ramsden's original, and it was this which had been accidentally laid on a wet plate on the table and when he held it to the fire to dry, had been slightly burnt at the bottom, rendering the last eight lines illegible.

The copy 'B' was accepted for probate on the affirmations of Matthew Berge and James Allan. (Punctuation is added here; wills, being legal documents, were not punctuated.)

> In the name of God Amen. I Jesse Ramsden of Piccadilly in the Liberty of Westminster, mathematical instrument maker, being at this time in a very debilitated state of Body but sound in mind and intellect do make this my last Will and Testament, that is to say I give and bequeath unto my son John Ramsden the Sum of ten pounds. I give and bequeath unto Mr Thomas Holroyd of Gerrard Street ten pounds and to his two daughters Mary and Eleanor Holroyd nine hundred pounds each. I give and bequeath to John Hill, Samuel Pierce, [small space in copy] Curtis, George Pope, and Edward Pritchard fifty pounds each and also to the last mentioned Edward Pritchard my small Dividing Engine together with the double Barr'd hand lathe now in the garratt. I give and bequeath to William Pye and to Gotlieb Hampandahl ten pounds each. I give and bequeath to [blank] Allan my shopman twenty pounds, to Peter Lealand fifty pounds, and to Elizabeth Tutt two hundred pounds. I give and bequeath to Isaac Dalby of Lisson Green one hundred and fifty pounds. I give and bequeath to Doctor John Hunter of Charles Street two hundred pounds which I beg him to accept merely as a token of my respect and esteem and to Mrs Hunter my watch. I give and bequeath to Samuel Pollard living near Hallifax five hundred pounds. I give and bequeath to Thomas Leicester of Salter Hebbel near Hallifax one hundred and fifty pounds, fifty of which for his own use and the remaining hundred pounds to be disposed of as he pleases among my relations. I give and bequeath to Joseph Pollard two hundred pounds. I give and bequeath to my old servant Matthew Berge the lease of my house and workshops N° 199 Piccadilly next St James's Church with my working tools and stock except those before mentioned on condition that he pays the remaining debts for building and repairing left at my death and I do so constitute and appoint the said Matthew Berge together with the above named Edward Pritchard to be joint executors of this my last Will and Testament and do decree this to be my last Will and Testament

In witness thereof I have hereunto set my hand and seal this sixteenth day of March in the year ~~of our Lord~~ [*sic*] one thousand eight hundred [signed] Jesse Ramsden [seal] in the presence of Joseph Green John Rooker

Some of the legatees are identifiable: Dr John Hunter (1754–1809) and his wife Elizabeth, *née* LaGrand lived in Charles Street, Westminster, a short distance from Ramsden. Hunter may have been his physician. He was unrelated to the Dr John Hunter mentioned in connection with Everard Home's concern with the human eye.[2] We met Isaac Dalby in Chapter 8; he was well known to Ramsden, who in 1785 had recommended Dalby as an assistant on the triangulation. Samuel Pollard of Halifax cannot be identified as there are too many of that name, but a family link is assumed as Joseph Pollard (d.1840), a Westminster optician, bequeathed his tools and goodwill to his son, John Ramsden Pollard.[3] Elizabeth Tutt remains mysterious: was this generous bequest in gratitude for her care of Ramsden after Sarah departed? Neither the will nor the death duties documents indicate her status or where she lived.

Berge and Pritchard legally transferred the duty of executing the will to Peter and John Dollond, who undertook to carry out Ramsden's instructions and settle his financial affairs:

> On the twenty fourth day of March in the year of Our Lord one thousand eight hundred and one administration with the Will contained in two paper writings marked A and B annexed of all and singular the Goods Chattels and Credits of Jesse Ramsden late of Piccadilly in the Parish of St James Westminster in the County of Middlesex deceased was granted to Peter Dollond and John Dollond the uncles and lawful attornies of John Ramsden Esquire the natural and lawful son and only child of the said deceased for his use and profit now on a voyage to the East Indies, having first sworn duly to administer. Matthew Berge and Edward Pritchard the Executors named in the said Will having first renounced the Execution thereof and no Residuary Legatee being named in the said Will or further [illeg.] decree having been first made and interposed for the force and validity of the said Will as by the Acts of Court app.[4]

The alleged events befalling the original will were not challenged. The Dollond brothers worked through the contents of Ramsden's house and workshops, trying to calculate his debts and the moneys owing to him. The probate inventory was registered on 27 March 1801.[5] To summarize the legally verbose document, the Dollond brothers found £10 cash in the house: they valued the lease of 196 Piccadilly at £200 and the lease of 199 Piccadilly and the workshops at £50. There was stock in trade to the value of £1026, 13s, 4d, and working tools to the value of £628, 9s, 6d. Ramsden's personal possessions – books, prints, plate, clothes and furniture – they valued at £170. There was £800 in annuities in the Bank of England, with £90 half-yearly dividend due, and £708, 0s, 3d cash with his [unnamed] bankers. His account books disclosed numerous debts owing to him, many of which 'appear to be desperate and irrecoverable but the Declarants believe they shall be able to recover other parts of the said debts to the amount of the sum of one thousand three hundred pounds or thereabouts'. They were able to raise £4983, 3s, 1d, more than adequate to meet his bequests totaling £1820, and Death Duty payable of £69, 4s.[6] John Ramsden was to receive a bequest of £10, but as the legal residuary legatee, he would have received an additional £3000 or thereabouts. Unfortunately we do not know the value of the irrecoverable debts due to Jesse Ramsden, nor his own debts: wages to his workmen, payments to his subcontractors for work and materials, and to his domestic suppliers of such essentials as food, fuel and clothing.

Notes

1   TNA, MS PROB 11/1355 ff.251ʳ–252ᵛ.
2   Wilkinson (1981–82).
3   TNA, MS PROB 11/1928 ff. 79 ʳ⁻ᵛ.
4   TNA, MS PROB 11/1928 ff.252ᵛ.
5   TNA, MS PROB 31/929/337.
6   TNA, MS IR 26/49, pp. 131–2. Ramsden's dutiable legacies.

Chapter 12

Sarah and John Ramsden

Sarah Dollond was aged 23 when she married Jesse Ramsden in 1766. When they parted company, Sarah took her son John back to the Dollond house at 35 Haymarket. A letter from Giuseppe Poli, instructor at the Naples Military Academy, to Joseph Banks, writing from Naples on 2 March 1784, suggests that the Ramsdens had parted by that date:

> In answer to your kind letter dated the 12th of January, I must acquaint you first of all that my gold medal of Captain Cook is to be given to the Duke of Marlborough as he desired me to subscribe for him. I subscribed likewise for a silver one, which I beg you will be so good as to give it to Mrs Ramsden, who will pay the money for me. ...[1]

There seems to have been no subsequent contact between Ramsden and his wife, and the bad feeling between Ramsden and Peter Dollond in 1789, delicately remarked on by de Rochon in 1790[2] (see Chapter 10) is testimony of their relationship. Sarah assisted her brothers with both their practical work and their accounts. In later years she was writing to Matthew Boulton, acting as agent to transmit a request from Poli who was asking that Boulton and Watt supply a steam engine to be erected there.[3] Oddly, steam engines were not prohibited machinery under the various Acts banning export of technology.

John Ramsden joined the East India Company Navy in 1780. He was only twelve years old, but following the practice of the times, presented a certificate signed by the priest at St Martin in the Fields stating that he had been baptised in September 1766 and was therefore fourteen, the legal minimum age for going to sea.[4] After five voyages to India, the last as Chief Mate of the ship *Sulivan*, on his return home on 5 July 1795, the Dollonds dined on board. On 9 September that year he was approved for Commander.[5] On 6 April 1797 he sailed in *Princess Amelia*; later that month this ship was lost by fire at Pigeon Island.[6] When his father died, John was on a voyage to China. Their relationship is not known. After payment of the bequests specified in Jesse's will, the remainder, unspecified in the Death Duty Registers but around £3,000, went to John Ramsden as residuary legatee. On 9 April 1806 John resigned his commission, 'with leave to reserve'.

Sarah Ramsden, and John when he was on home leave, joined in Dollond family life, either staying with Sarah's sister Susan, now married to the optician George Huggins and living at Lambeth, or at the St Paul's Churchyard house.[7] On 23 August 1796 Sarah Ramsden went to stay with Susan at Hercules Buildings, off Hercules Road, Lambeth. She died there a few days later, in the afternoon of 29 August 1796, aged 56, having left no will or other testament. She was buried in the churchyard of St Mary Lambeth, where her brothers John and Peter Dollond were later interred.

Sarah's niece Elizabeth Dollond, being of an age with John Ramsden, had probably grown up with him and they remained close friends thereafter. In 1789 Elizabeth married Timothy Tyrell (1755–1832), Remembrancer to the Corporation of the City of London.[8] John was a frequent visitor to their house at Kew, coming down from wherever he had taken rooms in London during his leave. In 1809 he travelled from his lodging at 8 Cecil Street, Strand, just to dine with them. On other occasions he stayed for several days at a time, joining the family on boat trips up the river Thames (they often took to the river for the journey between Kew and Guildhall where they attended functions). John and Elizabeth would play chess together in the evenings, a game at which they seem to have been fairly well matched. In November 1809, when John Ramsden was engaged to be married, it was Elizabeth who accompanied him to Kew to inspect houses that were on the market. It is likely, though not certain, that she attended his marriage on 30 November at St Helen's Bishopsgate to Mary Simmonds of Bishopsgate.[9]

During this time other characters come and go in the Tyrell circle. Besides George Huggins (who in 1805 had taken the name of Dollond), we hear of the numerous Gilbert family whose women and children often stay at Kew, their menfolk arriving from time to time. The family christian names and the fact that they arrive from or depart to Woodford and Leadenhall Street, where Elizabeth often calls on them when she and Timothy are overnighting in town, confirm that this is the instrument-making family of William and Thomas Dormer Gilbert, indeed Timothy Tyrell was one of the creditors at the Gilbert's bankruptcy, being owed £179.

When John Ramsden returned to sea – by 1811 he was commanding *Phoenix* taking troops to Batavia – Elizabeth and Susan were regular companions of his wife. The Ramsdens' first son was born on 22 November 1813 but lived for only three days; by 1818 they had moved, with their second son, John George, to Ivy Lodge, a house in London Road, Twickenham. John died on 6 April 1841, a month before his son John George married at Twickenham.[10]

In his will John left his property, worth nearly £50,000, to be divided between George Dollond, his own wife, and Richard Clark.[11] John George Ramsden took possession of the house, and lived there till his death in 1862, after which his widow continued in residence. There were no children of this marriage. Although his name appears in the death duty register indexes, several registers for that period have been destroyed; however, as he had Bank of England stock, the register was abstracted in their journals, showing that among the legacies was the remaining shop at 196 Piccadilly (formerly 199) originally leased to Jesse Ramsden then transferred to Berge, and which on the expiry of that lease in 1835 must have been released to John. By this time the property was no longer part of the Crown Estates, having been conveyed to Bethlem Royal Hospital in exchange for its estates which the government wanted for the extension to Charing Cross Hospital. Nothing remains of the Ramsden premises. Numbers 190–195 were rebuilt in 1881–83 as the Royal Institute of Painters in Watercolours (their façade survives) while in 1925 a branch of the Midland Bank (now in other hands) was constructed on the site of No.196.

Notes

1   BL, MS Add. 8095 f. 313.
2   Rochon (1801) discretely refers to Ramsden's 'brouillerie avec son beau frère', without further amplification.
3   BPL, B&W Coll., Boulton Letters Box 4 Bundle P and Boulton 251, Letters Box R, 1, 91, 92.
4   OIOC, MS Marine Records A146 669 and 670, birth affidavit.
5   OIOC, MS L/MAR/C/656 p. 84.
6   Perhaps the island of that name at 14°1′ N, 74°16′ E, south of Goa.
7   MHS, MS Blundell 7, Susan Huggins's journal. The descendants of George Huggins continued the Dollond business.
8   The Remembrancer's duty was (and is) to represent the Corporation on Parliamentary and other government committees.
9   LGL, MS 14951/1–2 Elizabeth Tyrell's diary.
10 Marriage of John George Ramsden, also death of John Ramsden, noted in the personal columns of *The Times*.
11 TNA, MS PROB 11/1946 sig. 354, will of John Ramsden.

Chapter 13

The Craft Inheritors at Piccadilly

The only difference, albeit considerable, between Berge and his master, is that the first lacked the creative genius of the second.'
                    Joseph Banks, as reported by Castelcicala to De Medici, 1808. ASP

Berge Continues the Business at Piccadilly

Berge lost no time in substituting his own name for that of Ramsden on the letterhead, which was otherwise unchanged (see Figure 13.1). His one-page 'Catalogue of optical, mathematical and philosophical instruments, made and sold by Matthew Berge, successor to the late J. Ramsden, 199 Piccadilly, London', listing 85 items, was probably issued shortly after Ramsden's death to reassure customers that the business continued to offer the same wide range of instruments[1] (see Figure 13.2) Sadly, there is no known likeness of Berge.

Figure 13.1   Berge's letterhead, 1801. RGO 14/17. f.107

Berge had inherited the lease of 199 Piccadilly on condition that he paid the debts for the remaining repairs and building works. This had been Ramsden's house; at around this time Piccadilly Circus and Regent Street were built and Piccadilly was renumbered and this house became 196. Ramsden's personal possessions remained in the house until his son John returned from China, and it was during this period, a month after Jesse Ramsden's death, that Berge suffered a theft of household linen and other small goods, as well as three barometer frames, which happily were recovered.[2] The Crown Estates registers record the transfer of Ramsden's lease to Berge,[3] and his tenure of a lease from 4 June 1803, 41 years from 5 April 1820, rental £28, 15s, 6d.[4] Whether Berge held on to

A CATALOGUE

OF

Optical, Mathematical, and Philosophical Instruments,

MADE AND SOLD BY

MATTHEW BERGE,

SUCCESSOR TO THE LATE J. RAMSDEN,

No. 199, *PICCADILLY, LONDON.*

OPTICAL INSTRUMENTS.

Improved Achromatic Refracting Telescopes, mounted with brafs fliding tubes, ufually termed Military Telefcopes, to draw out one foot in length	—	1 16 0
Ditto, if mounted with filver plated	—	2 10 0
Ditto, eighteen inches, in brafs	—	2 12 6
Ditto ditto, filver plated	—	3 13 6
Ditto, two feet in brafs	—	3 13 6
Ditto, ditto, filver plated	—	5 0 0
Ditto, two and half feet, in brafs	—	5 0 0
Ditto, ditto, filver plated	—	6 16 6
Ditto, three feet, in brafs	—	6 6 0
Ditto, ditto, filver plated	—	8 8 0
Tanned leather cafes, with belts for each of the above fizes, at	—	
A brafs ftand, with horizontal and vertical motions to the above Telefcopes, packed together in a mahogany cafe, from 2l. 2s. to	—	2 12 6
Achromatic Telefcopes, mounted in mahogany tubes for fea, of one foot in length	—	1 1 0
Eighteen inches ditto —	—	1 '11 6
Two feet ditto —	—	2 0 0
Three feet ditto —	—	3 3 0
Two feet, improved ditto —	—	3 3 0
Two and half feet ditto ditto —	—	4 4 0
Three feet ditto ditto —	—	5 5 0
Ditto, with a larger object glafs, and feparate, eye tube for hazy weather	—	6 16 6
Improved night or day Telefcopes	—	4 14 6
Night ditto, from 36s. to	—	3 0 0
Achromatic Telefcopes, twenty inches long, mounted in brafs, with a pillar, ftand, &c. one eye tube for terreftrial, and one for aftronomical objects, packed complete in a mahogany cafe	—	8 8 0
Achromatic Telefcopes, thirty inches long, &c. with two eye tubes for terreftrial, and one for aftronomical objects, in a mahogany cafe	—	12 12 0
Ditto, three feet long, ditto	—	18 18 0
Ditto, four feet, with improved rackwork, and five magnifying powers, thirty guineas to	—	42 0 0
Small achromatic perfpective glafs, with two concave eye-glaffes, of different magnifying powers, from 12s. to —	1 10 0
Ditto, with four eye-glaffes, and a ftand, complete	—	2 10 0
Opera Glaffes, mounted in ivory, or plated, gold, &c. of various kinds, from 10s. 6d. to	—	5 5 0
Reflecting Telefcopes, fourteen inches long, packed in a mahogany cafe	—	5 5 0
Ditto ditto, eighteen inches, ditto	—	8 8 0
Ditto ditto, two feet, ditto	—	12 12 0
Ditto ditto, ditto, with a rack-work ftand, and different powers	—	21 0 0
Ditto ditto, of larger fizes, and very complete to	—	500 0 0
Wilfon's pocket Microfcope, for viewing tranfparent or opaque objects	—	2 8 0
Ellis's ditto, with improvements	—	3 3 0
Compound Microfcopes complete, from 5l. 5s. to	—	15 15 0
Solar Microfcopes ditto, 6l. 6s. to	—	21 0 0
Lucernal ditto ditto	—	21 0 0
Optical machines for viewing prints, 1l. 1s. to	—	1 16 0
Camera Obfcuras, portable, 14s. to	—	2 12 6
Large box Camera Obfcuras, with pinion, and rack-work for adjuftment	—	8 8 0
Magic lanterns, 1l. 1s. to	—	2 2 0
Concave and convex Mirrors, from 10s. 6d. to	—	16 16 0
Spectacles, filver double jointed, with glaffes	1 1 0

Ditto, ditto with Brazil pebbles	—	1 16 0
Ditto, fingle jointed, with glaffes	—	0 14 0
Ditto, ditto, with Brazil pebbles	—	1 9 0
Ditto, tortoifefhell, fteel ditto, from 2s. to	0 16 0
Spectacle cafes, from 1s. 6d. to	—	0 18 0
Reading Glaffes of all kinds, in various mountings	—	

MATHEMATICAL INSTRUMENTS.

Hadley's improved Sextants, in brafs, with Telefcopes of different magnifying powers, adapted for determining the longitude at fea, &c. &c. from 9l. to	16 16 0
Brafs ftands, for Sextants of various conftructions, 2l. 10s. 6d. to	—	7 7 0
Artificial Horizon, with parallel glafs roof, mounted in wood, or brafs, from 2l. 2s.	—	3 3 0
Ditto ditto, with glafs plane and fpirit level	2 12 6
Hadley's Octants in ebony, with ivory arches, &c. from 2l. 5s. to	—	5 5 0
Theodolites, with plane fights	—	5 5 0
Ditto, with one or two Telefcopes, complete, four inches diameter to nine inches, from 10l. 10s. to	—	42 0 0
Ditto for extenfive furveys, graduated to read two or each fecond, eighteen or thirty-fix inches diameter, 160l, to	400 0 0
Plain Tables	—	3 13 6
Ditto improved, with a telefcope complete	15 15 0
Circumferenters, from 3l. 13s. 6d. to	—	10 10 0
Meafuring Wheels, from 8l. 8s. to	—	10 10 0
Spirit Levels, portable	—	1 1 0
Telefcopic level	—	7 7 0
Ditto, with parallel plates, compafs, box, and tripod ftand, complete, 10l. 10s. to	14 14 0
Protractors, from 5s. to	—	4 14 6
Tranfit Inftruments, from 15l. 15s. to	250 0 0
Aftronomical Quadrants, of twelve inches radius	—	31 10 0
Ditto ditto of larger radii, to	—	420 0 0
New invented circular inftruments for aftronomical obfervations, both for altitudes and azimuths, and with greater precifion than any other inftrument yet made	—	
Meafuring Chains of iron, or fteel, from 9s. to	—	26 5 0
Proportional Compaffes, from 1l. 11s. 6d. to	—	2 12 6
Improved Azimuth Compaffes, from 5l. 5s. to	—	15 15 0
Steering Compaffes of different forts, from 10s. 6d. to	—	2 2 0
Horizontal Dials, from 10s. 6d. to	—	12 12 0
Globes, with the lateft difcoveries, three inches diameter to twenty-one inches, from 1l. 4s. per pair, to	—	17 17 0
All kinds of cafes of Mathematical Inftruments, either in brafs or filver, from 16s. to	—	21 0 0

PHILOSOPHICAL INSTRUMENTS.

Electric Machines, from 6l. 6s. to	—	15 15 0
Air Pumps, from 6l. 6s. to	—	15 15 0
Improved ditto	—	36 15 0
Portable Barometers, from 2l. 12s. 6d. to	4 14 6
Improved ditto, with Thermometer in front	—	6 6 0
Ditto ditto, with apparatus complete, for meafuring heights, and fhuts up in its own fupport	—	10 10 0
Improved Marine Barometers, with ftand complete, packed in a deal cafe, and with Thermometer, 8l. 18s. 6d.	—	9 16 6
Thermometers of all forts, from 12s. to	4 4 0
Pluviameters, from 1l. 11s. 6d. to —	2 12 6
Mercurial Level for afcertaining the trim of a fhip	—	3 3 0

Printed by Henry Reynell, No. 21, Piccadilly, near the Black Bear.

Figure 13.2   Berge's price list of 1801. Hampshire Record Office, 38M49/A4/7.

the rebuilt workshop behind 196 Piccadilly, in order to finish the Dunsink circle and other large apparatus, or was able to move them into the smaller workshop at number 199, is not clear from the records of payments of rates or from the Crown Estates documents.

Among the many small instruments found in Britain and Europe produced under Berge's management, the optical instruments, levels (see Figure 7.5), and sextants (see Figure 2.3) are generally signed 'Berge late Ramsden'. The instrument historian Alan Stimson identified a sextant in private hands signed 'Berge late Ramsden' and numbered 1352 as having been made about 1797, with the words 'Berge late' engraved in a different hand.[5] This form suggests that Berge is acknowledging – perhaps profiting from – Ramsden's designs. Some gunners' callipers, and two of the five known distinctive marine barometers are simply 'Berge' or 'M. Berge London'. During his Peninsula campaign of 1804–14 the Duke of Wellington had a marine barometer, signed 'Berge London late Ramsden'; at Waterloo, in 1815, the Duke had a 5-draw brass and mahogany telescope, signed – unusually – 'Matthew Berge, London'. Both items are now in the National Army Museum, London. A letter from India, written long after Berge's death, rates Berge's barometers as 'very good', and at 300 rupees the pair, cheaper than the 'excellent' barometers of Troughton (400 rupees) and Dollond (350 rupees).[6]

Berge's competence was recognized by the mathematician George Atwood who was involved with the civil engineers Thomas Telford and James Douglass on the design of a bridge over the Thames at Blackfriars. The architects had proposed a single-span cast-iron structure and a parliamentary enquiry was set up to ascertain the loading that could be put on such a novel design. To verify the designs by means of models, Atwood engaged 'Mr Berge of Piccadilly, whose skill and exactness in executing works of this sort are well-known to the public'.[7] Berge made two brass models based on the drawings prepared by Telford and Douglass. Although brass was less rigid than cast iron, Atwood believed that his experiments would indicate the pressures which each sector of the arch would bear under certain loads.[8] Berge also provided Telford with tape measures for his survey for Sweden's Gotha canal, commencing in 1808.[9] In 1806 Berge replaced Dudley Adams as sole supplier of mathematical instruments to the Board of Ordnance.[10] On 25 August 1807, a sale of the Duke of York's instruments was held at Berge's premises.[11] Berge was also one of the advertised makers of Brewster's kaleidoscope, patented in 1817.[12]

Completing the Outstanding Orders

The outstanding commission, long overdue, was the great circle ordered in 1785 by Trinity College, Dublin, for its observatory at Dunsink. The Trinity College authorities had made interim payments to Ramsden; they now appealed to Maskelyne as Astronomer Royal, who reassured them that they were in no danger of losing both their money and the instrument.[13] Berge may have been reluctant to deal with this troublesome artefact. After four years he promised the instrument in the following August but it did not come. Two years later (1806) the astronomer, John Brinkley, complained that he could get no answer from Berge. In 1807 he learned that Berge would send the telescope in a month. He did not; but in 1808, about twenty-three years after the great circle was ordered, it was erected at Dunsink[14] (see Figure 6.1). The fate of this apparatus is recounted in Chapter 6.

Also outstanding was a portable zenith sector for the trigonometrical survey. Berge postponed the dividing of its arc until better weather arrived. On 1 April 1802 Mudge wrote to invite Banks to inspect the finished instrument: '… the zenith sector, long since bespoken of the late Mr Ramsden, being finished … would be erected in the Tower.'[15]

Cassini IV retired from the Paris Observatoire in 1793 without ever having seen the meridian transit instrument, ordered early in 1788 and promised for August 1789. Correspondence between English and French savants during the war years continued unimpeded and in 1801 Méchain wrote to Banks expressing his hopes that Berge would complete it:

> … I take advantage of the present situation to ask you to use your influence on Mr Berge, Ramsden's successor, that he will agree to complete and deliver to you as soon as possible a meridian or transit instrument of 8 feet focal length which his master, according to a contract made on November 1787 with Cassini, was to supply to our Observatory at the beginning of 1789, for which he was paid on account £59, 18s on 30 December of that year. But because of Ramsden's tardiness, and the war, we have obtained nothing. For over two years while Citizen Otto was in England he made every effort to get Ramsden to fulfill his agreements; he repeated these to Berge, his successor. Sir [Henry] Englefield also aided us but so far, to no avail. There are problems regarding these sums previously paid although I have to hand the agreements in good order and I have sent certified copies to Sir Henry and Citizen Otto.
>
> If you would be so good, Sir, as to persuade Mr Berge of our concern to finish the matter, that would surely bring us the joy of soon possessing an instrument which our Observatory lacks. Being put in charge of this establishment, lately restored, and provided with two great mural quadrants, one by Bird and the other by Sisson, and which will be acquiring a reflecting telescope of 22 feet focal length and 22 inches aperture, I earnestly desire to complete its provision with the meridian telescope for which we have waited so long in vain. Ramsden had fixed its full price before his death, at 200 guineas, the rest of this sum will be paid as soon as it is delivered. Sir Henry sent me word that it was well advanced, if your intervention can persuade the artist to complete it and bring us the benefit in good time, this will be another service for which the sciences will be indebted to you. Our Bureau of Longitude, and myself in particular, will be very much obliged to you.[16]

On 26 December 1802 the French ambassador to England was recruited to assist in procuring the transit for the Bureau des Longitudes, whose responsibility it would become. Banks and Méchain kept in touch over progress at Piccadilly, for which Méchain was duly grateful:

> I have received the letter which you kindly wrote on the 11th of last month. I am infinitely grateful for the trouble that you have taken in order to see for yourself the

present state of the transit instrument which the late Mr Ramsden had undertaken for our observatory. If as a result of your kind efforts Mr Berge finishes it and delivers it at the end of the summer, or at the year's end, it will be you whom we should thank ...[17]

Méchain was extremely relieved to receive Banks's letter of 24 March 1803, advising him that the transit would be finished in three weeks time.

I have received your letter ... alerting me that the passage instrument for our observatory which Mr Ramsden had undertaken in 1787, and which Mr Berge is completing, will in three weeks be ready to pack and send to Paris ...[18]

On 26 April 1803 the Bureau was notified that the transit was on its way; it was erected in the Observatoire in September that year.[19] There remained some uncertainty over the payment due to Berge. On 11 October 1806 the astronomer Jean-Baptiste Joseph Delambre wrote to Banks, 'On a été quelque tems incertain de la somme qui restoit due à M. Berge',[20] but a letter from Berge received on 10 October, together with the papers left at Méchain's departure from the Observatoire, clarified the situation. Delambre had Banks' own letter of 25 March 1803 attesting to having seen the instrument complete, and that there remained outstanding £119, 10s, not including the packing which could be up to £10. However, according to Méchain's notes, this was not the sum due, but what had been paid in two stages, one of £59, 12s, 8d, the other of £59, 18s. There was also a letter dated 18 July 1800 from Sir Henry Englefield, stating that Ramsden had fixed the price of the transit at £200 plus £10 for packing. Thus £90, 9s remained outstanding. Berge was claiming £101, 10s, and Delambre did not think that the Bureau des Longitudes would fuss over the small difference; in fact Jean-Nicolas Buache de la Neuville (1741–1825), who had succeeded Méchain as administrator, was already trying to speed payment and an undated note in French, signed by Berge, acknowledged receipt of the entire sum due.[21] This instrument is now in the Old Toulouse Observatory.

The Cartographic Survey of Sicily

A costed list of apparatus which the Palermo astronomer Giuseppe Piazzi prepared on 1 February 1808 for a cartographical survey of Sicily was conveyed to Joseph Banks by the Neapolitan Minister Plenipotentiary resident in London, Fabricio Ruffo, Prince of Castelcicala (1763–1832).[22] Banks responded that the listed instruments were 'proper for the survey of the island of Sicily and the geometrical observations which the Padre is about to undertake' and that 'the proper person to execute these instruments is M. Berge, successor of Ramsden, who worked with him during the time he made similar instruments for Gen. Roy's and Col. Mudge's surveys and of course is in possession of the necessary tools as well as in the practice of using them'.[23] Sending Banks's letter and its translation into Italian, to the finance minister Luigi De Medici, the Prince included a long memorandum, explaining that he had sought to borrow or buy the two great

Ramsden geodetic theodolites, but had been told by Banks in no uncertain manner that one belonged to the King, the other to the Ordnance Office, and both were in continual use for national cartographic surveys. They were not for sale or loan, and would not be entrusted to a sea journey. Banks did however assure the Prince that Berge, having been closely involved with Roy's instrument, had both the knowledge and the necessary tools to make another. Berge, he explained, had been Ramsden's first pupil and had worked with him for a long time. Everything which Ramsden did, Berge could do to the same standards. The only difference – a considerable one, nevertheless – between Berge and Ramsden was that Berge lacked the creative genius of his master.[24] This perceptive remark by Banks may explain why Berge had been content to remain with Ramsden, rather than establish his own business.

A list of 15 items in Berge's hand, dated 2 May 1808, includes 'a three-feet improved theodolite made after the construction of the instrument used by the late General Roy £420, an eighteen-inch Ditto £262, 10s. NB this Colonel Mudge consider should be the second theodolite' (see Figures 13.3 a and b). This brought the total estimate to £927, 17s, almost double Piazzi's first estimate.

Piazzi wrote to Banks in 1810 that Berge seemed reluctant to undertake this commission, but this was not a problem, as the war had prevented Piazzi from making use of the instruments sent earlier.[25] On 16 October 1812 Castelcicala wrote from London that the two geodetic theodolites were almost ready; there was £650 to pay, plus packing and embarcation charges. Nothing further is heard of these theodolites; probably their high price and the worsening political situation led to their cancellation. Berge did send the lesser items: 12 signed instruments remain at Palermo: a portable transit (see Figure 13.4), three measuring chains, two of 100-feet and one 50-feet (see Figure 13.5), two mountain barometers (see Figure 13.6), small theodolite, an achromatic telescope, two beam compasses, a surveyor's compass, and a pantograph. According to surviving records, these reached Palermo over a period of years, from 1809 until 1827. Those which arrived after Berge's death must have spent the intermediate years elsewhere, possibly with the astronomer Niccolò Cacciatore, then at Naples.[26]

James Allan and his Son

Some of Ramsden's men continued with Berge and testified at the hearing to prove Berge's will. James Allan (or Allen – both forms appear) senior (1739?–1816) then of Blewitts Buildings, Fetter Lane, received a gold medal from the Society of Arts for improvements to his dividing engine, which he described in a letter to the Society, supported by testimonies from Stancliffe and Berge, published 20 November 1809.[27] Writing to the Society of Arts on 16 January 1811 Allan claimed to have assisted about 25 years previously to make Borda circles and had been in business on his own account for the last 12 years.[28] It is unclear whether he had been among Ramsden's workmen and independent after Ramsden's death. His son, the offspring of a liaison between Mary Smith and Allan senior, who took his father's name, had worked for three years as Ramsden's shopman, as declared

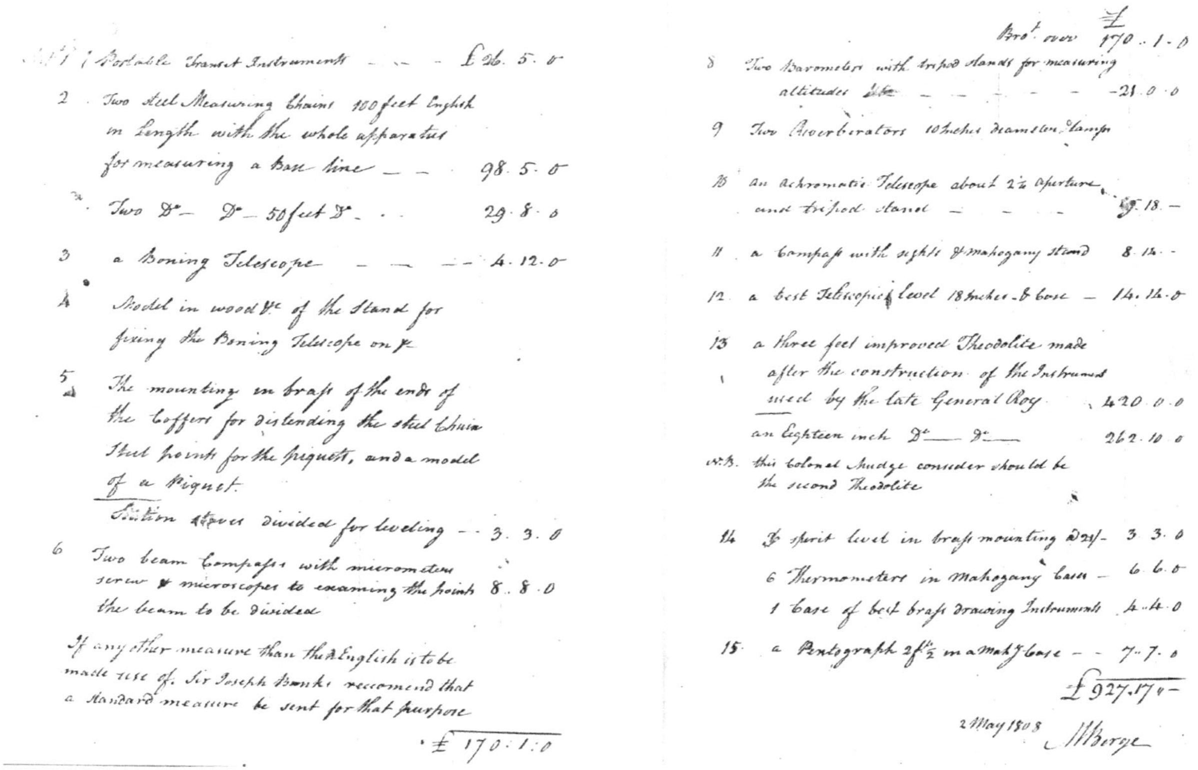

Bro't over £70 . 1 . 0

8 . Two Barometers with tripod stands for measuring
altitudes &c — — — — 21 . 0 . 0

9 . Two Accelerators Which Mountain Lamps — —

10 . An Achromatic Telescope about 2½ Aperture —
on a tripod Stand — — — ⅌ 18 . —

11 . a Compass with eight glmahogany stand 8 12 . —

12 . a best Achromatic level 18 inches & base — 14 . 14 . 0

13 . a Three feet improved Theodolite made
after the construction of the Instrument
used by the late General Roy — 420 . 0 . 0
an Eighteen inch D° — D° 262 . 10 . 0
N.B. this Instrument cannot be
the second Theodolite

14 . 6 foot level in brass mounting Do/ — 3 . 3 . 0
6 Thermometers in Mahogany case — 6 . 6 . 0
1 Case of 6 it brass Drawing Instruments 4 . 4 . 0

15 . a Pantograph 2½ in mahogany base — 7 . 7 . 0

£ 927 . 17 . 0
2 May 1808 Berge

I.... Portable Transit Instrument — — £ 26 . 5 . 0

2 . Two steel Measuring Chains 100 feet English
in length with the whole apparatus
for measuring a base line — — 98 . 5 . 0

⅏ . Two D° — D° — 50 feet D° . . . 29 . 5 . 0

3 . a Boning Telescope — — — 4 . 12 . 0

4 . People in wood D° of the Stand for
fixing the Boning Telescope on ⅌

5 . The mounting in brass of the ends of
the Coffers for distancing the steel Chain
steel knives for the Tripods, and a mould
of a Magnet
Iron steel Divider for leveling — — 3 . 3 . 0

6 . Two beam Compasses with microscopes
screw & microscope to examining the points 8 . 8 . 0
the beam to be Binch &c

If any other measure than the English is to be
made use of, six Scotch Banks recommend that
a Standard measure be sent for that purpose

£ 170 . 1 . 0

Figures 13.3 Berge's estimate to Piazzi for instruments for the proposed survey of Sicily, 1808. ASP, R.S.b.1480.

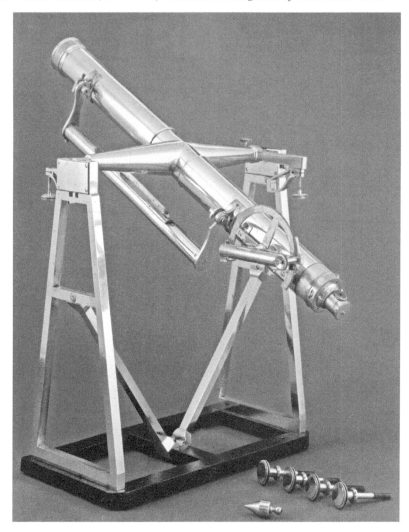

**Figure 13.4   Portable transit with 20-inch telescope signed *Berge*. Palermo
Observatory.**

in the probate of Ramsden's will;[29] at his father's death in 1816 he was declared
to be shopman to Berge of Piccadilly. Three pocket sextants and a 5-inch radius
reflecting circle survive, signed 'Allan' or 'Allen London'[30] (see Figure 13.7).

James Allan junior comes back into the picture in 1820. Before deciding
which of the applicants should be given custody of the Board of Longitude's
circular engine, the Board, prodded perhaps by Allan, investigated methods
of determining errors of dividing engines. A sub-committee of Captain Henry
Kater, Dr William Hyde Wollaston and a Mr Barrow, together with Ramsden's
former apprentice William Cary and a Mr Parsons, tested Ramsden's engine in
comparison with those of Stancliffe and Allan, and it was this last engine, which

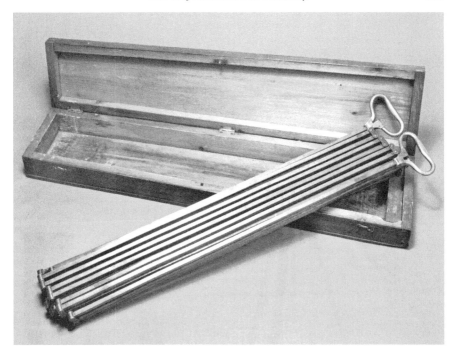

Figure 13.5   Surveyor's chain and box, signed *Berge London*. Palermo Observatory.

after all was the least worn, that received the Board's praise at their meeting of 3 February 1820, being deemed worthy of £100 recompense.[31] 'Mr Parsons' was probably a member of the Parsons family, makers and dividers of mathematical instruments, of Aldersgate. (They may have been related to Thomas Parsons of 14 William Street, Wilmington Square, Clerkenwell who in August 1846 applied to the Hydrographic Office for 'a copy of the publication on Ramsden's dividing engine made between 1759 and 1777 as he is about to make a dividing engine and would find it useful'[32] and even of the firm of W. T. Parsons who much later acquired the small Ramsden engine.)

On 7 June 1821 Wollaston published more on the comparison of dividing engines.[33] It seems that Allan allowed his advantage to lapse for the engine was sold to the instrument makers George and William Cook of Wapping.[34] A 'James Allen, 196 Piccadilly' attended the London Mechanics Institute from June 1825 to March 1826.[35] John or Jonathan Allan, son of James Allan, left England in 1807 and established himself in Baltimore 'at the sign of the mariner', advertising that all his instruments were graduated on his father's improved self-correcting engine.[36]

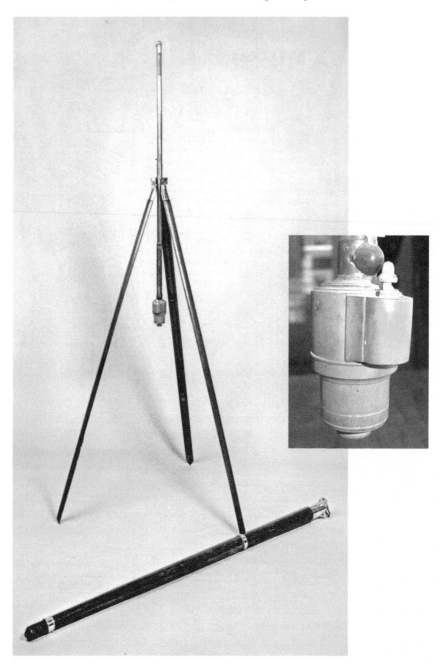

Figure 13.6   Mountain barometer, with tripod and case, signed *Berge London late Ramsden*. Palermo Observatory.

**Figure 13.7   Reflecting circle, radius 5 inches, signed *Allan*. Peter Ifland
Collection, Mariners Museum, Newport News, VA.**

Interlude – Réhé's Dividing Engine

Another dividing engine which was mistakenly taken for one of Ramsden's
manufacture. On 1 February 1803 Charles Blagden wrote to Banks from Paris
conveying a request from the astronomers at the Paris Observatoire:

> that you would have the goodness to assist General Andréossy in his negotiations
> for the purchase of a dividing engine now upon sale in London. This instrument is
> said to be the same with which Ramsden executed his divisions with some additional
> improvements by Troughton; it was to have been sold to Russia, but General Andréossy
> stopped the bargain by becoming a competitor for the purchase of it on the part of
> the French government. It is hoped, that you will inform the General whether it is as
> good a dividing instrument as those now in use for the same purpose in London, and
> also that you will give him such other instructions as may enable him to make a fair
> bargain with the proprietor.[37]

Antoine-François Andréossy (1761–1828) was ambassador to London during the brief period of Anglo–French peace, May 1802 to May 1803. From information supplied by the exiled Spanish astronomer Joseph Mendoza Rios (1761–1816), who wrote to Andréossy on 15 February, it is clear that this engine had no connection with Ramsden, but had been built by the engineer Samuel Réhé, or Rhee, (1735–1799) of Fleet Street, who had cut the screw for Troughton's dividing engine. Réhé's engine is in the Conservatoire des arts et métiers, Paris.[38] Two other letters from Blagden indicate that he was accused of being a spy of the English government, on the grounds that he and Banks endeavoured to prevent Andréossy from acquiring this dividing engine.[39]

Berge's Last Years

After his many years as a bachelor, Berge was married at St James's Church on or shortly after 22 July 1801 to Mary Ozard, widow of Robert Ozard (d.1792) of St Anne, Westminster.[40] In his will, dated October 1810, Berge bequeathed the premises in Piccadilly, 'wherein I now dwell' to his wife, all other bequests going to his or her relations. By 1815, when he came to write the first of two codicils, Berge had purchased land and property in Kent. His house, 'Mountains', just north of Tonbridge, is now a Grade II Listed Building and was in the early years of the twenty-first century on the market for £1¼ million.[41]

Berge continued to supervise the Piccadilly workshop; another court case, heard at the City of London Guildhall and subsequently at the Old Bailey, relates that on 1 January 1819 a naval officer had come into the shop and purchased a sextant for £16, tendering a £100 note which – as it turned out – had been stolen a few days previously. John Kimbell, Berge's shopman, did not have sufficient change in the till and called for Berge, who produced the necessary notes. Berge had recorded the numbers on the £100 note, and the £40 note given in part change, and was able to testify at the Old Bailey hearing on 30 April 1819. The guilty man's landlady produced the sextant and a considerable haul of jewellery, but as the latter was not part of the charge, only the sextant was seized by the prosecutor; it is not known if it was eventually returned to Berge.[42]

Berge died at 'Mountains' on 31 October 1819, 'in his 67th year'.[43] He was interred at St James's burying ground at Hampstead on 9 November, in the same cemetery as Ramsden.[44] James Allen and Samuel Pierce, who identified themselves as opticians of Piccadilly, testified to Berge's handwriting as the first codicil had not been witnessed.

Nathaniel Worthington

Berge's only registered apprentice: Nathaniel Worthington (1790?–1853), born at Gloucester, was bound from 16 August 1804 for seven years at a fee of £50. The rate books for 196 Piccadilly continue to acknowledge Berge or 'executors of Berge' as ratepayer, until 1831, when Worthington's name appears, and thereafter until 1845. A marine barometer on Berge's pattern by Worthington & Allen is in the Museo di storia della scienza in Florence and another is in private hands (see Figure 13.8). Worthington lived at 196 Piccadilly until shortly before 1851 when he moved to 13 Henrietta Street, Cavendish Square, where the Census described him as 'Optician, aged 61, with wife Elizabeth and three sons, 11, 8 and 1 – the two eldest born at Piccadilly – and a daughter, 6'. He died at Henrietta Street on 11 April 1853.

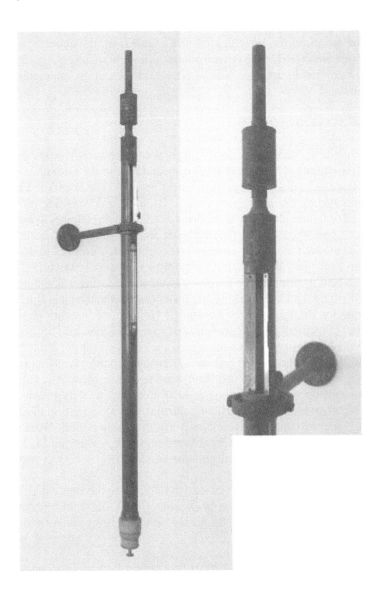

Figure 13.8   **Marine barometer signed** *Worthington & Allen London*

A fine altazimuth circle of 15½ inches diameter, signed 'Worthington London', survives at Palermo Observatory, although in 1904 it was said to be missing certain parts and therefore of no astronomical use beyond the instruction of students[45] (see Figure 13.9). This instrument was acquired by Niccolò Cacciatore, then director of Palermo Observatory, from the heirs of Giulio Tomasi (1815–1885), Prince of Lampedusa, whose passion for astronomy led him to erect in 1852 an observatory with a rotating dome, at Villa San Lorenzo Colli on the outskirts of the town. An engraving of this observatory figured on his cards and letterhead.[46] The Prince acquired the Worthington altazimuth from the astronomer Baron Ercole Dembowski (1812–1881), who in turn is thought to have possessed it from 1860 but its original purchaser, and where it had been in the intervening years, are unknown.

The Board of Longitude's Engines

When Berge died, other instrument makers sought to take over his privileges. Robert Brettell Bate (1782–1847) applied successfully to the Ordnance Office to succeed Berge as its mathematical instrument maker.[47] The Board of Longitude was approached by Thomas Jones, George Dollond and Worthington, each of whom applied to take charge of Ramsden's dividing engines. Jones, as mentioned in Chapter 3, had sought to take the circular engine after Ramsden's death. In April 1820 he was allowed to take the straight-line engine, having agreed to work to the same terms as Berge.[48] Worthington wrote to the Board on 3 March 1820, stating that he had been apprenticed to Berge for seven years, during which time his sole task was dividing, and likewise for years thereafter, and that he had since divided instruments for the East India Company.[49] His request for both engines was refused, but in October 1821 he was allowed to retain Ramsden's brass standard yard and the circular engine, having signed the required acknowledgement and obligation 'to keep the engine safe and in proper repair'[50] (see Figure 13.10). Worthington and Allan continued Ramsden's sequence of sextant numbers. A 12-inch sextant signed 'Worthington & Allan London', with its graduated scale marked 'I R', formerly at Bedford Museum, and several sextants in the National Maritime Museum and in private hands are from this period. Worthington's business is listed in trade directories as Worthington & Allen from 1822 to 1832, then on his own until 1852, this information conflicting with his recorded period as a ratepayer.

The Board of Longitude was abolished in 1828. Up to this time it had remained the legal owner of both engines. On 14 December 1829 Admiral Francis Beaufort wrote to Edward Sabine regarding the items which had meanwhile been transferred to the Royal Society:

> I have spoken to their Lordships about the books, papers and instruments of the late Board of Longitude. They have permitted me to give you up the two former to be kept at the Royal Society but they wish that you should carefully catalogue them as belonging to the late Board of Longitude, and therefore easily reclaimable by the Admiralty when necessary.

Figure 13.9 Portable altazimuth circle, diameter 15½ inches, signed *Worthington London.* Palermo Observatory

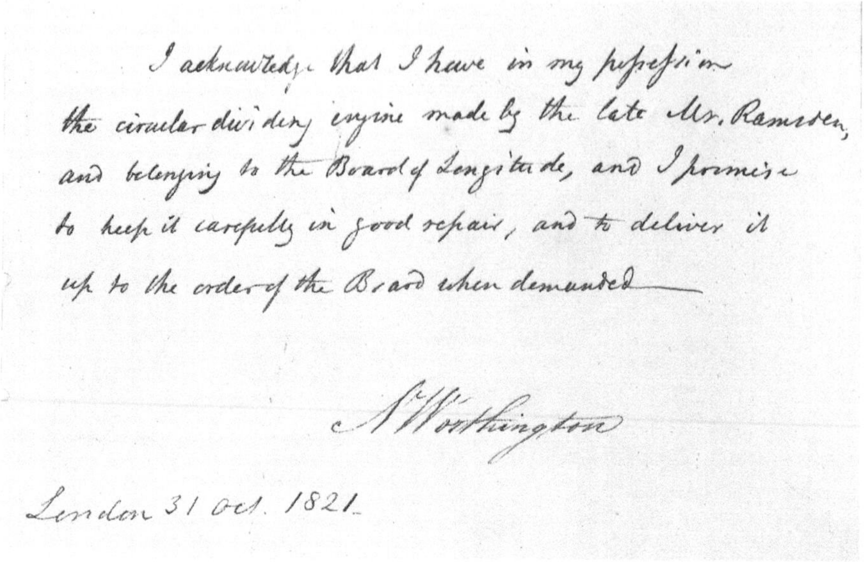

I acknowledge that I have in my possession the circular-dividing engine made by the late Mr. Ramsden, and belonging to the Board of Longitude, and I promise to keep it carefully in good repair, and to deliver it up to the order of the Board when demanded —

N Worthington

London 31 Oct. 1821.

Figure 13.10 Worthington's undertaking, 1821, to look after the circular dividing engine belonging to the Board of Longitude. RGO 14/7 f.224.

> Their Lordships entirely approve of your sending all your instruments to their office and they beg that you will examine them with me when [illeg] send at once separate the valuable part from the trash.[51]

Unfortunately no further mention of what was considered 'valuable' and what was deemed 'trash' has been located.

When Worthington left Piccadilly, Ramsden's dividing engine was around 75 years old, but it clearly had a recognisable value. The engine, and the apparatus with which Ramsden had cut its screw, was purchased by the American firm of Messrs Knox & Shain, of Philadelphia. In about 1880 both items were acquired by Professor Henry Morton, president of the Stevens Institute of Technology in New Jersey, and in March 1890 he presented them to the US National Museum – now the Smithsonian's Museum of American History, where they remain[52] (see Figure 13.11).

Notes

1   Winchester, Hampshire Record Office, 38M49/A4/7.
2   www.oldbaileyonline.org.
3   TNA, CRES 2/730.
4   TNA, CRES 39/65, second series of registers, 30.
5   Stimson (1976), 128.
6   'D' (1829), 314.
7   Atwood (1804), vii.
8   Atwood (1804), 35–8 and Plate III Figs 7 and 8.

Figure 13.11 Ramsden's circular dividing engine, formerly the property of the Board of Longitude, as it was when acquired by the Smithsonian Institution in 1890. National Museum of American History, Cat. MA*215518.

9   ICE, Telford papers T/GC.72/1–2.
10 Millburn (2000), 292–3, citing TNA WO 49/248, pp. 142–6.
11 Kitchiner (1825), 91. Frederick Augustus, Duke of York, did not die until 1827; the reasons for this sale are unknown.
12 Morrison-Low (1984), 18.
13 Ball (1895), 242–3.
14 Ussher (1787), 3–22.
15 BL, MS Add. 33981, f. 12. Mudge (1803). Pearson (1829), 2. 553.
16 BL, MS Add 8099 f187 Méchain to Banks, Paris, 31 October 1801.
    '… je profite de ces dispositions favorables pour vous prier d'employer votre crédit auprès de M Berge successeur de Ramsden, à fin de s'engager à terminer et vous livrer

le plutôt possible une lunette meridienne ou <u>transit instrument</u> de 8 pieds de foyer que son maitre, d'après un engagement pris en XI bre 1787, avec Cassini, devoit nous fournir pour notre Observatoire, au commencement de 1789; et pour le prix duquel il lui a été payé, en à compte, 59 livres 18 shillings le 30 décembre même anneé; mais par l'effet de la lentesse de Ramsden et de la Guerre, nous n'avons que rien obtenir. Depuis plus de deux ans que le Cit. Otto est en Angleterre, il a fait tous les efforts auprès de Ramsden pour tacher de le determiner à remplir ses engagemens; il les a réiterés auprès du successeur Berge, Sir Englefield a bien voulu s'employé aussi, mais tous cela était sans succès jusqu'à present. On élève des difficultés et presques les [?denes] sur les deux sommes déjà payées quoique j'en aie entre les mains les reconnaissances en bonne forme, et que j'en aie envoyé des copies certifiées à Sir Englefield et au Cit. Otto.

Si vous avez la bonté, Monsieur, de temoigner à M Berge quelque interest pour la fin de cette affaire, cela nous procureroit sans doute la jouissance prochaine d'un instrument qui manque à notre observatoire. Chargé de la Direction de cet établissement qui vient d'être restauré et muni de deux grands quart de cercle muraux, l'un de Bird et l'autre de Sisson, qui va posseder un téléscope de reflexion de 22 pieds de foyer et 22 pouces d'ouverture, j'ai fort à coeur de completer son ameublement par la lunette meridienne que nous attendons en vain depuis si longtems. Ramsden en avoit fixé le prix total avant sa mort, à deux cent guinées, on payeroit ce qui reste pour achever cette somme tout aussitôt que l'instrument seroit livré. Sir Englefield m'a mandé qu'il étoit fort avancé, si votre invention pouvait determiner l'artiste à le finir et à nous en faire jouir bientôt, ce serait un nouveau service dont les sciences vous servient redevables: notre Bureau de Longitudes, et moi en particulier, nous vous en aurions une grande obligation.'

17   BL, MS Add.8099, f.280. 'J'ai reçu la lettre que vous m'avez fait l'honneur de m'écrire le 11 du mois dernier. Je vous rends mille graces des peines que vous avez bien voulu prendre pour reconnoitre par vous même l'état où se trouvait le <u>transit instrument</u> que feu M Ramsden avoit entrepris pour notre observatoire. Si par l'effet de vos bons soins M Berge le termine et le livre à la fin de l'été ou de cette année, c'est à vous seul que nous le devrons …'

18   London BL MS Add.8099, f. 280, Méchain to Banks, Paris, 15 April 1803. 'J'ai reçu la lettre, datte du 24 mars dernier, pour laquelle vous m'avez fait l'honneur de me prevenir que l'instrument des passages destiné pour notre observatoire que M Ramsden avoit entrepris en 1787, et que M Berge s'est chargé d'achever, seroit pret dans trois semaines, à être mis en caisse et envoyé à Paris …'

19   Wolf (1905), 297–8.

20   de Beer (1960), 175–6.

21   RS, CB/1/1/209.

22   Foderà Serio and Nastasi (1985); Foderà Serio and Chinnici (1998).

23   ASP, RSV, file 1480 (unfol.) Castelcicala's transcript of Banks's undated letter.

24   ASP, RSV, file 1480, Castelcicala to De Medici, 1 May 1808.

25   BL, MS Add. 8100 f. 78. Piazzi to Banks 18 October 1810.

26   Foderà Serio and Chinnici (1997). Chinnici *et al.* (2000), 27. Foderà Serio and Indorato (1981), 217–24.

27   Anon (1810), 179–84. Brooks (1992), 101–35.

28   Allen (1811), 108.

29   TNA, PROB 11/1355 ff. 251r – 252v.

30   A box sextant signed 'Allan London' now in the Royal Engineers Museum at Chatham, was presented by the 4th Duke of Gordon to his son in 1813. Another is at Newcastle

Museum, a third at the National Museum of Scotland, in a box with label of John Allan. The circle is at Mariners Museum in Virginia.

31 RGO, MS 14/4, 112^r; RGO 14/10, f. 14r, 11 Feb. 1819: Allen explains the principle of his new dividing engine, 14/10 f. 17 ^{r}, 18^r, 17 June 1819: Kater and Wollaston declare it superior. See also 'Report of the Board of Longitude Committee for examining instruments and proposals, upon the mode employed for determining the errors of dividing engines', *Quarterly Journal of Science, Literature and the Arts* No. XVIII (1820), 347–9.

32 TNA, Taunton, Hydrographic Office Archives, Letters-in P 93.

33 RGO, MS 14/7, 364. Wollaston (1821), 381–8.

34 Stimson (1985), 103.

35 London, Birkbeck College Archives, Registers of the London Mechanics Institute.

36 I am indebted to Dr D.J. Warner of the National Museum of American History for the information on Jonathan Allen.

37 CU, Fitzwilliam Museum Library, Perceval Collection, H-206.

38 CNAM, MS C.32.

39 CU, Fitzwilliam Museum Library, Percevel Collection H-209 and 213.

40 Lambeth Palace Library, Faculty Office allegation, 22 July 1801.

41 Sale prospectus issued by Messrs Savills plc., 1991.

42 'Police. Guildhall, robbery at Inns', *The Times*, 13 March 1819, 3e.; 'Old Bailey. Extensive robbery', *The Times*, 1 May 1819, 3e.

43 *Gentleman's Magazine* 89 (1819), 475.

44 WA, St James Piccadilly, PR. Neither burial is recorded in the comprehensive manuscript survey of inscriptions in the Old Cemetery at Hampstead, made in 1883–88 by Frederick Snell. London, Society of Genealogists Library, MS F. Snell, 'Inscriptions in the chapel, and the Old and New Cemeteries, St James Hampstead', 2 vols.

45 Serio and Chinnici (1997), 77–8.

46 Vitello (1987), 254–65; Chinnici (1997).

47 TNA, OS 3/260, Ordnance Survey copy letter book, p. 128.

48 RGO, MS 14/4, 73^r.

49 RGO, MS 14/13, 140.

50 RGO, MSS 14/7, fol. 224.

51 RS, MS DM 4/128.

52 Watkins (1890) pt 1. 732. Miller (1998–99), 12–21.

Chapter 14

Tales of a Great Man

After Ramsden's death various stories circulated concerning his character – not always to his credit though generally affectionate in tone. The surgeon Everard Home, whose collaboration with Ramsden was described in Chapter 10, paid him a generous tribute in his Croonian Lecture of 1801, delivered, as it happened, on the first anniversary of Ramsden's death:

> The first of these experiments ... was made with the assistance of the late Mr Ramsden; and had not the death of that valuable member of this society deprived me of his further aid, the following observations would undoubtedly have been more deserving the attention of my learned audience.
>
> It is impossible for me to mention Mr Ramsden, from whom I received so much assistance in every pursuit connected with optics and mathematics, in which I have been engaged, without availing myself of this opportunity of paying that tribute of gratitude to his memory, which feelings of delicacy prevented me from offering to him while alive. It is unnecessary here to mention his genius, his merits, or his exertions for the promotion of science; these are equally well known to every member present, as to myself. It is only my individual obligations, in the prosecution of enquiries connected with the objects of this learned society that are meant to be taken notice of.
>
> To his friendly, and zealous assistance I am indebted for the information which was necessary to enable me to prosecute investigations upon the subject of vision: and without such assistance, I should have shrunk from the inquiry. It is also to his early friendship, and his readiness to communicate to me his knowledge, that I look back, as among the sources of early exertions, and love of philosophical pursuits.[1]

The Irishman Richard Lovell Edgeworth (1744–1817), who as a member of the Lunar Society knew Ramsden personally, related a story which has passed into many later accounts:

> I have mentioned that Ramsden, the celebrated optician, was of our society. Besides his great mechanical genius, he had a species of invention not so creditable, the invention of excuses. He never kept an engagement of any sort, never finished any work punctually, or ever failed to promise what he always failed to perform.
>
> The King [George III] had bespoke an instrument, which he was particularly desirous to obtain; he had allowed Ramsden to name his own time, but as usual the work was scarcely begun at the period appointed for delivery; however, when at last it was finished, he took it down to Kew in a postchaise, in a prodigious hurry; and, driving up to the palace gate, he asked if His Majesty was at home. The pages and attendants in waiting expressed their surprise at such a visit; he however pertinaceously insisted on being admitted, assuring the page, that, if he told the King that Ramsden was at the gate, His Majesty would soon shew that he would be glad to see him. He was right,

he was let in, and graciously received. His Majesty, after examining the instrument carefully, of which he was really a judge, expressed his satisfaction, and turning gravely to Ramsden, paid him some compliment upon his punctuality.

'I have been told, Mr Ramsden,' said the King, 'that you are considered to be the least punctual of any man in England; you have brought home this instrument on the very <u>day</u> that was appointed. You have only mistaken the <u>year</u>![2]

Edgeworth's daughter, the prolific writer Maria Edgeworth (1767–1849) introduced Ramsden's well-known delaying tactics into 'Conte Gervais le boiteux' a short story first published in French in 1808. The story concerns lame Jervas, a poor-lad-made-good, who becomes a scientific lecturer and is to emigrate to Madras where he will teach at the East India Company's orphans' home. Before departing for Madras Jervas is instructed by his patron to collect from Ramsden the apparatus ordered for him – two small globes, syphons, prisms, an air gun and an air pump, a speaking trumpet, 'a small apparatus for shewing the gases', and an apparatus for freezing water. Ramsden tells him that a small balloon and a portable telegraph in the form of an umbrella were expected to be sent next week. Ramsden is also to provide a set of mathematical instruments of his own making. 'But' adds he with a smile, 'you will be lucky if you get them soon enough out of my hands'. In fact, says Jervas, 'I believe I called a hundred times in the course of a fortnight upon Ramsden and it was the day before the fleet sailed that they were finished and delivered to me'.[3]

A more homely scene was depicted by William Kitchiner, writer of cookery books and light music, and who has figured in earlier chapters of this work as an enthusiastic collector of telescopes:

It was his custom to retire in the evening to what he considered the most comfortable corner in the house, and take his seat close to the kitchen fireside, in order to draw some plan for the forming a new instrument, or scheme for the improvement of one already made. There, with his drawing implements on the table before him, a cat sitting on one side, and a certain portion of bread, butter, and a small mug of porter placed on the other, while four or five apprentices commonly made up the circle, he amused himself with whistling the favourite air, or sometimes singing the old ballad of,

If she is not so true to me
What care I to whom she be
What care I, what care I, to whom she be!

And appeared, in this domestic group, contendedly happy. When he occasionally sent for a workman, to give him necessary directions concerning what he wished to have done, he first shewed the recent finished plan, then explained the different parts of it, and generally concluded by saying, with the greatest good humour, 'Now man, let us try to find fault with it;' and thus, by putting two heads together, to scrutinize his own performance, some alteration was probably made for the better. And, whatever expense an instrument had cost in forming, if it did not fully answer the intended design, he would immediately say, after a little examination of the work, 'Bobs, man! this won't do; we must have at it again:' then the whole of that was put aside and a new instrument begun. By means of such perseverance, he succeeded in bringing various mathematical, philosophical and astronomical instruments to perfection. The large theodolite for terrestrial measurements, and the equal altitude instrument for astronomy, will always be monuments to his fertile, penetrating, arduous, superior genius! There cannot be a lover (especially of this more difficult part) of philosophy, in any quarter of the globe, but must admire the abilities of Jesse Ramsden![4]

Louis Dutens, who contributed the personal description of Ramsden in Aikin's *Biography,* (see Appendix 1), regretted that Ramsden had not been better rewarded by his country, beyond being one of very few craftsmen elected to the Fellowship of the Royal Society, though he was clearly in no need of a civil pension. Dutens, himself a linguist and bibliophile, praised Ramsden for his love of poetry, prose and technical literature and his mastery of the French language which extended his reading. It is unfortunate that the circumstances of Ramsden's death did not call for a sale of his 'library' which might have revealed his tastes. Obedient to the 'delicacy' which was called for when speaking of the dead, Dutens nonetheless conveys a great affection for Ramsden, scarcely perturbed by what he terms Ramsden's 'few trifling failings'.[5]

There were of course those who considered Ramsden's sins of omission and commission to be more than 'trifling failings'; among them those who had purchased sextants with carelessly-divided graduation and those (very few) who, like William Roy and Giuseppe Piazzi, for whom Ramsden hastened his work and delivered an instrument with its defects uncorrected. Apologizing to Miklòs Vay for 'The ill treatment you as well as myself receive from my workmen ...', and that 'the men work so little' may have been just part of Ramsden's excuse for Vay's instruments not being ready on time.[6] Generally, his customers, notably Duke Ernst of Saxe-Gotha and Jean-Dominique Cassini, driven to despair by his endless delays, a sin made worse by his unreliable promises, admitted that they had little alternative. At the end of the day, his workshop was always busy, the waiting list of eager customers always a long one. His instruments continue to command a high price in the saleroom, and his reputation as England's leading scientific instrument maker of that period is undiminished.

Notes

1 Home (1802), 1–2.
2 Edgeworth, R.L. (1969), 1. 191–2.
3 Edgeworth, M. (1808), 387–414. English text published in Edgeworth collections, Pittsburgh, 1818.
4 Kitchiner (1815), footnote on pp. 87–9.
5 Aikin (1813), 8. 454b–457a. 'An elegant and interesting tribute of respect ... by a gentleman of considerable eminence in the scientific world ...' [i.e. Dutens].
6 Ramsden to Vay, 13 January 1788. Sarospatak Protestant College, Archive of the Vay family, Kii , Letter V-41.

Bibliography

Ramsden's Published Works

1770

Directions for using the New Invented Electrical Machine. As made and sold by J. Ramsden, Mathematical, Optical and Philosophical Instrument Maker, near the Little Theatre, in St James's, Haymarket. London [1770?] 12p., 8°

c.1772

Description of the Portable Barometer as made and sold by J. Ramsden

1773, 1774, 1779, 1791

Description of a New Universal Equatoreal, made by Mr J. Ramsden, with the Method of Adjusting it for Observation. ([London, 1773) 7 pp. + 1 pl. 4°. Another edition, 8 pp., 1774. Another edition, [15 pp. + 1 plate. (1779). (Some bibliographers list this as by James Stuart Mackenzie; the Institution of Civil Engineers' copy is annotated 'Mackenzie gave me this')
Nouvelle instrument appelée Equatorial Universel (London, 1773)
Another edition 'much improved by Mr Ramsden' also 'instructions for making observations' (January 1791) 28 pp. + plate
Description de l'equatorial universel, et de son nouvel appareil de refraction … avec la manière d'adjuster l'instrument pour s'en servir et les instructions pour en faire usage dans les observations (1791)

1775

Patent 1112 of 1775. Astronomical Equatorial Instrument
1775 *Description and Method of adjusting the Improved Hadley's Sextant, made and sold by J. Ramsden, Mathematical, Optical and Philosophical Instrument Maker, next St James's Church, in Piccadilly, London*
1775? *Directions pour l'usage de l'Octant de Hadley.* KCL MS 1493 refers to this document 'in lingua Gallica' but not otherwise seen
1775 *Description d'une lunette achromatique, faite et debitée par J. Ramsden. Faiseur d'instruments d'optique, de physique, et de mathématiques. Vis-à-vis de Sackville Street, Piccadilly, à Londres.* (1775) 4 pp. + 1 plate

1777 *Description of an Engine for Dividing Mathematical Instruments by Mr J. Ramsden, Mathematical Instrument Maker.* Published by order of the Commissioners of Longitude. (1777) 14pp. + 4 large plates

1790 *Description d'une machine pour diviser les instruments de mathématiques par M. Ramsden, de la Société Royale de Londres: publiée à Londres, en 1787, par order du Bureau des Longitudes; traduite de l'Anglois; Augmentée de la description d'une machine à diviser les lignes droites, et de la notice de divers ouvrages de M. Ramsden, [recte 1777] par M. de la Lande, de l'Académie Royale des sciences, de la Société Royale de Londres, etc. Pour faire suite à la description des moyens employés pour mesurer la base de Hounslow-Heath.* (Paris, 1790) 46pp. + 7 folding plates

1779 *Description of an Engine for Dividing Strait Lines on Mathematical Instruments.* Published by order of the Commissioners of Longitude. (1779) 16pp. + 3 large plates

1779 'The Description of Two New Micrometers', *Philosophical Transactions,* 69: 419–431

1783 'A Description of New Eye Glasses for such Telescopes as may be applied to Mathematical Instruments', *Philosophical Transactions* 73: 94–99

1791 *The Universal Equatorial and ... New Refraction Apparatus* see above. *Equatorial universal, et ... son nouvel appareil de refraction* see above.

1792 *An Account of Experiments to Determine the Specific Gravities of Fluids, Thereby to Obtain the Strength of Spiritous Liquors, together with Some Remarks on a Paper entitled The Best Method of Proportioning the Excise upon Spiritous Liquors, Lately Printed in the Philosophical Transactions.* By J. Ramsden. London, 1792. 33pp. + 1 plate.

1793 *Animadversions on a Paper entitled Supplementary Report on the Best Method of Proportioning the Excise on Spiritous Liquors. by Charles Blagden ... printed in the Philos. Trans. for the year 1792.* By J. Ramsden. London, 4°

Newspapers

Halifax: *Halifax Guardian* 10 January 1857
Halifax: *Courier and Guardian* 1935
London: *The Times* 1841
Palermo: *Giornale astronomico e meteorologico del Reale Osservatorio di Palermo* 1857
Westminster: *St James's Chronicle* 1800
Westminster: *Whitehall Evening Post* 1800

Abbreviations for Printed Sources

BAB   *Bibliographic Archive for the Baltic States,* A. Frey (comp.), mf., Munich 1999

DAB   *Dictionary of American Biography,* New York. 1999. 24 vols and online

DBE   *Deutsche Biographische Enzykloäpdie*, Munich 1995–99. 10 vols and English translation, 10 vols Munich 2001–06

DBF   *Dictionnaire biographique française*, Paris. 1933 – in production

EE   *Edinburgh Encyclopedia*, Edinburgh 1808–30. 18 vols

OxDNB   *Oxford Dictionary of National Biography*, Oxford 2004. 60 vols and online

PhTr   *Philosophical Transactions of the Royal Society of London*

Rees   Rees, A. *Cyclopedia, or an universal dictionary of arts, sciences and literature*, London 1819. 39 vols + 6 vols of plates

Adams, G. (1790), *A Short Dissertation on the Barometer, Thermometer, and other Meteorological Instruments*, London

Aikin, J. (1813), Article 'Ramsden' in *General Biography* (London, 1799–1815): vol. 8, 450–57

Allen, J. (1811), [Letter and Explanation of his improved reflecting circle] *Transactions of the Society of Arts*, 29: 106–13

Anderson, R.G.W., J.A. Bennett and W.F. Ryan (1993), *Making Instruments Count. Essays in Honour of Gerard Turner*, Aldershot

Andrewes, W.J. (ed.) (1996), *The Quest for Longitude*, Harvard, MA.

Andrews, J.H. (1975), *A Paper Landscape. The Ordnance Survey in Nineteenth Century Ireland*, Oxford

Andrews, W. (1887), *Famous Frosts and Frost Fairs in Great Britain*, London

Anon. [F.X. v Zach] (1798), *Allgemeine Geographische Ephemeris*, 1: 679

Anon. Article 'Cutting engines', *Rees*, vol. 2 (unpaginated)

Anon. (1772), 'Description du baromètre de Ramsden', *Introduction aux observations sur la physique, sur l'histoire naturelle et sur les arts*, 1: 509–12

Anon. (1778–79), 'Astronomia', *Antologia Romana*, 5/1: 1–3; 5/2: 9–12; 5/3: 17–21

Anon. (1788), 'Description d'une nouvelle balance, construite par M. Ramsden, de la Société Royale de Londres', *Observations sur la physique, sur l'histoire naturelle et les arts*. 2nd series, 33/2

Anon. (1789). *European Magazine*, 15: 91–6

Anon. (1800), *Monatliche Correspondenz*, 1: 68

Anon. (1803), 'Account of the life and labours of the late Mr. Ramsden, in a letter from Professor Piazzi, of Palermo, to M. De Lalande', 16: 253–62

Anon. (1810), 'Description of his mathematical dividing engine' and letter of 20 November 1809, *Transactions of the Society of Arts*, 28: 179–81

Anon. (1816), Article 'Graduation' in *EE*, Vol. 10: 313–84

Anon. (1820), *Das gelehrte Teutschland, oder Lexicon der jetz lebenden teutschen Schriftsteller*, 23 vols, 1796–1834. Lengo

Anon. (1822), 'Memoir of T. D. Pearse', *The British Indian Military Repository*, 1/2: 153–248

Anon. (1827), 'Memoir of the Life and Writings of M. Piazzi, Director-General of the Observatories of Naples and Palermo', *Edinburgh Journal of Science*, 6: 193–9

Anon. (1938), 'St James's Piccadilly', *Journal of the London Society*, No. 243 (May 1938): 69–73

Anon. (1998), *Sphaera* (Oxford Museum of the History of Science). No.18: 4–5

Anon. (1999), 'Poczobutt', *BAB* mf 275, frames 207–9

Archinard, M. (1980), *De Luc et la recherche barométrique*, Geneva

Atwood, G. (1784), *Principles of Natural Philosophy*, London

Atwood, G. (1804), *A Supplement to a Tract entitled 'A Treatise on the Construction and Properties of Arches, Published in the Year 1801'*, London. p. vii

Baiada, E., F. Bonoli and A. Braccesi (1995), *Museo della Specola*, Bologna

Balboni, A. *et al.* (2000), 'Gli strumenti dell'osservatorio astronomico di Brera', pp. 87–204, 94–5, in G. Buccellati (ed.), *I cieli di Brera. Astronomia da Tolomeo a Balla*, Milan

Ball, R.S. (1895), *Great Astronomers*, London & Bombay

Baring, C.A. (ed.), 1866 *The Diary of William Windham 1784–1810*, 2 vols

Bayley, W. (1769), 'Astronomical Observations made at the North Cape, *PhTr*, 59: 272

Beaglehole, J.C. (1961), *The Journals of Captain James Cook on his Voyages of Discovery, Vol. 2, The Voyage of the* Resolution *and* Adventure, *1772–1775*, Hakluyt Society, Extra Ser. 35. Cambridge

Beaglehole, J.C. (ed.) (1962), *The* Endeavour *Journal of Joseph Banks*, 2 vols Sydney

Beaglehole, J.C. (1969), 'Eighteenth century science and the voyages of discovery', *New Zealand Journal of History*, 3: 107–23

Beasley, A.W. (2002), *Home away from Home*, Wellington, NZ

Bedini, S.A. (2001), *With Compass and Chain. Early American Surveyors and their Instruments*, Frederick, MA

Beer, G. de (1960), *The Sciences were never at War*, London and New York

Bennett, J.A. (1987), *The Divided Circle: A History of Instruments for Astronomy, Navigation and Surveying*, Oxford

Bergman, G. (1952), 'Charles Apelquist. En pionjär inom den mekaniska verkstadindustrien I Sverige behandlas här av Fil. Kand.', *Dædalus. Tekniska Museets Årsbok*, 1952: 130–46. Stockholm

Bernoulli, J. (1771), *Lettres astronomiques ou l'on donne une idée de l'état actuel de l'astronomie practiqué dans plusieurs villes de l'Europe*, Berlin

Bernoulli, J. (1772), *Receuil pour les astronomes*, Berlin. 2 vols

Bernoulli, J. (1776–79), [on Ramsden's dividing engine], *Nouvelles littéraires de divers pays avec des supplemens pour la liste et la nécrologie des astronomes*, Berlin

Bernoulli, J. (1779), *Receuil pour les astronomes, Supplement*, Berlin

Berthaut, H.M.A. (1902), *Les ingénieurs géographes militaires 1624–1831*, 2 vols, Paris

Betts, J. (1993), 'The Eighteenth Century Transits of Venus, the Voyages of Captain Cook and the Development of the Marine Chronometer', *Antiquarian Horology*, 21: 60–69

Bickerton, D. and R. Sigrist, (2000), *Marc-Auguste Pictet 1752–1825: correspondance sciences et techniques; tome III: Les correspondants britanniques*, Geneva

Bigourdin, G. (1887), *Histoire des observatoires de l'Ecole Militaire*, Paris

Bird, J. (1768), *The Method of Constructing Mural Quadrants*, London

Blémont, H., Article 'Antoine-François Jecker' in *DBF*

Bozzolato, G., P. del Negro and C. Ghetti. (1986), *La specola dell'università di Padova*, Brugine

Bracegirdle, B. (1986), 'The First Automatic Microtome', *Proceedings of the Royal Microscopical Society*, 21: 145–7

Brazell, J. (1968), *London Weather*, London

Brissot, J.-P. (1911), *Memoirs 1754–93*, 2 vols, Paris

Brooks, J. (1992), 'The Circular Dividing Engine: Development in England 1739–1843', *Annals of Science*, 49: 101–35

Brooks, R.C. (1989a), 'The Precision Screw in Scientific Instruments of the 17th to 19th Centuries'. PhD thesis, Leicester

Brooks, R.C. (1989b). 'Gleaning Information from Screw Threads'. *Bulletin of the Scientific Instrument Society*, No. 22: 7–11

Brown, O. (1982), *The Whipple Museum of the History of Science. Catalogue 2: Balances and Weights*, Cambridge

Brown, J. (1979a), *Mathematical Instrument-makers in the Grocers' Company 1688–1800. With Notes on some Earlier Makers*, London

Brown, J. (1979b), 'Guild Organisation and the Instrument-making Trade, 1550–1830: the Grocers' and Clockmakers' Companies', *Annals of Science*, 36: 1–34

Bryden, D.J. (1992), 'Evidence from Advertising for Mathematical Instrument Making in London, 1556–1714', *Annals of Science*, 49: 301–336

Buczek, K. 'Perthees' (1980), *Polski Słownik Biograficzny*, XXV/4, 638–40. Warsaw

Buczek, K. (1982), *The history of Polish Cartography from the 15th to the 18th Century*, Amsterdam

Budde, K. (1993a), *Wirtschaft, Wissenschaft und Technik in Zeitalter der Aufklärung*, Mannheim

Budde, K. (1993b), 'Geschichte der Mannheimer Sternwarte im 18. Jahrhundert', Mannheim: Landesmuseum für Technik und Arbeit, *LTA-Forschung Diskussionsforum*, No. 12

Bugge, T. (1997), *Journal of a Voyage through Holland and England, 1777*, K.M. Pedersen and M. Dybdahl (trans.), Aarhus

Burnett, J.E. and A.D. Morrison-Low (1989), *'Vulgar and Mechanick'; The Scientific Instrument Trade in Ireland, 1650–1914*, Edinburgh & Dublin

Cacciatore, G. *et al.* (eds) (1874), *Corrispondenza astronomica fra Giuseppe Piazzi et Barnaba Oriani*, Milan & Naples

Cajori, F. (1921), 'Swiss Geodesy and the U.S. Coast Survey', *Scientific Monthly* 13

Calisi, M. (2000), 'Le specole romane del settecento', in L. Pigatto (ed.), *Giuseppe Toaldo e il suo tempo. Contributi alla storia dell'università di Padova 33.* Cittadella. pp. 423–45

Camilleri, S. (2001), 'Folders for Foreign Coins. …', *Equilibrium*, No. 2: 2561–8

Carter H.B. (1988), *Sir Joseph Banks*, London

Cassini, J.D. (1791), *Exposé des operations faites en France en 1787, pour la jonction des observatoires de Paris et de Greenwich par MM. Cassini, Méchain et Le Gendre* … [1790] Paris Cassini, J.D. (1810), *Mémoires pour servir à l'histoire des sciences*, Paris. The 'Eloge de M. le President Brochart de Saron' was also published as a monograph (1810). Paris

Chapman, A. (1983), 'The Accuracy of Angular Measuring Instruments used in Astromomy between 1500 and 1850' *Journal for the History of Astronomy* 16, 133–7

Chapman, A. (1993), 'Scientific Instruments and Industrial Innovation: the Achievement of Jesse Ramsden', in R.G.W. Anderson *et al.* (eds) (1993), pp. 418–30

Chatham, Royal Engineers Museum, 'Royal Engineers', printed leaflet

Chinnici, I. (1997), 'Gli strumenti del "Gattopardo"', *Giornale di astronomia*, 23/1: 24–9

Chinnici, I. *et al.* (2000), *Duecento anni di meteorologia all'Osservatorio Astronomico di Palermo*, Palermo

Chinnici, I. *et al.* (2001), 'The Ramsden Circle at the Palermo Observatory'. *Bulletin of the Scientific Instrument Society*, No. 71: 2–10

Christensen, D.C. (1994). 'Spying on Scientific Instruments: the Career of Jesper Bidstrip', *Centaurus*, 37: 209–44

Christensen, D.C. (2001). 'English Instrument Makers observed by Predatory Danes', *Mathematisk-fisiske Meddelelser*, 46/2: 47–63

Clarke, T.N., A.D. Morrison-Low and A.D.C. Simpson (1989), *Brass and Glass. Scientific Instrument Making Workshops in Scotland*, Edinburgh

Clifton, G. (1993a), 'The Spectaclemakers' Company and the Origins of the Optical Instrument-making Trade in London', in R.G.W. Anderson *et al.* (eds) (1993), pp. 341–64

Clifton, G. (1993b), 'Globe Making in the British Isles', in E. Dekker *et al. Globes at Greenwich. A Catalogue of the Globes and Armillary Spheres in the National Maritime Museum, Greenwich,* Oxford. pp. 45–57

Clifton, G. (1995), *Directory of British Scientific Instrument Makers 1550–1851*, London

Clifton, G. (2003), 'The London Mathematical Instrument Makers and the British Navy, 1700–1850', in P. van der Merwe (ed.) *Science and the French and British Navies, 1700–1850*, Greenwich. pp. 24–32

Clifton, G. (2004), Article 'The Dollond Family' in *OxDNB*

Close, C. (1969), *The Early Years of the Ordnance Survey*, Repr. Newton Abbott

Cotte, L. (1788), *Mémoires sur la météorologie*, 2 vols, Paris

Cox, T. (1879), *A Popular History of the Grammar School of Queen Elizabeth at Heath, near Halifax*, Halifax

Craik, G.L. (1881), *The Pursuit of Knowledge under Difficulties* (Reprint of 1830–31 ed.), London

Cranmer-Byng, J.L. (2000), *An Embassy to China, being the Journal kept by Lord Macartney during his Embassy to the Emperor Ch'ien-Lung 1793–1794*, vol. 6 of *Britain and the China Trade 1635–1842*, London

Cranmer-Byng, J.L. and T.H. Levere (1981), 'A Case Study in Cultural Collision – Scientific Apparatus in the Macartney Embassy to China', *Annals of Science*, 38: 503–25

Crawforth, M. (1979), *Weighing Coins: English Folding Gold Balances of the 18th and 19th Centuries*, London. pp. 47, 57, 60–61, 87, 115

Crawforth, M. (1985), 'Evidence from Trade Cards for the Scientific Instrument Industry', *Annals of Science*, 32: 453–554

'D.' (1829), On the most eligible Form for the Construction of a Portable Barometer', *Gleanings in Science*, No. 11: 313–19

Daumas, M. (1972), trans. M. Holbrook, *Scientific Instruments of the 17th and 18th Centuries and their Makers*, London

Daumas, M. (1973), 'Les méchaniciens autodidactes français et l'acquisition des techniques britanniques', pp. 301–31 in *L'acquisition des techniques par les pays non-initiateurs. Colloque du CRNS, 1970*, Paris

David, A. *et al.* (2001–04), *The Malaspina Expedition 1789–1794*, Hakluyt Society series III, vols 8, 11 and 13. London

Dawes, H. (1988), 'Marks, Doodles and Scratchings', *Bulletin of the Scientific Instrument Society*, No. 19: 14–15

Débarbat, S. (1989), 'Coopération géodesique entre la France et l'Angleterre à la veille de la Révolution Française: échanges techniques, scientifiques et instrumentaux', *114e Congrès national des Sociétés savantes*, Paris

Débarbat, S. (2002), 'Piazzi and his French colleagues', *Memorie della Società astronomica italiana*, Special Number 1: 17–25

De Luc, J.-A. (1778), 'An Essay on Pyrometry and Areometry and on Physical Measures in General', *PhTr*, 68: 419–533

Devic, M.J.F.S. (1851), *Histoire de la vie et des travaux scientifiques et littéraires de Jacques Dominique Cassini*, Clermont

Dollond, P. (1779), 'An Account of Apparatus applied to the Equatorial Instrument for Correcting the Errors arising from the Refraction in Altitude', *PhTr*, 69: 332–6

Dollond, P. (1789), *Some Account of the Discovery made by the late Mr John Dollond FRS*, London

Donnelly, M.C. (1964), 'Astronomical Observatories in the 17th and 18th Centuries', *Mémoires, Académie Royale de Belgique* (cl. sciences), 34/5: 1–37

Dunmore, J. (ed.) (1994–95), *The Journal of Jean-François de Galaup de la Pérouse 1758–1788*, Hakluyt Society, 2nd series, nos 180 and 181. 2 vols, London

Edgeworth, M. (1808), 'Conte Gervais le boiteux'. *Bibliographie britannique*, 38/7: 387–404

Edgeworth, R.L. (1969), *Memoirs of Richard Lovell Edgeworth Esquire*, London (Repr. Shannon.)

Edney, M.H. (1997), *Mapping an Empire*, Chicago & London

Englefield, H. (1802), 'On the Effect of Sound upon the Barometer', *Journal of the Royal Institution of Great Britain*, 1: 157–9

Estacio dos Reis, A. (1991), *Uma officina de instrumentos matemáticos e náuticos 1800–65*, Lisbon

Estacio dos Reis, A. (2006) *Gaspar José Marques e a Máquina a Vapor*, Lisbon

Everest, G. (1839), *A Series of Letters addressed to the Duke of Sussex*, London

Farington, J. (1978–98), *The Diary of Joseph Farington*, 13 vols, New Haven & London

Ferrighi, A. (2000), 'Toaldo, Cerato e la fabbrica della specola astronomica di Padova: un sodalizio esemplare tra astronomo e architetturo', in L. Pigatto (ed.) (2000). pp. 159–71

Fiorani, L. (1969), *Onorato Caetani, un erudito*, Florence

Flury, F. (1929),'Beitrag zur Geschichte der Astronomie in Bern', *Mitteilungen der Naturforschenden Gesellschaft Bern aus den Jahren 1927/1928*, Bern

Foderà Serio, G. and I. Chinnici (1997), *L'osservatorio astronomico di Palermo*, Palermo

Foderà Serio, G. and I. Chinnici (1998), *La collezione di strumenti topografici*, Palermo

Foderà Serio, G. and L. Indorato (1981), 'The Matthew Berge's [*sic*] instruments at the Palermo Astronomical Observatory', *Annali dell'Istituto e Museo di storia della scienze di Firenze*, 6: 217–24

Foderà Serio, G. and P. Nastasi (1985), 'Giuseppe Piazzi's survey of Sicily: the chronicle of a dream', *Vistas in Astronomy*, 28: 269–76

[Fontana, F.] (1783), *Opuscoli scientifici di Felice Fontana*, Florence

Forbes, E.G. (1980), *Tobias Mayer (1723–62), Pioneer of Enlightened Science in Germany*, Göttingen

Forbes, E.G. (1975), *Greenwich Observatory. Vol. I: Origins and Early History*, London

Fortescue, J. (ed.) (1927), *The Correspondence of King George the Third from 1760 to December 1783*, 6 vols, London

Gardiner, R.A. (1977), 'William Roy, Surveyor and Antiquary', *Geographical Journal*, 143: 439–50

Gascoigne, J. (1998), *Science in the Service of Empire*, Cambridge

Gaziello, C. (1984), *L'expédition de Lapérouse 1785–1788. Replique française aux voyages de Cook*, Paris

Golder, F.A. (ed.) (1922), *Bering's Voyages: an Account of the Efforts of the Russians to Determine the Relation of Asia and America*, 2 vols, New York

Graf, J.H. (1884), *Sammlung bernische Biographien*, 5 vols, 1884–1906. Bern, Vol. 1

Guijarro, V. *et al.* (1994), 'Los constructores británicos del siglo XVIII en la colección del Museo Nacional de Ciencia y Tecnología (Madrid)', *Llull*, 17: 25–59

Gumbert, H.L. (ed.) (1977), *Lichtenberg in England*, 2 vols Wiesbaden. Vol. 1, Tagebuch 1

Gunther, R.T. (1969a), *Early Science in Oxford*, 14 vols, 1923–35; repr. Oxford

Gunther, R.T. (1969b), *Early Science in Cambridge*, 1937; repr. Oxford

Hackmann, W.D. (1978), *Electricity from Glass: the History of the Frictional Electrical Machine 1600–1850*, Alphen aan der Rijn

Hall, M.B. (1991), *Promoting Experimental Learning. Experiment and the Royal Society*, Cambridge

Hammond, J.H. and J. Austin. (1987), *The Camera Lucida in Art and Science*, Bristol

Hargreaves, J.A. (1991), 'Religion and Society in the Parish of Halifax', PhD thesis, Hudderfield Polytechnic, Open University

Harley, J.B. and G. Walters. (1977), 'William Roy's Maps, Mathematical Instruments and Library: the Christie's Sale of 1790', *Imago Mundi*, 29: 9–22

Herbst, K.-D. (1991), 'Hans Moritz Graf von Brühl und seine Leistungen im Zusammenhang mit der Übergang zu den astronomischen Messinstrumenten mit Vollkreisen Ende des 18. Jahhunderts', *Die Sterne*, No. 676: 335–42

Hills, R. (2002–05), *James Watt*. Vol. 1: *His Time in Scotland, 1736–1774*, Vol. 2: *His Time in England, 1774–1819*, Ashbourne

Histoire de l'Académie royale des sciences, A. 1765 'Sur quelques moyens de perfectionner les instrumens [sic] d'astronomie' 1768, pp. 65–75

Histoire de l'Académie royale des sciences, A. 1765 'Une plate-forme à diviser les instrumens de mathématique proposée par M. le Duc de Chaulnes' p. 140 in 'Machines ou inventions approuvées par l'Académie en 1765' 1768, pp. 133–43

Holl, A. (2004), 'From the AGE to the Electronic IBVS: the Past and the Future of Astronomical Journals', in *The European Scientist. Symposium on the Era and Work of Franz Xaver von Zach (1754–1832)*, L.G. Balasz et al. (eds), Frankfurt am Main. pp. 224–32

Home, E. (1794), 'Some Facts Relative to the Late Mr John Hunter's Preparation for the Croonian Lecture', *PhTr*, 84: 21–7

Home, E. (1795), 'The Croonian Lecture. On Muscular Motion', *PhTr*, 85: 1–23

Home, E. (1796), 'The Croonian Lenture. On Muscular Motion', *PhTr*, 86: 1–26

Home, E. (1802), 'The Croonian Lecture. On the Power of the Eye to Adjust Itself to Different Distances when Deprived of the Crystalline Lens', *PhTr*, 92: 1–11

Howse, D. and B. Hutchinson, (1969), 'The Clocks and Watches of Captain James Cook, 1769–1969', London

Howse, H.D. (1975), *Greenwich Observatory, Vol. 3: The Buildings and Instruments*, London

Howse, [H.]D. (1989), *Nevil Maskelyne: The Seaman's Astronomer*, Cambridge

Howse, [H.]D. (1998), 'Britain's Board of Longitude: the Finances 1714–1828', *Mariners Mirror*, 84: 400–17

Hughes, T.S. (1820), *Travels in Sicily, Greece and Albania*, 2 vols, London

Hunter, J. (1778), 'On the Heat etc of Animals and Vegetables', *PhTr*, 68: 7–49

Ifland, P. (1998), *Taking the Stars. Celestial Navigation from Argonauts to Astronauts*, Newport News, VA & Malabar, FL

James, H. (1856), 'On the Figure, Dimensions, and Mean Specific Gravity of the Earth, as derived from the Ordnance Trigonometrical Survey of Great Britain and Ireland', *PhTr*, 146: 607–26

James, H. revised by D.A. Johnston (1902), *Account of the Methods and Processes adopted for the Production of the Maps of the Ordnance Survey of the United Kingdom*, London

Jenemann, H.R. (1997), *The Chemist's Balance*, Frankfurt am Main

Jeremy, D. (1977), 'Damming the Flood: British Government Efforts to check the Outflow of Technicians and Machinery, 1780–1843', *Business History Review*, 51: 1–34

Johnson, P. (1989), 'The Board of Longitude 1714–1828', *Journal of the British Astronomical Association*, 99: 63–9

Jones, T. (1807), 'Description of the Optigraph ...', *Philosophical Magazine*, 28: 66–9

Joost V. and A. Schöne (eds) (1983), *Georg Christoph Lichtenberg: Briefwechsel*, vol. 1, 1765–1779. Munich

Jungnickel, C. and R. McCormmach (1986), *Cavendish. Experimental Scientist*, Philadelphia

Kelly, P. (1813), *The Universal Cambist*, 2 vols, London

King, H.C. (1955), *The History of the Telescope*, London

Kitchiner, W. (1815), *Practical Observations on Telescopes*, London

Kitchiner, W. (1825), *Economy of the eyes*, Part II

Klüber, L. (1811), *Die Sternwarte zu Mannheim, beschreiben von ihrem Curator, dem Staats- und Cabinetsrath Klüber*, Mannheim & Heidelberg

Knoefel, P.K. (1982), 'The Astronomical and Meteorological Observatory of the Florentine Royal Museum of Physics and Natural History', *Physis*, 24: 399–422

Knoefel, P.K. (1984), *Felice Fontana. Life and Works*, Trento

Krüss, A. (1966), *Geschichte eines Hamburger Familian-Unternehmens, Krüss. 1796–1844*, Hamburg

Lalande, J.-J. de (1803), *Bibliothèque astronomique*, Paris

Lalande, J.-J. de (1790), *Description d'une machine pour diviser les instruments de mathématiques par M. Ramsden, de la Société Royale de Londres: publiée à Londres, en 1787, par order du Bureau des Longitudes; traduite de l'Anglois; Augmentée de la description d'une machine à diviser les lignes droites, et de la notice de divers ouvrages de M. Ramsden*, Paris

Lalande, J.-J. de., *Astronomie*, 1st ed. (1764); 2nd ed. (1771); 3rd ed. (1792), Paris.

Lambton, W. (1803), 'Account of a ... Survey across the Peninsula of India', *Asiatick Researches*, 7: 312–35

Lavoisier, A. (1862–93), *Oeuvres*, 6 vols, Vol. 3: 349–420. Paris

Law, R.J. (1971), 'Henry Hindley of York', *Antiquarian Horology*, 7: 205–21

Le Monnier, P. (1774), *Description et usages d'instruments d'astronomie* (Article 'Astronomie' in the encyclopedic *Descriptions des arts et métiers*, dated on article title page 1761)

Le Roy, [J.B.?] (1785), 'Sur une machine à electriser, d'une espèce particulière', *Observations sur la physique* ... 2nd series, 5: 53–9

Lefebvre, E. and J.G. de Bruijn (eds) (1969–76), *Martinus van Marum. Life and Work*, 6 vols, Haarlem & Leyden

London: Board of Longitude (1786), *Nautical Almanac for 1787*, Appendix, pp. 44–5

Lorenzoni, G. (1922), *I primordii dell'osservatorio astronomico di Padova*, Venezia

Maddison, R.E.W and F.R. Maddison (1956–7), 'Joseph Priestley and the Birmingham Riots', *Notes and Records of the Royal Society*, 12: 98–113

Magellan, J.H. de (1775), *Description des octants et sextants anglois ...*, Paris

Magellan, J.H. de (1779), *Description et usages des nouveaux baromètres pour measurer la hauteur des montagnes, et la profondeur des mines ...*, London

Magellan, J.H. de (1780), *Description d'une machine nouvelle de dynamique, inventée par Mr. G. Atwood, membre de la Société Royale de Londres; ... avec un précis des expériences rélatives à la premiere espece de mouvement, ... dans une lettre adressée à Monsieur A. Volta*, London

Mandell, R.B. (1960), 'Jesse Ramsden: Inventor of the Ophthalmometer', *American Journal of Optometry*, 37: 633–8

Mandrino A. *et al.* (1994), *Un viaggio in Europa nel 1786. Diario di Barnaba Oriani, astronomo Milanese*, Firenze

Mangani, G. (2000), 'Giovanni Antonio Rizzi Zannoni e i suoi rapporto con Giuseppe Toaldo', in L. Pigatto (ed.) (2000). pp. 173–90

Mansfield, John (ed.) (1906), *Dampier's voyage consisting of a new voyage round the world, a supplement to the voyage round the world, two voyages to Campeachy, a discourse of winds, a voyage to New Holland, and a vindication, in answer to the chimerical relation of William Fullell*, 2 vols, London

Martinez-Cañavate Ballesteros, L.R. (ed.) (nd ?1999), *La Expedición Malaspina 1789–1794, Vol. VI, Trabajos astronómicos, geodésicos e hidrográficos*, Madrid

Maskelyne, N. (1775), 'An Account of Observations made on the Mountain Schehallien, for finding its Attraction', *PhTr*, 65: 500–42

Matulaityte, S. (2004), *Senoji Vilniaus Universiteto astronomijos observatorija ir jos biblioteka*, Vilnius

McAdams, D.R. (1972), 'Electioneering Techniques in Populous Consistencies', *Studies in Burke and his Time*, 14: 25–53

McConnell, A. (1994), 'From Craft Workshop to Big Business: the London Scientific Instrument Trade's Response to Increasing Demand, 1750–1820', *London Journal*, 19/1: 36–53

McConnell, A. (1992), *Instrument Makers to the World. A History of Cooke, Troughton & Simms*, York

McConnell, A. and A. Brech (1999), 'Nathaniel and Edward Pigott, Itinerant Astronomers', *Notes & Records of the Royal Society*, 53: 305–18

McConnell, A., Article 'Chester Moor Hall' in *OxDNB*

McKay, C.G. (1979–80), 'Lichtenberg's Friend: the Progress of Genius in the Latter Half of the Eighteenth Century', *Lychnos*, 1979–80: 207–30

Memoires de l'Académie royale des sciences, A. 1765, 'Sur quelques moyens de perfectionner les instrumens [sic] d'astronomie', 1768, pp. 411–27

Millburn, J. (2000), *Adams of Fleet Street*, Aldershot

Miller, R.C. (1998–99), 'Circular Dividing Engines in the United States before 1900', *Rittenhouse*, 12/1: 12–21

Miniati, M. (1984), 'Le origine della specola fiorentina', *Giornale di astronomia*, 3–4: 209–20

Miotto, E., G. Tagliaferri, P. Tucci, (2000), 'La storia della specola di Brera dal 1762 al 2000', in G. Buccellati (ed.) *I cieli di Brera*, Milan. pp. 43–66

Modzalevskii, B.L. (1908), *Spisok chlenov Imperatorskoi Akademii Nauk 1725–1907*, St Petersburg

Montluzin, E.L. de. (2002), *Daily Life in Georgian England as Reported in Gentleman's Magazine*, Studies in British and American Magazines, vol. 14, Lewiston, New York & Ontario

Montucla, J.-E. (1799-1802), *Histoire des mathematiques*, 2nd ed. partly by Lalande, 4 vols, Paris

Morrison-Low, A.D. (1984), 'Brewster and Scientific Instruments', in *Martyr of Science. Sir David Brewster 1781–1868*, A.D. Morrison-Low and J.R.R. Christie (eds), Edinburgh. pp. 18–65

Morrison-Low, A.D. (1995), 'The Role of the Sub-contractor in the Manufacture of Precision Instruments in Provincial England during the Industrial Revolution', in I. Blanchard (ed.), *New Directions in Economic and Social History*, papers presented at the 'New Researchers' sessions of the Economic History Society conference, Edinburgh, 1995. pp. 13–19

Morrison-Low, A. and A. Simpson (1995), 'A New Dimension: a Context for Photography before 1860', in *Light from the Dark Room*, S. Stevenson (ed.), Edinburgh. pp. 15–28

Mudge, W. and I. Dalby (1799), *Account of the Operations ... 1787–96*, 2 vols, London

Mudge, W. (1803), 'Account of the Operations ...', *PhTr*, 93: 383–508

Nairne, E. (1771), 'Description and Use of a New Constructed Equatorial Telescope or Portable Observatory made by Mr Edward Nairne', *PhTr*, 61: 107–13

Neumann, J. (1994), 'Landesvermessung und Kartographie in der Kurpfalz in 18. Jahrhundert', Mannheim: Landesmuseum für Technik und Arbeit, *LTA-Forschung Diskussionsforum*, No. 19: 25–40

Nevskaya, N.I. (1977), 'Correspondence between Astronomers', in *USSR Academy of Sciences: Scientific Relations with Great Britain*, Moscow. pp. 177–216

Noorthouck, J. (1773), *New History of London*, London

O'Donoghue, Y. (1977), *William Roy, 1726–1790. Pioneer of the Ordnance Survey*, London

Odlanicka-Poczobutt, T. (1979), '250ᵉ Anniversaire de la Naissance de Marcin Odlanicki-Poczobutt', *Contes rendus de l'Académie des sciences*, Vol. 288, sect: Histoire des sciences, 9 April 1979: 127–8

Pearson, W. (1824–29), *Introduction to Practical Astronomy*, 3 vols, London

Pedley, M. (2000), *The Map Trade in the Eighteenth Century: Letters to the London Map-sellers Jefferys and Faden*. (Complete volume) *Studies on Voltaire and the Eighteenth Century*, 2000.06, Oxford.

Pellatt, A. (1849), *Curiosities of Glassmaking*, London

Phillimore, R.H. (1945), *Historical Records of the Survey of India*, 4 vols, Dehra Dun

Phipps, C.J. (1774), *Voyage towards the North Pole, 1773*, London

Piazzi, G. (1788), 'Lettre sur les instruments de M. Ramsden, de la Société Royale de Londres, adressée à de la Lande par le R. P. Piazzi, Théatin, professeur royal d'astronomie dans l'université de Palerme', *Journal des savants*, November issue: 744–52

Piazzi, G. (1792), *Della specola astronomica de' Regj Studj di Palermo*, 4 parts, Palermo

Pictet, M.-A. (1791), 'Considerations on Measuring an Arch of the Meridian', *PhTr*, 81: 106–17

Pictet, M.-A. (1797), 'Resumé of the Trigonometrical Survey, taken from the Philosophical Transactions of 1795, part 1', *Bibliographie Britannique* (Science & Arts) 4/4: 272–303

Pictet, M.-A., *Correspondence,* 4 vols. For vol. 3 see D. Bickerton and R. Sigrist (eds) (2000); for vol. 4 see R. Sigrist (ed.) (2004)

Pigatto, L. (2000a), *Giuseppe Toaldo e il suo tempo, nel bicentenario della morte. Scienza e lumi tra Veneto ed Europa*, Contributi alla storia dell'università di Padova, No. 33, Cittadella

Pigatto, L. (2000b), 'Giuseppe Toaldo, profilo biobibliografico', in L. Pigatto (ed.) (2000a), pp. 5–105

[Pigott, C.] (1792), *Jockey Club. A Sketch of the Manners of the Age*, 2 vols, London. 1: Preface

Poli, G.S. (1780), 'Memoria su di un nuovo micrometro di riflessione', *Opuscoli scelti*, 3: 111–18

[Pond, J.] (1813), Article 'Circle' in *EE*, vol. 6: 484–515

Porritt, A. (1970), 'Jesse Ramsden (An 18th century Halifax Inventor)', *Transactions of the Halifax Antiquarian Society* (no vol.): 15–27

Porter, R. *et al.* (eds.) (1985), *Science and Profit in Eighteenth-century London*, Cambridge

Portlock, J.E. (1833), 'Memoir of the Late Major-general Colby RE ... with a Sketch of the Origin and Progress of the British Trigonometrical Survey', *Papers on Subjects connected with the Duties of the Corps of Royal Engineers*, ns 3: i–lxiv. Reprinted as a monograph, 1869

Priestley, J. (1767–1775), *The History and Present State of Electricity*, 1st ed. 1767; 2nd ed. 1769; 3rd ed. 1775

Pritchard, E.H. (2000a), *The Crucial Years of Early Anglo-Chinese Relations, 1750–1800*, vol. 6 of *Britain and the China trade 1635–1842*, London

Pritchard, E.H. (2000b), *The Instructions of the East India Company to Lord Macartney on his Embassy to China and his Reports to the Company, 1792–94*, Vol. 7 of *Britain and the China trade 1635–1842*, London

Promies, W. (ed.) (1967–74), *Georg Christoph Lichtenberg, Schriften und Briefe*, 2 vols, Munich

Provost, M. (a), Article 'Cassini, J.D.', in *DBF* (1956), Vol. 7, cols 1331–2

Provost, M. (b), Article 'Chaulnes, Duc de' in *DBF* (1959), Vol. 8, cols 850–51

Proudfoot, W.J. (1868), *Biographical Memoir of James Dinwiddie LLD*, Liverpool

Quekett, J. (1987), *A Practical Treatise on the Use of the Microscope*, 1848 repr. Lincolnwood

Ragona, D. (1857, *Giornale astronomico e meteorologico del reale osservatorio di Palermo*, Vol. 2, No. 41 (n.p.)

Reid, J.S. (1990), 'Eighteenth-century Scottish University Instruments: the Remarkable Professor Copland', *Bulletin of the Scientific Instrument Society*, No. 24: 2–5; 3, citing Private coll.

Repsold, J.A. (1908), *Zur geschichte der astronomischen Messwerkzeuge von Purbach bis Reischenbach*, 2 vols, Leipzig

Richeson, A.W. (1966), *English Land Measuring to 1800*, Cambridge MA

Robinson, E. (1958), 'The International Exchange of Men and Machines 1750–1800. As Seen in the Business Records of Matthew Boulton', *Business History*, 1: 3–15

Robinson, T.R. (1828), 'On correcting Errors of the Astronomical Circle, by Opposite Readings', *Transactions of the Royal Irish Academy*, 15: 21–38

Robinson, T.R. (1871), 'Dublin Observatory', *Nature*, 3: 445–6

Rochon, A. (1801), 'Memoire sur les verres achromatiques adaptés à la mesure des angles', *Journal de physique, de chimie et d'histoire naturelle*, 53: 189–98

Rose, R.B. (1960), 'The Priestley Riots of 1791', *Past and Present*, 18: 68–88

Roy, W. (1785), 'An account of the Measurement of a Base on Hounslow Heath', *PhTr*, 75/2: 385–480

Roy, W. (1790), 'An account of the Trigonometrical Operation …', *PhTr*, 80/81: 111–270

Roy, W. (1777), 'Experiments and Observations made in Britain, in Order to Obtain a Rule for Measuring Heights with the Barometer', *PTr*, 67: 653–788

Saint Fond, B. Faujas (1799), *Voyage en Angleterre, en Ecosse et aux Îes Hebrides*, 2 vols, 1797, quotation from English translation, London

Santini, G. (1830), 'Descrizione delle principale macchine astronomiche, loro uso e verificazione', *Elementi di astronomia*, 2 vols, 1819–20, revised ed. 1830, Padova

Segnini, C.A. and R.V. Caffarelli (1990), *Antichi strumenti scientifici a Pisa (sec. XVII–XX)*

Setchell, J.R.M. (1972), 'Henry Hindley & Son. Clock and instrument makers and engineers of York', *Yorkshire Philosophical Society, Annual Report for the year 1972*. pp. 39–67.

Seymour W.A. (ed.) (1980), *A history of the Ordnance Survey*, Folkestone

Sharp, A. (1970), *The Journal of Jacob Roggeveen*, Oxford

Sheppard, F.H.W. (1960), *The Survey of London. XXIX The Parish of St James Westminster. Part One: South of Piccadilly*, London

Short, J. (1749–50), 'A letter from Mr J. Short FRS ... with the description and use of an equatorial telescope', *PhTr*, 46. 241–6

Shuckburgh, G. (1779), 'On the Variation of the Temperature of Boiling Water', *PhTr*, 69: 362–75

Shuckburgh, G. (1793), 'An Account of the Equatoreal Instrument', *PhTr*, 83: 67–128 + 6 plates

Sigrist, R. (2004), *Marc-Auguste Pictet 1752–1825: correspondance sciences et techniques; tome IV: Les correspondants suisses, italiens, allemandes et autres avec supplements aux trois precedents, postface et index generaux*, Geneva

Simms, W. (1839–43), 'On a Self-acting Circular Dividing Engine'. *Monthly Notices of the Royal Astronomical Society*, 5: 291–2

Smeaton, J. (1754), 'Description of a New Pyrometer, with a Table of Experiments made therewith', *PhTr*, 48: 598–613

Smeaton, J. (1776), 'Observations on the Graduation of Astronomical Instruments', *PhTr*, 76: 1–47

Sorrenson, R.J.(1993), 'Scientific Instrument Makers at the Royal Society of London, 1720–1780', PhD thesis, University of Princeton, NJ

Stevenson, J. (1992), *Popular Disturbances in England 1700–1832*, 2nd ed., London and New York

Stimson, A. (1976), 'The Influence of the Royal Observatory at Greenwich upon the Design of Seventeenth and Eighteenth Century Angle-measuring Instruments at Sea', *Vistas in Astronomy*, 20: 123–30

Stimson, A. (1985), 'Some Board of Longitude Instruments in the Nineteenth Century', in P.R.de Clercq (ed.), *Nineteenth-century Scientific Instruments and their Makers*, Amsterdam. pp. 93–115

Stimson, A. (1996), 'The Longitude Problem: the Navigator's Story', in W.J.H. Andrewes (ed.), *The Quest for Longitude*, Cambridge, MA. pp. 72–84

Stock, J.T. (1969), *Development of the chemical Balance*, London

Stock, J. (1973), 'Weighed in the Balance', *Analytical Chemistry*, 45: 974A–80A

Stothers, R.B. (1966), 'The Great Dry Fog of 1783', *Climatic Change*, 32: 79–89

Talbot, S. (2003), 'The First Telescope Dynameter as Designed and Constructed by Jesse Ramsden, London, c.1780', *Bulletin of the Scientific Instrument Society*, No. 77: 8–9

Taylor, E.G.R. (1968), *Mathematical Practitioners of Tudor and Stuart England, 1485–1814*. 1954, repr., 1968. Cambridge

Taylor, W.B.S. (1845), *The History of the University of Dublin*, London

Thiout A. *Traité d'Horologerie* (2 vols, 1741, repr. 1972), Paris

Thrower, N.J. (1982), *The Three Voyages of Edmund Halley in the* Paramore, *1698–1701*, 2 vols Hakluyt Society, second series Nos 156–7; Vol. 1, 268–9, London

Timmins, S. (1890), *Dr Priestley's Laboratory, 1791. Articles Reprinted from the Birmingham Weekly Post March to April 1890*, Birmingham

Triarco, C. (1999), 'La riforma scientifica di Leonardo Ximenes (1716–1786)', PhD thesis, Firenze: Università degli studi

Triarco, C. (2000), 'La Specola di Leonardo Ximenes a Firenze e la catalogazione dei suoi strumenti', in L. Pigatto (ed.) (2000a), pp. 381–409

Trechsel, F. (1819), 'Notice sur la triangulation execute dans le canton de Berne' *Bibliothèque universelle*, ns 10. sect: Sciences et Arts: 77–89

Troughton, E. (1809), 'An Account of a Method of Dividing Astronomical and other Instruments', *PhTr*, 99: 104–45

Turnbull, H.W. (1959), *The Correspondence of Isaac Newton. Vol. 1 (1661–1675)*, Cambridge

Turner, A.J. (1987), *Early Scientific Instruments. Europe 1400–1800*, London

Turner, A.J. (1989), *From Pleasure and Profit to Science and Security. Etienne Lenoir and the Transformation of Scientific Instrument making in France, 1760–1830*, Cambridge

Turner, A.J. (1998), 'Mathematical Instrument-making in Early Modern Paris', in R. Fox and A.J. Turner (eds), *Luxury Trades and Consumerism in Ancien Regime Paris: Studies in the History of the Skilled Workforce*, Aldershot. pp. 63–96

Turner, A.J. (2002), 'The Observatory and the Quadrant', *Journal for the History of Astronomy*, 22: 373–85

Turner, A.J. (2003), 'John Dee, Louvain and the Origins of English Instrument-making', *Musa MusaeiStudies on scientific instruments and collections in honour of Mara Miniati. Nuncius*, 49: 63–78

Turner, G.L'E. (1967), 'The Auction Sale of the Earl of Bute's Instruments, 1793'. *Annals of Science*, 23: 213–42

Turner, G.L'E. (1989), *The Great Age of the Microscope. The Collection of the Royal Microscopical Society through 150 Years*, Bristol

Turner, G.L'E. (2000a), *Elizabethan Instrument Makers. The Origins of the London trade in Precision Instrument Making*, Oxford

Turner, G.L'E. (2000b), 'The Government and the English Optical Glass Industry 1650–1850', *Annals of Science*, 57: 399–414

Udias, A. (2004), *Searching the Heavens and the Earth: the History of Jesuit Observatories*, Dordrecht & London

Uglow, J. (2002), *The Lunar Men: the Friends who made the Future*, London

Ure, A. (1832), *Dictionary of Arts and Manufactures*, Repr., R. Hunt (ed.) London, 1867; Repr. London, 1999

Ussher, H. (1787), 'An Account of the Observatory belonging to Trinity College, Dublin', *Transactions of the Royal Irish Academy*, 1: 3–22

Ussher, H. (1788), 'An Account of a New Method of Illuminating the Wires, and Regulating the Position of the Transit Instrument', *Transactions of the Royal Irish Academy*, 2: 13–26

Vancouver, G. (1984), W.K. Lamb (ed.) *A Voyage of Discovery to the North Pacific Ocean and Round the World 1791–1795*, London: Hakluyt Society, second series, Vols 163–6, Cambridge

Vargha, M. (2005), *Franz Xaver von Zach (1754–1832). His Life and Times*, Budapest: Konkoly Observatory Monograph No. 5

'Veritas' (1790), *Gentleman's Magazine*, 70: 1116

Vince, S. (1790), *Treatise on Practical Astronomy*, Cambridge

Vitello, A. (1987), *Giuseppe Tomasi di Lampedusa*, Palermo

Volta, A. (1949–55), *Epistolario di Alessandro Volta*, 5 vols, Bologna

Walckenaer, C.M. (1940), 'La vie de Prony', *Bulletin de la Société d'encouragement pour l'industrie nationale*, 139: 68–78

Walker, J., 'Our local portfolio' *Halifax Guardian*, 10 January 1857; portrait and article in *Courier and Guardian*, 24 March 1934

Wallis, P.J. and R.V. (1985), *Eighteenth-century Medics*, Newcastle upon Tyne

Wallis, R. (2000), 'Cross-currents in Astronomy and Navigation: Thomas Hornsby FRS (1733–1810)', *Annals of Science*, 57: 219–40

Warner, B. (1979), *Astronomers at the Cape of Good Hope*, Cape Town & Rotterdam

Warren, J. (1811), 'Account of Observations ... at the HEIC Observatory near Fort George', *Asiatick Researches*, 10: 513–25

Watkins, J.E. (1890), 'The Ramsden Dividing Engine', *Smithsonian Report, 1890 pt 1*, pp. 721–39

Watson, J. (1973), *The History and Antiquities of Halifax*, London, 1775. Repr. 1973. Manchester

Wayman, P.A. (1987), *Dunsink Observatory 1785–1985*, Dublin

Weld, C. (1848), *A History of the Royal Society*, 2 vols, London

Whittington, G. and A.J.S. Gibson. (1986), *The Military Survey of Scotland, 1747–1755: a Critique*, Lancaster: Historical Geography Research Group of IBG, No. 18

Whyte, L.L. (1961), *Giuseppe Ruggiero Boscovich SJ, FRS. Studies of his Life and Work*, New York

Wilkinson, L. (1981–82), '"The Other" John Hunter MD FRS (1754–1809)', *Notes and Records of the Royal Society*, 36: 227–41

Wolf, C. (1905), *Histoire de l'Observatoire de Paris*, Paris

Wolfschmidt, G. (2004), 'Zach's Instruments and their Characteristics', in L.G.Balázs *et al.* (eds). *The European Scientist. Symposium on the Era and Work of Franz Xaver von Zach (1754–1832). Proceedings of the symnposium held in Budapest on September 15–17, 2004*, Acta Historica Astronomiae 24. Frankfurt am Main. pp. 83–96

Wollaston, F. (1793), 'A Description of a Transit Circle, for Determining the Place of Celestial Objects as They Pass the Meridian', *PhTr*, 83: 133–53

Wollaston, W.H. (1807), 'Description of the camera lucida', *Philosophical Magazine*, 27: 343–7

Wollaston, W.H. (1821), 'On Comparison of Dividing Engines', *Quarterly Journal of Schience*, 12, 381–8

Woolf, H. (1959), *The Transits of Venus: a Study of Eighteenth-century Science*, Princeton, NJ

Yolland, W. (1847), *Measurement of the Lough Foyle Base*, London

Young, T. (1801), 'On the Mechanism of the Eye', *PhTr*, 91: 23–88

Zach, F.X. von (1818), 'Lettera XXIII', *Correspondence astronomique*, 1: 445–57

Zach, F.X. von (1823), 'Lettre I', *Correspondence astronomique*, 8: 3–39

Zschokke, E. (trans.) (1882), *Memoirs of Ferdinand Rudolph Hassler*, Nice

Appendices

Appendix 1

J Aikin, *General Biography* vol. 8 (1813), 450b–457a
   Article signed 'M' and probably by the Revd Thomas Morgan, but sourced as the Piazzi 'Life', plus original communications sent by the Revd L. Dutens.

Ramsden, Jesse, a very eminent English mathematical and astronomical instrument maker in the 18th century, was the son of an innkeeper at Salterhebble, near Halifax, in Yorkshire, where he was born in the year 1735. From the register of Halifax it appears that he was baptised on the 3rd of November, in that year, when he was, probably, about a month old. At nine years of age he was admitted into Halifax free-school, where he was instructed in the rudiments of classical learning during three years. His father then removed him from this school, and sent him to an uncle's in Craven, a district in the northern part of Yorkshire, where a clergyman of the name of Hall kept a school, and had acquired [a] reputation by his success in teaching the mathematical sciences. Under this gentleman's tuition young Ramsden became proficient in geometry and algebra and was proceeding with delight in studies for which his genius was particularly suited, when his father sent for him home, and put him an apprentice to a clothier in Halifax. After he had followed this occupation three years, he was placed in the capacity of clerk with another manufacturer in the same town, in whose service he continued till he was about twenty years of age. At this period of his life he went to London, where he became clerk in a wholesale cloth warehouse. This situation he retained for two years and a half, when his inclination for the sciences revived; and as he possessed at the same time a strong mechanical turn, he resolved to qualify himself for some business which should prove suitable to the bent of his mind. With this determination he bound himself an apprentice for four years to Mr Burton, who lived in Denmark Court near the Strand and was one of the best workmen of his time in making thermometers and barometers, and in engraving and dividing mathematical instruments. Not long after the expiration of his apprenticeship, he became a partner with a workman of the name of Cole, under whom he was at first a journeyman, with no higher wages that twelve shillings a week. This partnership, however, [p. 451a] did not long continue; and after its dissolution Mr Ramsden opened a workshop on his own account, where he soon recommended himself to employment by some of the most eminent and mathematical instrument makers, particularly Sisson, Dollond, Nairne and Adams. In the course of that employment, his repeated examination of the instruments which were sent to him to be engraven or divided, led him to notice their defects, and his genius suggested to him the means of removing them, or of constructing better instruments. Having resolved to attempt that task, he soon made himself master of the file and the lathe, and even of the art of grinding glasses. He now formed the design of examining every astronomical instrument in use, with the view of correcting those which, being founded on good principles, were faulty only in the construction, and of proscribing those which were defective in both these respects. About this time, by his marriage with Miss Dollond, he became possessed of a part of Mr Dollond's patent for achromatic telescopes.

In the year 1766 Mr Ramsden opened a shop in the Haymarket, and continued there till 1774, when he removed to another in Piccadilly, which he retained as long as he lived.

Before his settlement in the first of these shops he had brought the sextant to its present improved state, and he had invented, though he had not brought to perfection, his famous dividing machine. Until his time Hadley's sextant, though so much employed in the British navy, and so useful an instrument, was in a very defective state. The essential parts were not of sufficient strength; the centre was subject to too much friction; the index could be moved several minutes without any change being produced in the position of the mirror; the divisions in general were very coarse; and Mr Ramsden found that the Abbé de la Caille was right, when he estimated at five minutes the error which might take place in the observed distances of the moon and stars, and which might occasion in the longitude an error of fifty nautical leagues. Mr Ramsden therefore changed the construction in regard to the centre, and made these instruments so correct as never to give more than half a minute of uncertainty. His sextants of fifteen inches he warranted to be correct to within six seconds. From that size he made them to an inch and a half radius, and in the latter the minutes can be clearly distinguished; but he recommended for general use those of ten inches, as being more easily managed, and [451b] susceptible of the same exactness. The methods for dividing mathematical instruments before his time were very inaccurate. Graham and Bird made use of beam compasses; and the latter kept his method a secret till it was purchased from him by the Board of Longitude, in order to be published. Mr Ramsden, however, had already discovered a method of his own, which in accuracy exceeded that of Bird. For large works he still continued the use of beam compasses; but as it is necessary in the greater number of instruments to save time, he contrived a dividing machine, in which ease and expedition are so happily united, that a sextant can be divided with it in the space of twenty minutes. Having spent ten years in bringing to perfection this machine, which reflects high honour on his inventive genius and superior talents, Dr Shepherd made it known to the Board of Longitude, who gave him a premium of a thousand pounds, and caused a description of it, with a plate, to be published in 1777. This edition was unfortunately burnt by accident. Mr Ramsden also constructed an instrument for dividing straight lines, a description of which has been printed. While he was employed on his dividing machines, he made very important improvements in other instruments. The theodolite before his time consisted merely of a telescope, turning on a circle divided at every three minutes, by means of a vernier; but in the hands of Mr Ramsden it is become a new and perfect instrument, which serves for measuring heights and distances, as well as for taking angles. The largest and most admirable theodolite ever constructed was made by him for General Roy, for the purpose of measuring the series of triangles in England, which at present join those in France. Though this is only of eighteen inches radius, its accuracy is so great as not to admit an error of a single second. It is furnished with two telescopes, each of which turns on a horizontal axis, and by which the angles between objects more or less elevated are reduced to the horizon and measured. With this instrument General Roy measured the angle between the pole-star and the sides of his triangles, in order to have the convergence of the meridians such as it is actually, in our oblate spheroid, and he has shewn, that the difference between the meridians of the two observatories of Paris and Greenwich is 9' 20".

Mr Ramsden greatly improved the barometer for measuring the height of mountains. His method of marking at the bottom the line [p.452a] of the levels, and of looking at the top to the contact of the index with the summit of the mercury, renders it possible to distinguish the hundredth part of a line, and to measure heights within a foot. He shewed M. de Luc that it was the summit of the column, and not the part which touches the glass, that ought to be observed; and he caused to be engraved a table to accompany his barometers, which, without the trouble of calculation, gives the heights of places according to the height of the barometer, and even for different degrees of heat. He simplified also, in the most ingenious manner, the apparatus for the conveyance and support of this portable barometer. Various other instruments for philosophical purposes were executed by Mr Ramsden, and always with new improvements; such as an electrical machine; a manometer for measuring the density of the air; an instrument for measuring inaccessible distances, which renders it unnecessary to measure a base; assaying ballances which turn with a ten thousandth part of the weight used; levels exceedingly sensible; the optic rectangle, prismatic eyeglasses where much fewer rays are lost than by the reflection of an inclined mirror, when it is necessary to look on one side; and the dynameter, for measuring the magnifying power of a telescope, by applying before the eyeglass a small scale divided into hundredths of a line to measure the pencil or image of the object glass. The pyrometer, for measuring the dilation of bodies by heats, also exercised the talents of Mr Ramsden; And with the happiest success, as may be seen in the "Philosophical Transactions" for 1785. On examining the pyrometer then in use, he observed the radical defect of that instrument, in which the bodies subjected to experiment were not sufficiently separated. But with this microscopic pyrometer he found means to compare the natural state of a body with the same body exposed to any degree of heat or of cold, and by a micrometer adapted to the microscope he measured these variations with an exactness before unknown, and which furnished the measure of a base with a precision ten times greater than in any of those ever before measured. On this occasion, as on all others, Mr Ramsden shewed a natural sagacity in discovering the essential faults of an instrument, and in inventing the most simple and exact methods of correcting them. The science of optics was no less indebted to him. He invented a method of correcting, in a new and perfect manner, the aberration of sphericity and refrangibility in compound eye-glasses [452b] applied to all astronomical instruments. Opticians had imagined that the purpose might be accomplished by making the image of the object glass fall between the two eye-glasses; which was attended with this very great inconvenience, that the eye-glass could not be touched without deranging the line of collimation, and the value to the parts of the micrometer. To remedy this inconvenience Mr Ramsden set out from a very simple experiment, namely, that the edges of an image observed through a prism are less coloured according as the image is near the prism. This led him to attempt placing the two eye-glasses between the image of the object-glass and the eye, without failing to correct the two aberrations, which he did by changing the radii of the curves, and placing the glasses in a manner altogether different from that commonly employed. Mr Ramsden also invented a reflecting object-glass micrometer; he points out the defects and inconveniences of that of Bouguer, invented in 1748, in which the different positions of the eye, with respect to the pencil of light, cause the two images to appear sometimes to touch each other, sometimes to be separated, and sometimes alternately by a kind of oscillation. He found also, that the aberration of the rays, which renders the object badly defined, increased the inconvenience of that instrument. He thought, therefore, that it would be necessary to abandon the principle of refraction, and to substitute that of reflection. This instrument, not less simple than ingenious, contains no more mirrors or glasses than what are necessary for the telescope; and the separation of the two images depends solely on the inclination of the mirrors, and not on the focus. He turned his attention, however, to the improvement of the refracting micrometer, and conceived the happy idea of placing this micrometer not towards the object-glass, but exactly in the conjugate focus of the first eye-glass. By these means the contrary refraction of the two plano-convex lenses, and the convex lens, corrects the error which takes place in object-glass micrometers where the image depends only on the focus of the two plano-convex lenses; and the image, being already considerably magnified before it falls on the refracting micrometer, the imperfection of the glasses can occasion only an insensible error in the masurement of angles. It is true, indeed, that by this position the field of the micrometer will be smaller that what it [p.453a] would be were the micrometer near the object-glass; but Mr Ramsden has contrived means for making the images to be uniformly illuminated in every part of the field. With this micrometer, the diameter of the planets may be measured in every direction; it may be adapted to all kinds of achromatic telescopes; it may be brought near to or removed from the object-glass at pleasure, to render vision distinct; and it may be taken from the tube of the eye-glasses, that the telescope may be used without a micrometer. The merit to which Mr Ramsden was entitled in consequence of the inventions already mentioned, rendered his friends desirous that he would consent to have his name hung up among the candidates for admission into the Royal Society; but his great modesty would never suffer him to yield to their wishes. However, without his knowledge, they proposed him, and he was elected a fellow of that body in the year 1786.

But the objects hitherto mentioned are not the most important of Mr Ramsden's works. The equatorial, the transit instrument, and the quadrant, received new improvements in his hands. The equatorial, first constructed by Sisson, and somewhat improved by Short, was much further improved by Ramsden. In the first place, he rejected the endless screw, which by pressing on the centre destroyed its precision; he placed the centre of gravity on the centre of the base, and caused all the movements to take place in every direction; he pointed out the means of rectifying the instrument in all its parts; and he applied to it a very ingenious small machine for measuring or correcting the effect of refraction. This invention is considerably prior to that given by Mr Dollond in the Ph.Tr. Mr Ramsden had a patent for this kind of equatorial, of which a description was printed, written by the Hon. Stewart Mackenzie, brother of the Earl of Bute. But Mr Ramsden did not always strictly adhere to this description, his inventive genius rarely allowing him to construct the same instrument many times in the same manner; and it often happened that he brake to pieces instruments which had cost him a great deal of labour, if they proved not so correct as he wished. The greatest equatorial instrument ever attempted is that which he constructed for Sir George Shuckburgh, on which he was employed at least nine or ten years. In this instrument the circle of inclination is four feet in diameter, so that observations can be made nearly within a second; [453b] the telescope is placed between six pillars, which form the axis of the machine; and the whole turns round two pivots resting on supporters of mason work. The transit instrument is made use of in all the large observatories of Europe; but Mr Ramsden has added to it several improvements. He invented a method of illuminating the wires, by making the light pass along the axis of the machine. The reflector is placed in the inside, and obliquely in the middle. He did not lessen the aperture of the object-glass; and as the light passes through a coloured prism, which may be moved at pleasure, the light may be increased or diminished. In order to adjust this essential instrument, Mr Ramsden invented a method which superseded the use of a spirit-level, on which he set no value, because it does not give that exactness which it was always his aim to obtain. His method is, to suspend a thread and plummet before the telescope placed vertically. This thread passes over two points, which are marked on two pieces fixed one above and the other below the telescope, and one of which has a small motion. The thread is quite detached from the telescope, and when it corresponds on the same points in the two different situations of the telescope, the observer is certain that the axis is horizontal. What is most new and ingenious in this method is, that the thread and plummet sometimes pass over the images only of the points which are formed in the focus of a lens, because the observer is sometimes obliged to remove the thread to a considerable distance from the instrument and from the points; but the exactness is not lessened and there is no parallax. Mr Ramsden's meridian telescope, which he made for Blenheim, Manheim [sic], Dublin, Paris and Gotha, are also remarkable for the excellence of their object-glasses.

In the mural quadrant Mr Ramsden has distinguished himself by the exactness of his divisions, and by the manner in which he has finished the planes, by working them in a vertical position. The thread and plummet are placed by him behind the instrument, that there may be no necessity for removing it when observations are made near the zenith. His methods of illuminating [p.454a] the object-glass and the divisions at the same time, and of suspending the telescope, are also new, and improvements deserving of notice. In those of eight feet, which he made for the observatories of Padua and Vilna, and which Dr Maskelyne examined, the greatest error does not exceed two seconds and a half. He also made one of the same size for the observatory at Milan. The mural quadrant of six feet, which he made for the Duke of Marlborough at Blenheim, is a most admirable instrument. It is fixed to four pillars which turn on two pivots, so that it may be placed north and south in a minute. For this instrument, which is as beautiful as it is perfect, Mr Ramsden invented a method of rectifying the arc of 90 degrees, respecting which an able astronomer had started some difficulties; but with a horizontal thread and plummet forming a kind of cross which does not touch the quadrant, he shewed him that there was not an error of a single second in 90 degrees; and that the difference arose from a mural quadrant of Bird in which the arc of ninety degrees contained several seconds too much, and which had not been verified by so exact a method as his. But the quadrant is not the instrument which Mr Ramsden principally valued. It is the whole circle: and he demonstrated to M. de la Lande, that to attain to the utmost degree of precision of which observation is susceptible, we must renounce the quadrant entirely. His principal reasons are, 1st, The least variation in the centre is perceived by the two points diametrically opposite. 2dly, As the circle is turned the plane is always rigidly exact; which cannot possibly be the case in the quadrant. 3dly, Two measurements can always be had of the same arc; which serve for verifying the accuracy of the observation. 4thly, The first point of the division can be verified every day with the greatest ease. 5thly, The dilation of the metal is uniform, and can produce no error. 6thly, This instrument is a meridian telescope, as well as a mural. 7thly, It becomes a moveable azimuth circle by adding a horizontal circle below the axis, and then gives the refractions independently of the measure of time. Mr Ramsden made a circle of five feet for the observatory at Palermo; and one of twelve feet for that at Dublin. One of seven or eight feet, however, is sufficient to give precision within a half second, as in the zenith sector, which is employed for the most rigorous observations in regard to the figure of the earth. So wide as we have seen was the field in which [454b] the inventive genius and superior talents of Mr Ramsden were exercised.

That every part of his instruments might be fabricated under his own inspection, he collected in his workshops men of every branch of trade necessary for their construction. The same workman was always confined to the same branch, and by that means arrived at the greatest correctness and nicety in executing it. But, notwithstanding the perfection of his instruments, which ought to have secured to Mr Ramsden a large fortune, he sold them cheaper than any other artist in the same line in London; sometimes even one third below the usual price. Such was the demand for them, from every part of the world, that though he employed near 60 men, he was not able to execute all the orders which he received. He was indefatigable, however, in his endeavours for that purpose, till at length, by too intense application to his business, without allowing himself the necessary relaxation, he so far injured his health, that he was obliged to quit London and to visit Brighthelmstone for the benefit of the sea air. He died at that place on the 5th of November 1800, in the 65th year of his age. He had been admitted a member of the Imperial Academy of Sciences at Petersburg, in 1794 [*recte* 1793], and in 1795, he was honored with the annual gold medal adjudged by the Royal Society to persons distinguished for their excellence in the sciences.

[pp.454b-457a] An elegant and interesting tribute of respect ... by a gentleman of considerable eminence in the scientific world [ie Louis Dutens].

In person he was above the middle size, slender but extremely well made, and to a late period of life possessed of great activity. His countenance was a faithful index of his mind, full of intelligence and sweetness. His forehead was open and high, with a very projecting and expressive brow. His eyes were dark hazel, sparkling with animation but without the least fierceness. His nose aquiline and very handsome. His mouth rather large, but in speaking it had an expression of cheerfulness and a smile the playful benevolence of which will not easily be forgotten by his friends. His tone of voice was singularly musical and attractive, and his whole manner had a character of frankness and good humour which he well knew to be irresistible. When he attempted a bow to persons of rank his air was bashful [455a] and awkward, but when at ease with his friends, his motions and attitudes were in an uncommon degree graceful. To attempt in any degree to describe a mind like his, will be acknowledged by all who knew him to be a most arduous task. He was by nature endowed with uncommonly strong reasoning powers, and a most accurate and retentive memory, but with a quickness of penetration unequalled by any person I even saw, which enabled him, as it were at a single glance, to view in every light the subject on which he thought, and adopt at once the most advantageous mode of considering it. That quality of the mind, which is emphatically styled elegance, which in the abstract sciences leads to clearness, simplicity and precision, as in the fine arts it gives the last polish to genius, and is more generally known by the apellation of taste, he possessed in so exquisite a degree that it was perhaps the most leading and prominent feature in his character. In conversation it was most delightful; in the few treatises which he committed to paper it was eminently conspicuous; and in those immortal works which proceeded from his shop, and were the wonder of the whole scientific world, it is almost equally to be admired with the genius which invented them. This feeling for perfection [here Dutens inserts the equivalent Greek term] led him, in the most minute and insignificant parts of his instruments, to a polish and grace which sometimes tempted those to smile who did not perceive that the same principle which enabled him to carry the essential parts of his instruments to a degree of perfection unknown, and considered as impossible before his time, induced him to be dissatisfied if a blemish of some sort, even the most trifling, appeared to his exquisite eye. To these uncommonly strong natural endowments he added all that the most constant and intense study could bestow. Temperate to abstemiousness in his diet, satisfied with an extremely small portion of sleep, unaquainted with dissipation of amusement, and giving but very little time even to the society of his friends, the whole of those hours which he could spare from the duties of his profession were devoted either to meditation on further improvements of philosophical instruments, or to the [455b] perusal of books of science, particularly those mathematical works of the most sublime writers which had any connection nwith the subjects of his own pursuits.

Mr Ramsden's only relaxation from these constant and severe studies was the occasional perusal of the best authors both in prose and verse; and when it is recollected that at an advanced age he made himself so complete a master of the French language as to read with peculiar pleasure the works of Boileau and Moliere, he will not be accused of trifling even in his lighter hours. Short and temperate as were his repasts, a book or a pen were the constant companions of his meals, and not seldom brought on a forgetfulness of hunger; and when illness broke his sleep, a lamp and a book were ever in readiness to beguile the sense of pain, and make bodily sickness minister to the progress of his mind. Of the extent of his mathematical knowledge he was always from innate modesty averse to speak, though I have heard him say that he never was at a loss when his profession required the application of geometry. His knowledge in the science of optics is well known to have been perfect; and when we add that the works of Bouguer and the great Leonard Euler were his favourite study, we shall not lightly rate his proficiency in mathematics. Of his skill in mechanics it is unnecessary to speak. Nor let it be supposed that his science in his profession was limited to the higher branch of invention and direction of the labours of others. It is a well known fact that such was his own manual dexterity, that there was not any one tool in any of the numerous branches of his profession, which he could not use with a degree of perfection at least equal to that of the very best workman in that particular branch; and it is no exaggeration to assert, that he could with his own hands, have begun and finished every single part of his most complicated instruments. It is scarcely necessary to remark, that this practical knowledge of the minutiæ of his profession must have been extremely conducive to the perfection of his instruments. It may not be foreign to this part of his character to observe, that his drawings were singularly neat and accurate, and his handwriting so beautiful, that when he chose to exercise his skill, few writing-masters could equal it.

Of the qualities of his heart a friend may be allowed to expatiate with a peculiar and melancholy pleasure; but he will be acquitted of exaggeration when he appeals to the numerous list of those who, for a long series of years, loved and cherished him, and who, without a single exception, having once known, even continued to feel for him the most warm and sincere regard. In this respectable list will be found many of those men whose high rank has added lustre to their learning and abilities; and [456a] many of those whose powers of mind have raised them, like himself, from obscurity to splendour. To name those of the living who now lament his loss, may be offensive to the delicacy of some of them; but among those admirers of his talents and virtues who are now no more, I feel pride in placing here the honoured names of Stuart Mackenzie, Col. Calderwood and John Hunter. To have been loved and honoured by such men as these, is a sanction to the truth of the eulogium of the warmest friend. In fact, such was the candour and disinterestedness of his mind, such was the benevolence of his heart, that his few trifling failings could scarcely be contemplated; and the facility of temper which induced him to give to the person present what had been ten times promised to his absent friends, and then disarm their ill humour by such irresistibly good-tempered excuses as convinced them, in spite of experience, that he meant all he said, was rather the subject of laughter than spleen to his friends, as they knew that it proceeded in a great degree from his total indifference to pecuniary concerns, and were sensible that no views of profit or advantage could have engaged him to a dishonourable act. He, in truth, considered money only as the means of making further improvements in science. Misers have often been as temperate in diet, as simple in dress, and as frugal in expence as he was; but in hime those qualities proceeded from an utter disregard of everything (if I may be allowed the expression) material; and in the prosecution of a new idea, expence, while he possessed a farthing, never came into his mind, nor did he feel that it was a loss if he threw by an expensive instrument half finished, when a plan for its improvement struck his mind. Such was the placability of his temper, that although for the moment extremely sensible to the unmerited injuries which he sometimes met with, he was very soon induced even to lay aside his own vindication, rather than bear hard on the reputation of others.

Conscious of his own merit, he possessed in the highest degree an independent spirit, and he was roused to indignation by the bare mention of injustice or opression of any sort. But although he could not for a moment have borne the imputation of paying court to the most exalted man for any advantage that could be derived to himself from it, no man was more sensible of the necessity of subordination in society, or more punctiliously and unaffectedly exact in a respectful demeanour [456b] to those persons of high rank in whose society he often lived. Humble without meanness to his superiors, he was of course kind and gentle to those below him. His servants, his workmen unite in this most flattering testimony to the good temper and humanity of his soul, and his last will is a faithful picture of his mind in this respect. His workmen are there all remembered in proportion to their merits and services, and amongst them the greater part of his fortune is divided. The smallness of the sum of which he died possessed, might justly surprise those who knew the great extent of his commerce, and the ascetic frugality of his life, did they not at the same time know that he, in fact, was never paid for his great works but in reputation, for that the time consumed on them rendered every one of them a real loss to him as a trader. This, however, is not peculiar to himself. Several others among those artists whose abilities have done honour to their country, and most materially advanced the cause of science, have barely obtained a maintenance by their superior talents, while the steady plodding trader in the same line has, without fame, acquired great wealth. Mr Bird, second only to Ramsden in the art of constructing astronomical instruments, died worth rather less than him; and Shelton, whose time-pieces were long unrivalled, and have perhaps never been excelled, ended his life in absolute want.

It is not possible here to suppress a regret which everyone must feel at those instances of national neglect of high and rare merit. The encouragement of individuals cannot properly compensate men of this class of powers. Government can only bestow such reward as as may be at the same time substantial and flattering, and which would at once place the receiver in such a situation as would enable him to make vigorous exertions, and, by gratifying his homest and laudable pride, attach him by the strongest ties to the hand which protects and crowns him. In the case of Ramsden, it is scarcely to be imagined how much this country has lost by not having thus claimed for her own this her most distinguished son. While Europe in every corner repeated his name with respect, it was to a great proportion of his countrymen scarcely known but as that of a very idle spectacle maker and he of course worked for every foreign nation with a marked predilection over his own countrymen. The excellent and most ingenious works which the patronage of the Duke of Richmond, while Master-General of the [457a] Ordnance, engaged him in that department, and which the removal of His Grace from that office unfortunately suspended, are a shining proof that his abilities were so extensive, that they might with signal advantage have been directed to every purpose where mechanics as a science are applicable. And the navy of this country will now, with unavailing regret, hear that many most original and ingenious ideas on subjects of the highest importance to their profession, are, for want of that species of encouragement which could alone engage him to mature and perfect them, irrecoverably lost to the world. Among these are I fear to be reckoned a very improved log, and an instrument for measuring leeway; a most simple and accurate instrument for trimming a ship, or ascertaining her line of floating; and another, equally new and ingenious, for measuring the angle of her inclination either from the pressure of her sails or the action of the waves, or, in other words, her heel and roll.

Appendix 2

'The most perfect instruments for a complete astronomical observatory'
Ramsden to Fontana, undated but probably 1778/9. Florence: Archivio di Stato di Firenze, MS Misc. Finanze 438, 14pp. within a large bundle, unfoliated. The letter employs an odd mix of French and English.

1st. Two Quarts de Circle Mural of 8 feet radius, such as are describ'd by Mr Bird, one to be placed on the meridian towards the North, and the other towards the South. The following additions have been made to this instrument since.

Mr Aubert hath applied a very simple piece of mechanism, whereby the whole weight of the Telescope, or very nearly, is taken off the centre of this Quadrant which is of the greatest importance particularly when achromatic telescopes are applied to them. I have also had reason to suspect, that some of the inconsistencies which frequently happen in the observations which happen with this instrument, might very probably arise from different expansions from the degree of heat not being the same in different parts of the instrument. This difference is perhaps encreased by arising from the effect of the wall against which it is supported, the form of the observatory, &c. In some experiments I made on the Quadrant at Greenwich the difference amounted sometimes to 6 degrees Farenheits scale. To correct this defect I applied to different parts of the Quadrant thermometers constructed in a peculiar manner, which give the heat of the part of the Quadrant to which they are applied without being affected by the temperature of the air.

These Quadrants are divided by the micrometer screw to every second of a degree and have two arcs divided on their limb. One of these arcs is divided into degrees, minutes, &c, and the other is divided by continual bisection; as these arcs are graduated independent on [sic] each other, the exactness of the dividing will be shown by the correspondence of the two verniers, and will answer that the difference of these two arcs will nowhere exceed 3 seconds. Although in the Quadrant at Greenwich the difference of the two arcs in some places amounts to 4 seconds.

I have lately finished one of these instruments for the University of Padua, and at present am making one for His Majesty the King of Poland.

The price is 400 Guineas each and can be finished in 12 months.

2nd. A complete circle of twelve feet diameter, which will serve the purpose of both the Mural Quadrants; it is constructed on an entirely new principle: its radii instead of being solid is [sic] made with conical tubes, by this method the circle is made inconceivably strong and so very light that it has not the least tendency to alter its figure by its own weight, even if made of twenty feet diameter.

Astronomers must readily see the immense superiority of a Circle to that of a Quadrant; in the latter there being no means of detecting errors of the construction and executing. The instrument is taken for perfect tho: perhaps very erroneous; with the Circle it is not the case, every imperfection of the centerwork, errors of dividing &c will be shewn double and may thence be corrected, and its very advantageous construction obviates many difficulties to which the Quadrant is unavoidably subject.

The telescope applied to this circle is twelve feet long. It is fix'd on an axis of six feet, which turns in a conical tube fix'd on the centre of the circle and at right angles to the plane of the limb, by this means the line of collimation may with certainty be made to describe a true plane, and thereby obviates a very complicated error to which the quadrants are subject, for if the limb of the quadrant be not in a true plane, which is impossible to be done in an instrument of any size, the distance in altitude between the line of collimation and the first division of the nonius will vary; the ingenious Mr Mayer (of Mannheim) was the first who discovered this error, which in his quadrant of six feet radius made by Mr Bird amounted sometimes to eleven seconds; but the circle being turn'd in a path on its own axis must have its limb a true place yet let be ever so irregular the telescope moving on an axis independant of the limb, cannot be affected by any irregularity of it, and therefore the distance in altitude between the line of collimation and the first division of the nonius must constantly be the same.

By the application of a nonius to each end of the telescope, that is to say, at 180 degrees from each other, the accuracy of the instrument so far as respecting divisions may be examined and any irregularity, whether arising from the errors of the centre work, expansion, etc, will be detected which cannot be detected in quadrants.

The different temperature of the different parts of the quadrant are in great degree caused by the difference of temperature between the wall against which it is suspended, and the temperature of the air. The circle has no need of a wall it is fix'd to a strong vertical axis and turns round with it, so that the face of the circle may be set either east or west; by this means any error in the adjustment in the line of collimation will be of no effect, for if the altitude of any object be taken with the face of the instrument towards the east, and also with the face towards the west, the sum of those altitudes divided by two will be the true one, altho the line of collimation be not adjusted.

From this imperfect description we see that as every imperfection of execution etc. will be immediately shewn in this instrument, it will therefore require the greatest attention and care to execute it, which enhances the price of it. Such an instrument will at least cost 600 guineas, and will require about twelve months to make.

3rd. A Secteur d'Aberration of 12 feet radius, the length of its telescope 12 feet, with an aperture of 4 inches. The arc of this instrument extends 10 degrees on each side the zenith, and by means of the adjusting screw, angles may be taken to the exactness of the fraction of a second of a degree.

I have lately improv'd this instrument by making the axis on which the telescope turns much longer than usual, and in a different manner; in those hitherto made, the plumb line was suspended by a noose on a small pin on the end of the axis, but this has been found very liable to error. If the telescope was set to the same object, by moving it to the right and towards the left, the plumb line would sometimes be found to vary fifteen seconds. This error is rectified by suspending the plumb line so that it passes over a point or dot at the center of the axis, whereby the friction is avoided.

The method of reversing the instrument in order to adjust the line of collimation has been found exceeding inconvenient, being obliged to take the telescope off the point of suspension to reverse it made the line of collimation liable to be displaced. This is avoided by suspending the whole instrument on a vertical axis, by which the face of the instrument may be turn'd towards the East or West by turning the vertical axis. The price of this instrument made in the most complete manner is 120 Guineas and will require about 6 months to finish.

4th. An Instrument de Passage having its telescope 5 feet long and an aperture of at least 3½ inches diameter. The magnifying power is about 100 times. By this telescope the motion of a star &c in the Equator is visible in the one-fortieth second of time. The transverse axis on which the telescope turns is 4 feet long. It rest[s] in angular notches, and has proper adjustment whereby it may be moved both in the horizontal and vertical directions, and in order to prevent the pivots from wearing, the whole weight of the instrument, or very nearly, is suspended on a counterpoise. The axis is set horizontal by means of a spirit level suspended on two hooks; I have added an apparatus furnished with a plumb line, whereby the transverse axis may be set horizontal, to the exactness of a quarter of a second of a degree.

It has been found very difficult to reverse these instruments when made on so large a scale, which is absolutely necessary in order to set the line of collimation, or misplace some other part of the instrument. I have for that purpose contrived a stand whereby the instrument may be reversed with the greatest facility, without endangering it. The diameter of the semicircle applied to this instrument is two feet, and is divided by the nonius to every 30 seconds. I am at present making one for His Majesty the King of Poland of this size; the price is 100 Guineas.

5th. An instrument equatorial of a new construction, of the greater use for determining the path of a comet, or the Right Ascension or declination of any star, or planet. This instrument also at all times, shews the quantity of refraction either in altitude, Right Ascension, or declination, with many other very important uses in astronomical observations. I am at present making one of these instruments, for the Royal Observatory at Greenwich. The declination and Equatorial Circles have each 5 feet diameter; the telescope is 5 feet long, and has an aperture of 4 inches. It turns on a transverse axis of 30 inches. The exactness of the Equatorial Circle will be to every ¼ of a second of time, and the declination Circle in the same proportion.

This instrument has the peculiar property of shewing any error in the observation, as well as those arising from the imperfection of construction and of making of it. It appears in any respect by much the most perfect instrument yet used for astronomical observations. Since my invention of this instrument, I have had many of them bespoke, particularly by Sir Geo: Shuckburgh who wishes to have one made on a yet larger scale. The price allow'd me by the Royal

Observatory is 500 Guineas, but I flatter myself to be able to make another of the same size for 400 Guineas and it could be executed in about 9 or 10 months.

6th. A Quart de Circle mobile pour les hauteurs correspondants, de trois pieds de rayon. It is placed on a vertical axis or pillar; with it altitudes may be determined to the exactness of about 3 seconds of a degree. There are three telescopes applied to this instrument. In order to render it complete for the purpose of measuring angles on a horizontal plane, I have also contriv'd a method to adjust the instrument by means of a ground spirit level, which is of very great use, particularly when the instrument is expos'd to the air, which every portable instrument must necessarily be. The price of this instrument with every requisite to render it complete both for astronomical and geodesical purposes is 180 Guineas.

7th. An Equatorial Instrument of a construction I lately invented, which besides its application to the common purposes of astronomy, is particularly adapted to determining longitudes at [sic] land. It is very portable and exceeding proper from taking plans of a country, and by a peculiar method of using this instrument horizontal angles may be measured with greater exactness than can be done by any other instrument of triple its dimensions. It has every property in common with the large ones, such as correcting the effects of parallax and of refraction. I have also added my new Micrometer to the telescope, whereby the diameter of a planet may be measured to the exactness of a second of a degree. The price of this instrument without the micrometer is 80 guineas. The diameter of the azimuth equatorial & declination Circles are 11 inches each. It may require about 6 months to finish one.

8th. Une morse a tenir un telescope, which is of 3 feet length and hath an aperture of 2¾ inches. This telescope is furnished with a micrometer whereby the difference between the Right Ascension and declination of the Moon or a Comet & Star &c may be measured with great precision, when the distances do not exceed 5 degrees. The price is 50 Guineas and is an instrument which can be executed in about 3 months.

9th. A Sextant de Hadley of use for determining the distance of the Moon from the Sun, or a star, or measuring terrestrial angles; they are furnished with a telescope for the more accurately determining the contact of the two images. The index is furnished with an adjusting screw and many other improvements. As a reward for my improvement of this instrument and the method of dividing the Board of Longitude were pleased to honor me with a reward of £600.
  The price of this instrument 15 inches radius is 10 Guineas.

10th. A theodolite on the same principle as that I had the honour to make for the Royal Society which was used by Dr Maskelyne to determine the dimensions of the mountain whereon he made his experiment, to determine the attraction of the plumb line. Since that time I have added many improvements, such as making an axis to the horizontal index which takes off the friction of the two horizontal plates, and prevents the center from wearing so as to shake. This instrument is furnish'd with two telescopes, one is fix'd on a vertical semicircle which is attached to the horizontal index and the other is attached in such a manner to the horizontal circle, as that it may either move with or without it. Both the telescopes describe vertical planes. The lower telescope having an horizontal motion is of great service when angles are required to be measured with great accuracy. The price of this instrument when the circles are 9 inches radius is 24 Guineas.

11th. Une Telescope de Reflection made according to the construction of Cassegrane [sic] which admits of a much larger aperture and consequently magnifying power than any other construction. These telescopes are mounted on a Polar axis with graduations whereby to find a star or planet in the daytime. I have lately invented a new Micrometer applicable to reflecting telescopes which separates the images in the same manner as the object glass micrometer and without any additional glasses or mirrors to those used in the telescope. By this means the telescope is free from those additional aberrations which are unavoidable when the object glass micrometer is applied and the distance of the image depending on the angular inclination of two mirrors and not on the focal length of any lens or mirror. It therefore cannot be affected by any alteration in the eye of the observer. By these advantages the diameter of a planet &c may be measured with a much greater precision than can be done with the object glass micrometer.

The price of a telescope of this construction of 9 feet long and having an aperture of ten inches will be 200 Guineas and it will bear a magnifying power of 500 times.

The price of the Micrometer to such a telescope will be 15 Guineas and will require 12 months to make it.

12th. Une Telescope de Refraction having its object glass 4 feet focus with an aperture of $3 \, {}^8/_{10}$ inches. It is mounted on a stand of a new construction which prevents the trembling of the telescope. It is mov'd horizontally & vertically by means of endless screws.

I have also invented a new Micrometer applicable to refracting telescopes. It is in some degree similar to that applied to the reflecting telescope as it separates the images without the addition of any glass to those necessarily used in the telescope. The price of this telescope with its different eye-tubes & Micrometer made in the complete manner is 50 Guineas and may be done in six months.

13th. Une lunette cometere of 30 inches long having an aperture of 3 inches. This telescope mounted on a stand with proper movements so as to sweep parallels of latitude or declination for the conveniency of finding comets, new stars &c will cost 9 Guineas.

14th. Deux Barometre [sic] de mon dernier construction avec leur supports qui servira aussi a measure les hauteurs de montagne.

15th. 3 or 4 Thermometres avec les Echelle de Farenheit et de Reaumur, one graduated on each side the tubes and the other [here Ramsden's pen runs dry and he does not complete the sentence.]

16th. Une manometer which serves to give the densite of the air very proper to correct the tables of mean refraction.

The price of these last articles together may amount to about 25 Guineas.

17th. 2 Pendule astronomique having their pendulums properly corrected for the differences of temperature 40 Guineas.

18th. Une porteur du temps price 6 Guineas.

Index